PIONEER IN MODERN MEDICINE

David Linn Edsall of Harvard

PIONEER IN MODERN MEDICINE

David Linn Edsall of Harvard

BY

JOSEPH C. AUB, M.D.

AND

RUTH K. HAPGOOD

FOREWORD BY PAUL DUDLEY WHITE, M.D.

Harvard Medical Alumni Association

1970

Library of Congress Catalog Card Number: 78–145896

Manufactured in the United States of America

Foreword

Having known David Edsall in his pioneering days in medical teaching, research, and administration at the Massachusetts General Hospital and Harvard Medical School at the beginning of my own medical career, for the direction of which he was largely responsible, and being myself intensely interested in the details of medical history which this book exemplifies, I am very appreciative of the invitation to read this manuscript and to comment upon it.

Those of our readers who are too young to have known him and to have seen and experienced the primitive state of medical science which he did so much to change, cannot do more than imagine the transformation for which he and a very small band of patient, dedicated, and courageous physicians were responsible. Undaunted, though temporarily discouraged, by his failures in Philadelphia and St. Louis, he finally succeeded in Boston despite the many complications of the Boston scene. I suspect that by the time he arrived to take his Jackson Professorship at the Massachusetts General Hospital he had himself become wiser and more patient, and that the revolution in medical teaching was already in the air. It was to change the daily grind of lecture after lecture before massed medical classes into the current personal participation by the individual students in laboratories and in clinics and in the homes of the patients themselves. In any case Edsall was a great leader, perhaps the most important one, to put theory into practice. He was also an important pioneer in the fields of industrial and of preventive medicine, and as an administrator he vigorously and successfully promoted, often against strong opposition, a part-time clinical faculty of teachers and investigators to share equal responsibility with the full-time faculty.

This book presents in considerable detail not only the background and the life of David Edsall but the history of the struggles of the faculties of the medical schools of three universities: Pennsylvania in Philadelphia, Washington in St. Louis, and Harvard in

Boston. It was an exciting time of change and it was natural for these pioneers in their impatience often to disagree with each other. Dr. Aub and Mrs. Hapgood have done for us an important service in presenting the complexities of the life of our own hero David Edsall and we are grateful.

Dr. Joseph C. Aub, a longtime colleague and friend, and like myself a pupil of David Edsall, has himself made a significant contribution to medical education by proving that his original and important researches in metabolism, lead poisoning, and cancer greatly enhanced the impact of his teaching on both undergraduates and graduate students. His own pupils, scattered widely, have continued to spread his example throughout the country. His intimate association with Dr. Edsall through many years has made him an ideal biographer of that great pioneer in medical education.

PAUL DUDLEY WHITE, M.D.

Acknowledgements

THE AUTHORS wish to take this opportunity to acknowledge their great debt to the many people who have helped them in elucidating the history of this period.

First and foremost their thanks go to the Edsall family, Dr. John Edsall and Dr. Geoffrey Edsall, the late Mrs. David L. Edsall, Mrs. Dorothy Roberts and Mrs. Euena Snook;

To Dr. Carl Binger, the late Dr. Stanley Cobb, Dr. Edward D. Churchill, Dr. Francis R. Dieuaide, Dr. Edwin Locke, the late Dr. Dr. James Howard Means, the late Dr. Thomas Pellett, the late Dr. Borden S. Veeder—all good friends of Dr. Edsall—who shared their special knowledge of him:

To historians Dr. George Corner, Dr. Jean Curran, Mr. Saul Benison, and the late Dean Sidney Burwell, for their help in locating materials;

To Mr. Richard J. Wolfe, Rare Books Librarian, his assistants Miss Ruth Linderholm and Miss Helen Olney, and the staff of the Countway Library of Medicine; to Mr. Kimball C. Elkins, Senior Assistant in the Harvard University Archives, to Mr. David Bailey and Mr. Sargent Kennedy, Secretaries to the Harvard Corporation, and to the Harvard Corporation itself; to Miss Genevieve Cole, Librarian, and Mr. George Jacobsen, Archivist, at the Massachusetts General Hospital; to the Dean's Office and the Schlesinger Library of Radcliffe College; to Dean Edward Dempsey and Dean M. Kenton King at the Washington University School of Medicine and to Mr. William A. Deiss, Archivist, Washington University Libraries; to Mr. Leonidas Dodson, Archivist of the University of Pennsylvania, and to Mr. W. B. McDaniel, 2nd, Curator, Library Historical Collections of the College of Physicians of Philadelphia; to Miss Esther Stamm of the Rockefeller Foundation; to Miss Margo W. Brown of the Harvard University Health Services;

To the following for gracious permission to quote from unpublished material:

Dr. Arlie V. Bock | Mrs. Jean Fitz Buffum

Dr. P. S. de Q. Cabot
Dr. Bradford Cannon
Mr. Camillus Christian
Mr. James B. Conant
Dr. Francis R. Dieuaide
Dr. Anne S. Drinker
Mr. Charles W. Eliot
Mr. James Thomas Flexner
Mr. Raymond B. Fosdick
Mrs. Alan Gregg
Mr. Hugh M. Hamill
Mrs. Amy Washburn Hamilton
Mr. Richard Hocking
Mrs. Jean Flexner Lewinson
Mrs. Bertha Rosenau Ilfeld
Mrs. Aatos Ketro
Mr. Lincoln Kirstein
Mrs. Linda Pellett Lannin
Mrs. Janet Longcope
Mrs. Roger I. Lee
Dr. Roger I. Lee, Jr.
Mrs. George R. Minot
Mr. James J. Minot
Mr. Henry W. Minot
Mrs. Natalie Cabot Neagle
Dr. Eugene L. Opie
Mrs. Richard M. Pearce
Dr. Willard C. Rappleye
Dr. John P. Riesman
Mr. Robert Brookings Smith
Miss Madeline Stanton
Miss Elizabeth Thomson
Dr. Paul D. White
Mrs. T. North Whitehead
Mr. Ray L. Wilbur, Jr.
Dr. Hans H. Zinsser

To Mr. Howard J. Sachs, Harvard classmate of J. C. Aub, and to Dr. Dana L. Farnsworth, Dr. James M. Faulkner, and Dr. Joseph Garland for their aid and encouragement in bringing this book into being.

JOSEPH CHARLES AUB
RUTH KNOTT HAPGOOD

Table of Contents

List of Illustrations

Illustrations facing page 194

CHAPTER 1

Home Ground, 1869–1890

WRITING the life of David Edsall has also meant writing the medical history of an era — a very important era. Every age seems to produce a few outstanding personalities who typify the dominant characteristics of their time. These are the leaders who direct the activities of their own decades and so teach subsequent leaders, and who thus affect the attitudes of more than one generation.

Such a leader was David Edsall. He came into American medicine just before the beginning of the modern investigative approach, the period in the early decades of this century when medical knowledge was undergoing careful scientific reexploration. Edsall was one of the earliest leaders in this movement. He acquired the scientific point of view early in life, even though in college he was rebuffed by the professor of chemistry who held that chemistry was a waste of time as training for medicine. In medical school, research laboratories were just starting in an occasional institution, and Edsall found these opportunities open to him at the University of Pennsylvania, though like most of his generation, he owed much to years of study in Europe.

But Edsall was fortunate and able, and he helped create the environment he wanted for himself and for his disciples. Because his scientific ability was united with the other gifts of the healer, he became nationally known as a clinician. He was a highly regarded professor in three of the country's leading medical schools. Perhaps the quality which has made the most enduring imprint on medicine was his ability to search out talented young men and find ways for them to develop the best that was in them — a quality shown in his teaching, in his leadership of the medical services and the research at the Massachusetts General Hospital, and as dean of Harvard's schools of medicine and of public health at a critical time in their history. Further, as a trustee of the Rockefeller Foundation and advisor of other national organizations, he had a part in the wider flowering of medicine which took place in the '20's and '30's.

Six-foot-four, deep-voiced, he had a breadth of view to match his physical proportions. He had a phenomenal memory and the ability to marshal many facts in a balanced way. These powers of mind were combined with an almost boyish idealism, tempered but not blunted in the course of a difficult career. He remained a doer, a builder, to the end of his life.

This was a man, not a collection of virtues. He made some mistakes and met some resounding defeats. While he showed a calm serenity to the world, his devoted friends knew that he was at times depressed and had need of their friendship. At play he often sought out the company of young people. To his younger colleagues, he was never too busy, never superior, one's teacher but always one's friend.

A personal note enters this story. I am a physician and never expected to be anyone's biographer.* Edsall was my professor of medicine at the Harvard Medical School, my chief when I interned at the Massachusetts General Hospital, my friend and mentor when I returned to Harvard after serving in France. Those of us who grew up under his eye — Paul Dudley White, J. Howard Means, Carl Binger, Stanley Cobb, Edward Delos Churchill, to name a few among many — knew him for one of the makers and shapers of those stirring times. From a perspective of fifty years in medicine, Edsall's achievements are all the more remarkable — both for his advances, which were revolutionary, and for the balanced way in which he used innovation to strengthen the best of the old institutions.

David Linn Edsall was born in Hamburg, New Jersey, on the 6th of July, 1869, sixth son of Richard E. Edsall and his wife Emma Everett Linn. Edsalls had been living in New Jersey for six generations, both Edsalls and Linns being of English descent with an admixture of Dutch and Scotch. David Edsall's forebears had settled the little towns and carved the farms out of the wilderness. His mother's grandfather, Martin Ryerson, a land surveyor, was one of the first settlers of Hamburg.[1]

* As J.C.A. is the only one of the authors to have had a personal relationship with David Edsall, he is the only one to make use of the first person singular in the text.

The first Edsall to land in the New World was Samuel Edsall,[2] who left Reading, England, in 1648 and settled in New Amsterdam where he became a successful merchant, being admitted as a burgher of the city in 1657. On his many trading expeditions to Long Island and along the Hudson and Delaware Rivers, he learned the Indian languages, and acted as interpreter between the Dutch and English and the Indian tribes. It was he who negotiated with the Indians for the sites of Elizabeth Town and Newark.

Samuel Edsall married four times, his first wife and the mother of most of his children being Jannetje Wessels, a belle of New Amsterdam straight from Holland, as the rumor runs. The mother of the line which led to David Edsall was English, his third wife Ruth Woodhull.

Surviving two shifts in sovereignty from Dutch to English, Samuel Edsall was appointed to the council of Jersey governor Philip Carteret in 1668, and was president of Bergen and a member of the Court of Judicature. Another shift in politics nearly caused him to lose his life in supporting Governor Leisler of New York, on whose council he served in 1689 and 1690.

When James II fled the throne of England and word came to the New World of the Glorious Revolution of 1688, the enraged citizens of Boston threw the unpopular Governor Andros into prison, leaving New England, New York, and New Jersey without a head. In New York, Leisler took over the government and declared for William and Mary,[3] but by the time a new legitimate governor arrived in 1690, opposition to Leisler was so strong that Leisler and his son-in-law were condemned for treason, and other members of his council narrowly escaped death. Edsall himself was "under bail" until the arrival of a new governor, and did not return to public life until the attainders were reversed by Parliament in 1698.

His son Richard moved that branch of the family to New Jersey and made his living as a surveyor. Richard's grandson, also named Richard Edsall (1750–1823) was a captain in the 2nd New Jersey Volunteers during the Revolution. His son Joseph (1783–1833) was a quartermaster in the War of 1812 and a member of the New Jersey legislature. It was he who brought the Edsalls to Hamburg, where he was a forge owner, employing seventy men and producing five tons of pig iron a day in the 1820's.[4]

David Edsall's father Richard was born in Vernon township in 1813, the son of forge-owner Joseph Edsall and Sarah DeKay.[5] At seventeen he left home, a little like his ancestor Samuel, to make his way in the world. After a number of business ventures which took him to New Orleans and to Iowa, he started a general store in Hamburg which prospered for years and was eventually taken over by his sons.

In 1855 Richard Edsall was elected sheriff of Sussex County, and in 1863 the voters sent him to fill an unexpired term in the state legislature. This was followed by six years as state senator, where he was chairman of the committee on railroads and canals and a member of other important committees. His political career was the more remarkable since Richard Edsall was staunchly Democratic and the New Jersey attitude toward Democrats was expressed in the phrase "sand Spaniards," the "sand" referring to New Jersey and the "Spaniard" being a way of saying that anyone fool enough to be a Democrat must be a foreigner.[6]

An old Hamburg story tells how when he served as postmaster in Hamburg, he used to hand out the *New York Tribune* on a shingle because he wouldn't touch such a Republican sheet.[7]

David Edsall kept up this Democratic tradition, to the occasional astonishment of staid Republican friends in Boston; in fact his sister-in-law Edith Tileston thought his politics "peculiar."[8] The story went around Boston at one time that Dr. Edsall was a Jew, for what seemed three obvious reasons: one, his name was David; two, he was an outspoken supporter of Justice Brandeis who had just been appointed to the Supreme Court; and three, he was a Democrat.[9]

Edsall's father waited until he was forty-six to marry and then had seven sons in the following fifteen years. Sixth of the seven boys was David Edsall, born in the big old family house on the main street of Hamburg (since torn down to make way for the Hardyston National Bank). Richard Edsall not only sent three of his sons to college, he sent them abroad for further study, in days when a trip abroad must have been a rather unusual event for a young man from a small town like Hamburg. David went abroad for the first time while he was still in school.

Two of the boys went into medicine, perhaps in part because of traditions in their mother's family. Emma Linn Edsall's grandfather was a Dr. Andrew Linn; another doctor was her brother Theodore who died in 1852 after practicing for only two years. There was also a cousin, Dr. Alexander Linn, a physician of prominence in New Jersey with an extensive country practice. A graduate of Jefferson Medical College in 1836, Dr. Alexander Linn settled at Deckertown, now named Sussex, in New Jersey. His fellow members of the Sussex County District Medical Society seem to have regarded him as "the most brilliant star in their circle," and the Sussex hospital is named in his honor. Theodore Linn studied with him before going to Jefferson Medical College. Dr. Linn died the year before David Edsall was born.[10]

Edsall loved his old home and the rolling hills and fertile valleys that lay around the little village. The Jersey countryside around Hamburg bears no resemblance to the flat and sandy south Jersey which for most people typifies the state. It is like a gentler Vermont, high, rugged, hilly country with many little lakes, and, like Vermont, a notable dairying region. The long, shaggy, forested ridges have kept the country open even to this day. In Edsall's time there were plenty of upland pastures covered with blueberries, rocks and mines to explore, and occasional limestone quarries for swimming holes. Edsall was never much of a swimmer, and indeed almost drowned in the Wallkill River during his childhood, but the love of tramping and mountain climbing stayed with him throughout his life. In his Harvard days he liked nothing better than to go off tramping with a few congenial companions, friends of the out-of-doors like Walter Cannon and Carl Binger, or a group of young people from the medical school or hospital. The Blue Hills with their lovely scenery were always available at the times he lived in Milton near Boston, or he would visit Fitzwilliam, New Hampshire, in the Monadnock region, or the White or Green Mountains. Eventually he settled in northern Vermont itself in spite of the severe climate, rebuilding a farmhouse in Greensboro where there are perhaps four or five shirtsleeve days in a summer. This suited him perfectly because next to a brisk walk across the hills the thing he probably liked best was a blazing fire and a well-filled pipe and a chance for storytelling and anecdotes with his family or friends.[11]

Hamburg ranks as the principal hamlet of the township of Hardyston, and when Edsall was growing up there in the early

1880's it boasted two general stores in addition to Edsall, Chardavoyne & Co., a drugstore kept by Charles H. Linn, three blacksmith shops, two wheelwrights, a harness shop, a coal and lumber yard, two hotels, and the creamery around which much of the enterprise of the region centered. The creamery was said to consume the milk of five hundred cows.[12]

Edsall's niece, Dorothy Roberts, daughter of his older brother Thomas, recalls the old Edsall home as it was in the days before the First World War. The house had an old-fashioned well in front, with a balanced windlass and a bucket on each end of the rope. This well was used by most of the townspeople until a dead dog was found in it one winter. The Linn Realty Company then drilled a well which supplied the Edsalls, Dr. Pellett, and several other homes, and a nearby neighbor would stop the pump when the water overflowed from the tower and splashed on her house.

There were five bedrooms in the Edsall house, a large parlor, a large living-dining room, and a small library. When a tower room was added, it became the smoking room and was filled with old magazines, to the delight of generations of children. The original kitchen was in the basement, with a hand pump for water from a cistern, and a dumbwaiter.

There were three barns, one for horses, one for carriages, and a wood barn. One barn housed the wagons used for deliveries by the store across the street. Edsall's brothers Thomas and Robert were running the store in those days, and it sold "furniture, groceries, and everything from barrels of sugar, flour, apples, crackers and other groceries that came in barrels . . . to wall paper, shoes, dress material, notions and everything one would need in operating a home."[13] There was a large yard and garden enclosed by a picket fence.

According to his own account, David Edsall's schooling followed a rather checkered pattern. In his teens he suffered from what he later thought must have been an attack of tuberculosis, though he made a complete recovery. While he was out of school he was tutored. One of his tutors was the Reverend Joseph Smith, who lived next door and was the rector of the local Church of the Good Shepherd.[14] Edsall's son John remembers that his father also had a tutor named Magee, "apparently quite a remarkable man with a real love of learning. Father found him very inspiring, much more so than school would have been at that time, so that he really took to reading the classics with enthusiasm. He read a lot in the original," particularly Cicero and other Latin authors. This explains why Edsall intended to follow a career in the classics when he entered Princeton.[15]

Edsall had not been long at Princeton, however, when he had so much trouble with his eyes that he left college temporarily. He spent most of that year in Hamburg, forbidden to read, and finding solace in long tramps across the countryside. The rest and proper glasses completely restored his eyesight, and the enforced idleness may not have been time wasted. It is curious how often an interval of this kind occurs in the early life of a person who later makes his mark in the world.

David Edsall stood well over six feet and was built to scale, which caused inevitable comparisons with Barnum's famous elephant Jumbo, and "Jumbo" he was promptly called by his Princeton classmates. It may have been his size which prompted the publication in a Princeton annual of a photograph of the enormous college water tower with the caption, "Edsall's only rival." On the other hand, the reference may be to a rather notable capacity for beer. One waggish tale recounts how Edsall had been crossing the campus with twenty-four bottles of beer when he tripped and fell. "Heavens," said the hearer, "what a lot of beer must have been spilled." "No, there wasn't any lost," was the reply, "the beer was all inside him."[16]

Edsall shifted direction quite radically in his first years at college. A serious-minded young classicist when he entered, he gave up his ambitions in this field when he encountered a Princeton professor more interested in the letter than the spirit of Greek. As for history, years later in thanking the historian Samuel Eliot Morison for the pleasure he had had in reading his Oxford History, he described his encounter with the Princeton history department:

> I had the misfortune when a student in college to be under an historian who was then, and still is, a very distinguished man but who had the unhappy gift of making history an extremely repulsive thing . . . Two or three years ago, at a luncheon of the Foreign Policy Association . . . the chief speaker was introduced and his name mentioned and suddenly I had an extraordinarily vivid and intense feeling: "You are

the wretched person that made me hate history over thirty years ago and it took me about ten years to recover from you." [17]

Still enrolled in the academic department, Edsall in his sophomore year probably took the required course in zoology and botany.[18] Professor John S. Schanck gave lectures on anatomy and physiology in the first term and Professor George McCloskie taught elementary zoology in the second term. Presumably this was the "year's work in biology" to which Edsall referred later, describing his first taste of research:

"That little piece of work, on the development of the chick's ear, was one of the really significant things in my training as it gave me some idea, early in my life, as to what it meant to work and think independently and to make an effort in an elementary way to arrive at a scholarly judgment on some subject, even though it was a small one." [19]

Equally memorable and a source of pleasure and inspiration to him through life was the astronomy course given by Professor Charles A. Young, a well-loved teacher whose students called him "Twinkle." [20]

In senior year, 1889–90, he took Professor William Libbey's course in histology, in which he valued the laboratory work. "[It] taught me methods with the microscope and something about the structure of tissue and made it much easier for me to get on when I entered the medical school than for the men who had not had this experience."

He also had a year of chemistry, "entirely lectures and demonstrations," and a year in physics. "The latter was extremely valuable, the former was largely a variety show." [21]

The lack of serious training in chemistry troubled him for years. Edsall said in a speech later:

> It has proved that most of the research that has especially engaged my personal interest has been in more or less definitely chemical lines. Through reading and through consulting persons trained in chemistry, I gained some knowledge and I blunderingly learned some methods. Years afterward in this country and abroad I got a little of the training that I wanted to get in college. But the efforts were stumbling and insecure and always with consciousness of lack of the power and insight that might have been given by sound and continued basic training . . .

He went on to describe what happened in college when he tried to go further with his chemistry.

I had a strong desire to know more about the subject and a rather vague but insistent idea that it might help me to understand medicine better and perhaps aid me if I wished to attempt investigations in the vast area of the unknown in medicine. With the consent of the dean, I took the then radical step of asking the professor of chemistry in the School of Science to allow me to take the elementary laboratory course and perhaps other courses with the Bachelor of Science group. He asked me what I was headed for, and I said medicine. He replied with entire disapproval, "A doctor does not need to know any chemistry. You would be wasting your time" – and looked at the door. I left with the feeling that I had been merely stupid and impractical . . .

Until very recently, speaking in the usual historical sense, serious devotion to research in medicine was looked at askance by very many of the medical profession itself. Weir Mitchell told a group of us once that after he had graduated in medicine and had a service in a hospital, his father, then a prominent professor of medicine, asked him what he intended to do next, and he said that he thought he would spend a few years in research in an intensive way. His father threw up his hands in astonishment and almost anger, and replied, "Weir, if you do that, people will look upon you as they would if you joined a circus." [22]

Years later I had an analogous experience when I was coming home from France at the end of the First World War. (I was twenty-nine then.) I wanted to leave clinical medicine and go back into a physiology laboratory for a time and get better training with Walter Cannon, the outstanding physiologist of his day. I asked two men who could, I thought, advise me better than most: Richard Cabot, who had done a great deal of work studying blood smears, and Sir James Mackenzie, the pioneer in cardiac irregularities. I asked them both, "What will happen to me if I study physiology for a couple of years instead of continuing with clinical medicine?" I had had more than my fill of clinical medicine during the war. Both men said it would ruin my life; I would lose touch with people; wouldn't know how to handle them well; one could not break away from clinical medicine so readily and remain very good. These two men were the best I could consult, I thought. They both gave me the same advice. They both told me not to do it. I did it anyhow, and frankly, I could not see then nor later that it did me anything but good.

I tell this story to indicate how precocious was Edsall's desire to study fundamental science nearly twenty-five years before me—and even then the real change did not come about until the relatively large group of young doctors who, following the Second World War, returned to the physiological and chemical laboratories in preparation for careers in "academic medicine."

Edsall's older brother Frank Hynard Edsall, seven years David's senior, had already gone into medicine. Graduating from the medical school at the University of Pennsylvania in 1885, Frank went abroad to study in Heidelberg, and about 1890 he settled in Pittsburgh, where he specialized in eye, ear, nose and throat. In 1898 he moved to Madison, Wisconsin, later returning to New Jersey and going into public health work. It is possible that David visited his brother abroad before going on to medical school in his turn.

Looking back on his own preparation for medicine, Edsall wrote: "If I had to do it over again perhaps the chief things I would change would be in the college course. I would take more science in college than I had, especially exact methods and definite work on problems. This not in order to be able to do better research but chiefly because it is so valuable as early training in preparation for directly clinical work and the problems of which this consists. I think the amount now required is, however, entirely adequate for the usual student [this was 1925], though more is wise for those who have a bent for actual science."

He was also a strong advocate of the study of general literature both in college and in later life. This he felt of value not merely for the pleasure and relaxation to be derived from it but for the understanding of human nature which a good physician would need. Without such reading of good fiction, poetry, and biography, he felt that "little is learned in ordinary living except by long continued experience . . . Generous reading of these constitutes a valuable psychological training and is of more value in learning to understand human beings than any courses in psychology I had or am familiar with, though I do not wish to disparage psychology."[23]

CHAPTER 2

Medical Student, 1890–1895

DESPITE the eminence of the College of Physicians and Surgeons in New York, and the increasing reputation of the medical school at Harvard under the first twenty years of President Eliot's leadership, the University of Pennsylvania's medical school was still, in Osler's words, "the premier medical school of the United States."[1] Osler had left Pennsylvania the year before Edsall arrived, after five years there as professor of clinical medicine, to join the staff of the Johns Hopkins Hospital as physician-in-chief. The Johns Hopkins Medical School did not accept its first student until 1893, and had not yet begun to dazzle the medical world where it reigned supreme for decades. In 1890 Edsall settled on Pennsylvania, as had his brother Frank before him.

Almost all the famous luminaries who had been his brother's professors were still there: white-whiskered Joseph Leidy, professor of anatomy since before the Civil War, whom Edsall called "the greatest Anatomist of his time in this country; also a great Anthropologist and an authority on precious stones"[2] *; precise-minded John Ashhurst, professor of surgery, whose skill and vigilant care of his patients gave him as good a record as the younger men whose aseptic principles he refused to adopt; George E. de Schweinitz, regarded by Edsall as "the leading ophthalmologist of the country; great as an investigator and a practitioner, and as a personality"[3]; Edward Reichert in physiology; William Goodell in the diseases of women and children; Louis Duhring in diseases of the

* Weir Mitchell's biographer tells a delightful story of Leidy, who was one of the members of the Biological Club in Philadelphia, where the members dined in gourmet style. One day a member asked Leidy, "How did you find the dozen terrapin I sent you a month ago?" Leidy replied enthusiastically, "Magnificent! I dissected the entire twelve and discovered in the twelve intestines, three parasites previously unknown!" To the friend's scandalized protest, Dr. Mitchell said, "Never give Leidy anything that is edible and worth dissecting. We all know where it will go!" (Anna Robeson Burr, *Weir Mitchell: His Life and Letters*. New York: Duffield, 1929, pp. 136–37.)

skin; Theodore Wormley in materia medica; Juan Guitéras in pathology. Dr. James Tyson filled Osler's vacant chair of clinical medicine.

At the head of this group stood the remarkable Dr. William Pepper, senior professor of medicine, provost, and ex-officio president of the department of medicine. In the reforms of the late nineteenth century through which every leading medical school struggled, Pepper was the acknowledged leader at Penn. In the '70's and '80's he fought for the raising of entrance requirements, including a preparatory examination; the lengthening of the school year; the grading of the course so that one subject led to another; clinical and laboratory instruction; and the end of the old fee system.[4] Formerly, the only medical school income had been students' fees, divided among the professors at term-end — a system which resulted in continued pressure for high enrollment and consequent low standards.

While Edsall was in school, Pepper was fighting to carry through the four-year course. Having provided the university with a new teaching hospital, he also started a movement to endow a proper new laboratory. He himself gave $50,000, and the laboratory, named in honor of his father the William Pepper Laboratory of Clinical Medicine, opened in 1895.[5] Not content with raising the standards of the medical school and working hard for the university, Pepper also made basic contributions to the life of Philadelphia. In 1891 he brought about the creation of Philadelphia's first free library, and he was a moving spirit in founding the Museum of Science and Art, and the Commercial Museums. He was Edsall's professor of medicine, then and always a powerful influence in his life.

Professor Tyson wrote of Pepper after his death:

> His greatest ability was shown in teaching clinical medicine. He attracted students and patients from all parts of the country, and his Saturday clinics were often made up of cases who had thus come to seek his opinion. He never hesitated to take up any case, however difficult, and generally succeeded in unfolding it to the edification of the class and satisfaction of the patient . . .
>
> Dr. Pepper's conception of the office of the medical teacher was a very broad one. He would have him broadly educated in letters and arts as well as learned in medicine, an associate of men and interested in

public enterprises — in a word, a man of affairs, not a mere pedagogue in the narrower sense of the term.[6]

Edsall used to tell a story, probably from one of these same Saturday clinics, which demonstrates Dr. Pepper's extraordinary ability to make lightning diagnoses. He had already started to lecture on cirrhosis of the liver when his resident announced that a case suitable for clinical demonstration had just appeared in the outpatient department. While the patient was being brought in, Pepper was talking about cirrhosis, and started to say, "One of the outstanding things about cirrhosis is . . ." In mid-sentence the patient was wheeled in and after one glance Pepper continued, ". . . the difficulty in differentiating it at times from Addison's disease, which, as you'll see in this patient, is in some ways strikingly similar."[7]

In spite of Pepper's reforms, the entrance requirements on Edsall's arrival in the fall of 1890 were still what we would consider rudimentary. His A.B. from Princeton was more than enough to guarantee his admission. The catalogue stated:

> Candidates for admission are required: *First,* to write an *Essay,* of about three hundred words in length, as a test of Orthography and Grammar; *second,* to pass an examination in *Elementary Physics* (Part I. of Fownes's *Chemistry*). A candidate who has received a collegiate degree, or passed the matriculate examination of a recognized college, or who has a certificate covering the *required* subjects from a recognized normal or high school, or a duly organized county medical society that has instituted a preliminary examination — such as that adopted by the Medical Society of the State of Pennsylvania — may enter without examination.[8]

The catalogue went on to call attention to the "unusual advantages of the Course in Philosophy in the College Department of the University."[9] Two years of that course would exempt students from the preliminary examination. Even these requirements were not rigidly enforced. It was not until 1896 that the equivalent of high school graduation was required, with serious repercussions on the enrollment.[10]

The medical course when Edsall entered required three years, though this was raised to four shortly after he left, when Pepper's reform was carried through. First-year courses included chemistry, pharmacy, osteology, histology, and dissection, and the privilege of attending lectures in general medicine and surgery. The second year added medical chemistry, pathological histology, and physical diagnosis, a great deal more anatomy, and medical and surgical clinics at the Philadelphia and University Hospitals. Third-year men were divided into sections for bedside instruction in clinical medicine, surgery, and gynecology, and there was work in special departments at the University Hospital. A fourth year was optional and devoted to the various specialties.

The necessary textbooks had largely been written by the Pennsylvania professors—Ashhurst's *Principles and Practice of Surgery*, Pepper's *System of Practical Medicine*, Goodell's *Lessons in Gynaecology*. The day was packed solid with lectures from 9 to 2 P.M., beginning again with more lectures at 3:30, and laboratory work in the evenings. By the third year, the pattern shifted to clinics and ward rounds in the hospitals, but the work still ran straight through from 9 A.M. to 6:30 in the evening or later.

Said Edsall, looking back on those days:

> When I was a medical student the medical sciences were still largely taught by clinicians, who gave only a fraction of their time to these subjects; and they did this chiefly because of the prestige that teaching gave them in practice, not because they had any deep interest in the subject or had the slightest thought that it was a worthy life work. The school where I graduated was, I think, at the time that I entered it, the most prominent in the country, but like many others it encouraged large numbers of students chiefly because, with the methods then in vogue the costs were low enough so that with a large group of students the fees exceeded the costs of the teaching and at the end of the year the professors made suitable division of the excess and added it to their salaries. Medical faculties were in those days considered by the general university faculties to be inferior and in large degree mere vocational members of the university body.[11]

Large requirements for laboratories, equipment, and specially trained teachers were still to come. The medical school had had a laboratory building since 1878, but it was not until 1885 that they had enough microscopes for each second-year student in pathologic histology. At the University Hospital, when William

Osler was clinical professor there in the '80's, he owned the only microscope in the hospital and the first blood-counting apparatus in the city of Philadelphia.[12]

The new scientific medicine was only beginning to make its appearance in Edsall's student days. He recalled:

> The resistance of professional opinion to change of established beliefs is shown by the fact that as a medical student, twenty-five years after Lister's most essential work, I was taught by our distinguished professor of surgery that pus was often laudable. We did to be sure feel as students somewhat confused and cautious about accepting that because the Prof. of Clinical Surgery had some disturbing modern ideas and he taught that pus is always an evil, and since he examined us separately the situation appeared to a student, to be somewhat dangerous.[13]

On another occasion he recounted:

> Surgical antisepsis had, of course, been in use for some time and with great advantage, but that it was in its infancy is indicated by the pictures that I can still see of one of our Professors of Surgery operating day after day in the same unwashable cloth operating jacket, always the same jacket, and while he told us to use antiseptic methods, it was difficult to see in what efficient ways he used them himself, while the other Professor of Surgery, who was already an ardent advocate of antisepsis, would regularly scrub his hands thoroughly and after that would reach into his trousers' pocket for his pocket-knife, scrape beneath his finger nails, replace the knife, and proceed to operate.[14]

Edsall's senior professor of surgery was John Ashhurst, Jr.; the rising young rebel was J. William White.

Although the teaching may have represented the best of the old nineteenth century style of lecturing to students, it was soon to be superseded. "A most pernicious system," Osler called it, "bad for the teacher, worse for the pupils."[15] Didactic lectures, a theater clinic or so, and ward classes in physical diagnosis were not enough — "the students had no daily personal contact with patients."[16] Getting the students into the hospital, and giving them responsibilities for patients — the clinical clerkship, which owed so much to Osler's methods at Hopkins — was yet to come.

In general Edsall had little to say about the medical school of his student days. In his Autobiographical Notes he recounts what meant more to him: some outside work he did during a summer holiday and during his final year.

> One summer I spent six weeks in a Surgical Out Patient Department doing simply routine work, but this was useful in teaching something of the handling of patients, technique of minor surgery and the progress of minor ailments.
>
> The other voluntary work was offered to two or three of us by one of the ablest young men in Clinical Medicine in the University at that time, a man who has since become conspicuous in Clinical Medicine. This consisted essentially of opportunities to meet him in the hospital and hospital laboratories, where we learned much about gross pathology and many aspects of the newer methods of clinical and laboratory investigation of patients. The clinical laboratory methods were scarcely taught at all then to students and were only beginning to be recognized as important in Medicine. They had really been taken up by only a very few persons.
>
> Along with this work, we had, with the same man, evening conferences for the whole evening once a week or oftener, that were a sort of seminar. The whole thing was closely similar to what the British mean by tutorial instruction. I look back upon it as a really quite exceptional opportunity which was beyond all question the most valuable thing to me in the whole undergraduate medical period, because of the point of view of independent thought, reading, etc., which it gave. I selected Clinical Medicine very largely because of its offering opportunity for concentrated work without narrow restrictions, but also in considerable part because of the above mentioned contact.[17]

At this late date, there seems to be no way to establish for certain the name of this man so important in Edsall's life, and responsible in large part for the tutorial system instituted at the Harvard Medical School during Edsall's administration. The likeliest person seems to be Dr. Alfred Stengel, who graduated from the Pennsylvania medical school four years ahead of Edsall, became professor of clinical medicine in 1899 and succeeded Edsall as professor of medicine in 1911. Stengel became assistant in medicine to Dr. Pepper in 1891, and was quiz master and clinical clerk before becoming instructor in medicine in 1893–1894. Looking back from 1926, Edsall might well have remembered him at this time as "a young instructor."[18] Edsall and Stengel were close friends and colleagues in Philadelphia for many years and the breaking of their friendship later was a sad blow to Edsall.

This type of personal teaching was much on Edsall's mind when he came to Boston, and he liked particularly the way Dr. Harry Newburgh instituted a review course in the evening which was analogous to this. Edsall watched the work with great interest. Unfortunately, Dr. Newburgh could not develop it further at Harvard, as he went to the University of Michigan as professor of medicine, but Edsall carried it out later in a system which is still being used.

As a result of his extra work in clinical medicine as a student, Edsall later stated:

> When I graduated in medicine, my roommate and I were, I think, the only students among the six or seven hundred then in the whole of that school who had ever counted blood or examined stained specimens of blood. We had, but only because we had by chance close personal relations with a young instructor who had become interested in these new methods that were just coming in and he kindly taught us. There were no methods then of measurement of blood pressure in patients, the x-ray was unknown, bacteriological methods were, to be sure, beginning to be known, but they were little understood and hence little used . . .
>
> The diagnostic use of bacteriological methods was already possible in a few ways, but was little comprehended or employed. For instance, with that great disease, tuberculosis, ever about and in many cases suspected to be the cause of obscure symptoms, my able chiefs never once asked for an examination for tubercle bacilli during the whole time that I was a hospital intern, though that was already feasible.[19]

Edsall was already developing the habits of mind that were to take him so far. This is revealed in the way he handled his studies. He once wrote: "As a medical student I worked fewer hours than most of the other men and had more diversion. On nights before examinations my room-mate and I usually quit work very early, or did none, and played billiards, took a walk, or the like, with, I think, advantage to both of us in knowledge as well as in standing, since we both had done reasonably hard and concentrated work throughout the individual courses."[20]

There is another element important to mention in Edsall's

medical education, and this harks back to the old apprentice system in which many intangibles were transmitted on a personal basis from teacher to disciple. Edsall apparently had very close contacts with the two doctors of Hamburg after he reached an age to be interested in medicine. He said that he "spent a fair amount of time going about with two physicians in my home village, both of whom were exceptionally able men. In this way I learned their methods in the homes of patients, which were very illuminating, and I got many very sane, mature and wise ideas from them as to the general approach to the practice of medicine and the handling of people, and how to carry on when isolated from special expert advice. After entering the medical school I continued to do this at times through the holidays and also after graduating. When I had learned a little pathology, I did a number of autopsies for them."[21]

The town's doctors were Dr. Jackson B. Pellett and Dr. Joseph Couse. It is probably Dr. Pellett of whom Edsall wrote, "My contact with one of these men, who is still living, has continued and, although he is now nearly eighty years old, I look upon him still as one of the most stimulating and wise influences in my life."[22]

Dr. Pellett's son (also a doctor in Hamburg) remembers Edsall coming to call on his father whenever he returned home, dressed in white duck pants and carrying a cane. Dr. Jackson Pellett had read medicine with a Doctor Allen before going to the College of Physicians and Surgeons in New York, and he was one of the oldest graduates of the Columbia school. Apparently he envied Edsall his opportunities of study in Europe, although he poked fun at his continental medical year as "mostly a beer drinking course."[23]

On graduation from medical school, Edsall could have interned at either of the two well-known hospitals in Philadelphia, but instead he went to Mercy Hospital in Pittsburgh. His brother Frank was then practicing in the city, but that seems not to have been a deciding point. Edsall says he made the decision largely upon the advice of certain older men as to the advantages of the large and active clinical service to which he would be going. "This was I think absolutely wrong advice and my choice was one of my greatest mistakes," Edsall wrote later, and went on to explain:

The large clinical experience and the handling of great numbers of patients gave me much responsibility and the sense that I was learning a great deal, but the staff of the hospital was mediocre or poor, except in a few instances, as was the instruction in the course of the service. It led to very bad habits of superficial and rapid work that required afterwards a good deal of time and effort to overcome.

I would unhesitatingly recommend the choosing of a hospital on the basis of the thoroughness of work it demands, the character of the staff, and their familiarity with and success in teaching, as against the relatively slight utilitarian consideration of whether it is in a locality where the man expects to practice or the *amount* of experience and responsibility it gives.[24]

Edsall told me that it was unwise to think that because one saw a lot of cases one necessarily became a better clinician. "All you do," he used to say, "is get to be sloppy. It's a great deal better to see a few cases well than to see a vast number of cases of all sorts and not learn very much." In his later life he thought it was a great deal better to go to a place like the Massachusetts General Hospital where cases were very carefully studied.

In one respect his work in Pittsburgh had a value not perceived at the time. Among the many cases of all sorts were ailments directly connected with the taxing work of the steel mills, and in particular work under conditions of intense heat, so that a patient would come into the hospital with his shoes incrusted with salt from his sweat. Here Edsall saw his first cases of heat cramp, not then differentiated from heat stroke, heat prostration, and the like, but which came in later years to bear the name "Edsall's disease." Here also he became aware of the intimate relation between occupation and disease, which was to bear fruit in later years in his researches in industrial disease, a field which in the United States was making its first tentative beginnings and in which Edsall was to be a pioneer.

He was an intern at the Mercy Hospital in Pittsburgh and at the Rosalia Foundling Asylum until 1894, when he went abroad, studying in London, Vienna, and Graz (a noted Austrian medical center) for a year. If the Mercy Hospital experience had been a mistake, the year abroad was something else again. In his Autobiographical Notes he called it "perhaps the most important single thing, for my own development, that I ever did."[25]

It was during this period abroad that he saw the beginning of the new treatment for diphtheria in Vienna.

One of the most dramatic and affecting experiences that I have ever had was when I was working, a year after I had graduated, in Vienna and was daily going the rounds in the Children's Hospital, including the diphtheria wards. At that period diphtheria antitoxin came into use. We had seen in the same hospital wards a series of cases of diphtheria treated without antitoxin, and then the use of antitoxin began and we saw the marvelous difference. Old Professor Widerhofer, accustomed to the previous conditions all his life, day by day would stop amazed by the beds of children that a few hours before had been prostrate and dull and terribly ill, who were sitting up in bed, playing with toys and utterly transformed. His warm heart was always overwhelmed by the extraordinary change and he could only exclaim each day "Wunderbahr, meine Herren, wunderbahr." The reduction since then in the deaths from diphtheria is one of the happiest stories in medicine.[26]

When on his way home from Europe in June 1895, Edsall wrote to his mother about the achievements of that year.

I shall be very glad to start for home but very sorry to leave London for some reasons. However I find that the somewhat unsystematic way they have of teaching here is growing somewhat tedious now and I think I have spent quite all the time here that I could now to advantage. I hope I may be able, sometime in the dim future to come back here and to Vienna when I have been in practice some years and learn of some definite special things I may want to work up. As for general work I feel quite well satisfied that I have stayed long enough.

When I left home I felt that I knew too little to go into practice and take the management of lives upon myself without many qualms of conscience. My opportunities over here have been so very advantageous that I now feel that I have a right to go to work. No one can ever know enough of medicine but I feel quite satisfied that, speaking in all modesty, I know a good deal more from my five years study and experience than most men do from their two or three years and that I shall not, at any rate, be likely to deliberately kill many people without knowing it. Even though it has been a large expense I believe I shall be well repaid for it in manners, morals & general capacity.[27]

Perhaps more revealing than anything Edsall wrote on the subject of what the German influence could mean to a young American doctor, is a letter from Osler to his house officers at Johns Hopkins, which dates from 1890.

It may interest you to know the general impression one gets of the professional work over here. I should say that the characteristic which stands out in bold relief in German scientific life is the paramount importance of knowledge for its own sake. To know certain things thoroughly and to contribute to an increase in our knowledge of them seems to satisfy the ambition of many of the best minds. The presence in every medical center of a class of men devoted to scientific work gives a totally different aspect to professional aspirations. While with us – and in England – the young man may start with an ardent desire to devote his life to science, he is soon dragged into the mill of practice, and at forty years of age the "guinea stamp" is on all his work. His aspirations and his early years of sacrifice have done him good, but we are the losers and we miss sadly the leaven which such a class would bring into our professional life. We need men like Joseph Leidy and the late John C. Dalton, who, with us yet not of us, can look at problems apart from practice and pecuniary considerations.

I have said much in my letters of splendid laboratories and costly institutes, but to stand agape before the magnificent structures which adorn so many university towns of Germany and to wonder how many millions of marks they cost and how they ever could be paid for, is the sort of admiration which Caliban yielded to Prospero. Men will pay dear for what they prize dearly, and the true homage must be given to the spirit which makes this vast expenditure a necessity. To that *Geist* the entire world to-day stands debtor, as over every department of practical knowledge has it silently brooded, often unrecognized, sometimes when recognized not thanked.

The universities of Germany are her chief glory, and the greatest boon she can give to us in the New World is to return our young men infected with the spirit of earnestness and with the love of thoroughness which characterize the work done in them.[28]

Edsall returned to Pittsburgh and opened his office. Then followed that tedious dull period which plagues a beginning practice and which caused Dr. Conan Doyle to invent Sherlock Holmes.

"I found it impossible to do anything, in Medicine," said Edsall, "but sit in an office and wait for patients to come to me. There was no opportunity to do any out-patient or in-patient hospital work, and no suggestion even of laboratory opportunities in relation to clinical medicine. In fact quite against my own judgment I was drifting into Surgery because of a brilliant chance in practice and a good hospital opening in Surgery. However, I was so dissatisfied that after six months I moved to Philadelphia."[29]

CHAPTER 3

Young Doctor Edsall, 1895–1900

PHILADELPHIA had the advantage of being a live medical center and home ground to the Pennsylvania graduate, although at first Edsall found opportunities hard to come by. Nevertheless, he soon had a great piece of good fortune, which followed his landing an assistantship in medicine at the medical school. One day he was quizzing in medicine when Dr. William Pepper was there to give a lecture. They left the building together and Dr. Pepper said, "Edsall, drive in town with me." As the horse carried them along he asked, "Tell me about yourself." Next morning Pepper told him, "I need an assistant and I want to get you." [1]

Pepper paid his new assistant the compliment of working him hard. Edsall became Pepper's recording clerk.

> In this position [Edsall wrote] I had to prepare the clinical and laboratory records of the patients that he showed in his clinic, and had the run of his wards. He expected absolute thoroughness and accuracy and knowledge of the literature. The experience was extremely valuable for these reasons as well as for the extraordinary opportunity it gave for association with, I think, the most brilliant and the ablest clinician, and perhaps the most intelligent man I have ever known. It also gave me the opportunity, under excellent auspices, for working in the wards of the most carefully and thoroughly run hospital in the city. [2]

It was indeed his "rare fortune" as he called it that brought him in contact with such a man. He used to tell a story that showed what Osler thought of Pepper, representing as it did his own view as well. Osler had, of course, known Pepper well during his Philadelphia years.

> On one of my visits to Johns Hopkins [said Edsall] I made the ward rounds with Osler and his students. We spent a few moments at the bedside of an advanced gastric carcinoma, and as we walked on, Osler with characteristic mixture of whimsicality and seriousness sud-

denly caught the arm of the student and said: "Tell me what can put forty pounds on a patient with cancer of the stomach?" Hopelessly puzzled, the student said he did not know. "An optimistic consultant," said Osler. "William Pepper could do it every time, couldn't he, Edsall?"

. . . I have never seen a man live under such driving pressure, a pressure that killed him early; but with all the distractions of his life I never knew Dr. Pepper to neglect to give every one who needed it a word of cheer and a stiffening of courage, and with a gracious charm that no one who met it ever forgot.[3]

Pepper maintained his extraordinary pace on little sleep but frequent small catnaps. At the opera, or before a consultation, he would disappear for five minutes and return refreshed and seemingly inexhaustible.

Edsall became an associate in clinical medicine at the William Pepper Laboratory before 1895 was out. Here he made many of his closest friends. Among them was Alfred Stengel, who had begun in the same manner a few years before, then went on to marry Pepper's daughter and, rising fast, became professor of clinical medicine by 1899. Another dear friend was Sam Hamill, also a member of the Pepper Laboratory at this time and an instructor in clinical medicine, who became a distinguished Philadelphia pediatrician.

Another of this lively group at the laboratory was Alonzo E. Taylor, nicknamed "Coxey" after the radical leader of the time. In those days, Taylor and Edsall shared a common enthusiasm for studying the urine, and Taylor was analyzing every unusual sample he could get hold of. One day Edsall brought him a clear yellow specimen that he said deserved study and Taylor spent two days working it over and finally returned in desperation. "I can't make head or tail of this specimen," he said. "It isn't like anything I have ever gone over before." Edsall asked him what tests he had done on it and Taylor listed a long number. Finally he asked, "Have you tasted it?" Taylor replied, "Naturally not." Sticking his finger into it, Edsall licked a drop, and said, "Taste it. Maybe that will help you find out what its origin is." The answer, of course, was beer.

Taylor did not have to wait long for his revenge. One day Edsall brought in a large and handsome bath sponge, which he had bought in spite of its having a brown stain in one section. He intended to bleach out the stain in the laboratory, and then use the sponge for his morning cold bath. But before he could get to work on it, Taylor abstracted it briefly and fixed the brown stain with as permanent a dye as he could discover. Edsall spent two weeks bleaching that sponge before his friends let him in on the secret.[4]

Taylor and Edsall also shared a love of music, and Edsall used to tell how they once attended an especially fine concert but had been unable to sit together. When the performance ended, Edsall went in search of his friend whom he found still in his seat in the rapidly-emptying hall, head back and eyes closed, conducting over again his favorite passages.[5]

Edsall's first published paper, written with Dr. Pepper, came out in the *American Journal of the Medical Sciences*, in July of 1897, "Tuberculous Occlusion of the Esophagus with Partial Cancerous Infiltration." It is an interesting case report, showing as it does the results of examination by Pepper's informed fingers and senses, together with Edsall's analyses of gastric contents (the patient was being fed through a fistula in his stomach and the opportunity was not to be wasted), and in addition Edsall's encyclopedic collection of relevant cases from the medical literature all the way back to the Middle Ages.

This was followed in September, 1897, in the *University of Pennsylvania Medical Magazine* by a paper "On the Estimation of Hydrochloric Acid in Gastric Contents" by Edsall alone, and on January 22, 1898 in the *Philadelphia Medical Journal* by a further chemical study, "On the Estimation of HCl: Note on the Gastric Condition in Anemias." These papers dealt largely with the accuracy of the systems for determining the hydrochloric acid in gastric contents, which had only been studied since 1891 following the new use of the stomach tube. One had to differentiate between acid phosphate and hydrochloric acid in cases where no free hydrochloric acid could be found. This problem had occupied chemists for ten years, but even in this simple chemical problem Edsall had to devote much time to methodology.

This was the beginning of a series of fruitful investigations. The papers dealt with patients, using the relatively new techniques of the chemical laboratory to study their diseases. This represented a new approach to clinical medicine; in America, not even counting of blood was yet indulged in by pioneers like Richard Cabot. It was an active and exciting period.

As Edsall described his start in this field:

> I drifted almost completely for a couple of years without accomplishing anything immediate. It was only when I fell into close association with a man who had training in advanced methods of research that I began to make any progress. Obviously the most important thing to do is not to work alone or in casual contact with some other person, but to work in immediate association with a man of ability and excellent training and thus to have daily contact with his methods of thought. At the time I am speaking of this was difficult, for there was almost no one of that sort in this country doing research in clinical medicine. Really serious investigative work, other than straight clinical or pathological casuistic, was even thought by many of my seniors to be likely actually to interfere with a man's career and mark him as unpractical.[6]

This was a time when there was only one clinical laboratory in the University of Pennsylvania hospital and when very few people did chemistry. Chemistry in hospitals was just a new-fangled idea in spite of the fact that as early as 1855, for example, the Massachusetts General Hospital had a "Chemist and Microscopist."[7] Actually, beyond some pharmacological work, chemists did little in hospitals. In later years Edsall used to regale me with stories of the character of the laboratory work which was done in this early period at the University of Pennsylvania hospital. He had a tale of one man who didn't seem to do much work in the laboratory but kept publishing very interesting experiments with wonderful results. The discrepancy between work and results intrigued the other members of the laboratory, particularly when it was noticed that the several chemical bottles that he was supposed to be using in his observations were piled high with laboratory dust. A plot was laid to see if his fingerprints were on the bottles and before long it became clear that this young man wasn't doing any experiments. His associates then began to read his published articles carefully and found that they were translations from fairly recent German articles. In fact, he was acquiring a reputation for very little work.

The practice of medicine, too, was very different from what it is today. In a reminiscent mood, Edsall said:

When I entered medicine, diagnostic skill really meant chiefly ability to designate accurately what would be found in the body if the patient died, rather than what effect was actually being exerted by his disorder upon his life functions at the moment. As to prevention, we were so powerless in the vast majority of instances that hygiene meant for the most part quarantine of dangerously contagious disease, suitable plumbing installations, and the control of nuisances. The doctor's campaign at that time was largely a defensive one . . .

With all this, there has been a still more fundamental change in the mental atmosphere in which the doctor works. When I entered medicine, and for a considerable period after, the practice of medicine depended almost entirely upon the trained senses, especially sight, hearing, and touch, and the exercise of judgment, experience, and imagination in interpreting these results . . . Only if death occurred and there was an autopsy could it be actually determined whether many of our observations were correct . . .

Anyone who is vain of his powers of observation and their accuracy can easily have his vanity shaken by such simple things as, for example, comparing his estimates of the extent of anemia, made from merely looking at patients, with accurate blood examinations in a series of cases. And yet in even such simple things, the doctor had rather recently no methods of checking up . . . We were, for example, taught earnestly as students to estimate the tension of the pulse by feeling it. I have often, in later times with a group of able and highly trained physicians, had the individuals of the group estimate the blood pressure by touch, and then check it with a blood pressure apparatus, and have at times found the individual estimates to range 100 points wide of the actual blood pressure.[8]

Edsall never forgot the importance of the postmortem, with its final answers, a point of view which made him sympathetic to the exercise of the "Clinico-pathological Conference" which Cabot had originated in Boston, in which all the clinical means available were employed to arrive at a diagnosis, the pathologist then having the final word.

Edsall wrote in 1926 in his Autobiographical Notes: "During this time I took advantage of many opportunities to do post-mortem work in the hospital, and took much interest in securing autopsies on interesting patients who left the hospital to die at home. As compared with conditions at that time, I am much struck by the neglect many young clinicians show now of persistent training in gross pathology and autopsy work. The natural reaction against purely descriptive and diagnostic pathology has swung many so far as to make them overlook those ways in which it is essential to continue clinical training."

It is remarkable to see how in those early days Edsall was building for the future. A basis can be found for his remarkable knowledge of the literature and of rare and obscure diseases, for instance, when we realize that for seven years he did the reading for and the writing of much of the article on medicine in the Saunders yearbook, *The American Year-Book of Medicine and Surgery Being: A Yearly Digest of Scientific Progress and Authoritative Opinion in All Branches of Medicine and Surgery, drawn from Journals, Monographs, and Text-Books of the Leading American and Foreign Authors and Investigators.* The article on medicine had been written by Dr. Pepper and Alfred Stengel, and after Pepper's death it was signed Alfred Stengel and David L. Edsall.

> This required me to read literally all I could [Edsall wrote] of the available literature on clinical medicine and on related work in pathology, biochemistry, physiology, etc., in English, French, German and what little I could pick up in one or two other languages that I knew slightly. This had to be done as thoroughly as possible and, each year, when finished, tightly compressed into an article of one hundred and twenty thousand words, written in the form of a connected critical review. At the same time I did a good deal of abstracting, reviewing, proof reading, and other editorial work as well as the accepting or rejecting of articles for the American Journal of Medical Sciences and one or two other journals.* I mention this in some detail because the training it furnished was really quite striking in the power it gave of rapid, critical surveying and judgment of literature and the general information in medical literature that necessarily resulted. It was one of the most valuable parts of the training I had. I mention these things because I believe a fair number of young men could make a simple living in ways that give mental training instead of in ways that merely provide some income, if they seek such.[9]

By 1900 he was writing to his Princeton class secretary: "Be-

* *Philadelphia Medical Journal, University of Pennsylvania Medical Magazine.*

side practice I have been interested in hospital and research work, spending most of my spare time in the William Pepper Laboratory of Clinical Medicine and the Hospital of the University of Pennsylvania; in the former I am an associate; in the latter, assistant physician. In the past year I have been teaching in the Medical Department of the University of Pennsylvania as Instructor in Clinical Medicine. The other positions that I hold are physician to the Home for Incurables, to St. Christopher's Hospital for Children, and pathologist to the Methodist Hospital." [10]

Ten years later he could confess, looking back on this period: "The plan which I outlined for myself in the beginning of my career and which was chiefly to lead up ultimately to a teaching position, and consulting practice, made the first ten years of my work exceedingly difficult and of very doubtful outcome." [11]

Nevertheless, he seems to have had a clear idea of where he was headed, and he got there with remarkably few detours along the way.

"I did a fair amount of out-patient work in the earlier days," he wrote later, "and I have always looked upon that work, as then often done, as very hasty and superficial and largely wasted time except for the work done under the one able man mentioned above. Later, opportunities for hospital positions came rather rapidly and I soon found the chief thing I had to do was to restrict closely the number I took advantage of and resign the previous ones, in order not to spread out too thin. I have seen much damage done to many careers by taking on too many posts, with the feeling that each had something in it too good to let pass." [12]

"I saw a few private patients during these ten years," he noted, "but never had many to see . . . My time there was pretty consistently occupied for ten years afterwards with either the mornings given to hospital work and the afternoons given to investigative work or, after I got a major appointment in a hospital, most of the day given to the hospital for six months of the year and practically all of the day given to laboratory and clinical study for the other six months of the year." [13]

Another element in the progress he made was the schedule he followed. "I worked for many years usually from breakfast time until approximately midnight with little time given to anything else but some exercise. This was done not because of a sense of pressure but because practically all the work I had to do was interesting and was done as a pleasure. Distinctly less intense work would have accomplished much less, I think, at that stage of independent effort." [14]

This picture of a happy man should be balanced by the knowledge we gain from Edsall's Philadelphia friends that he was also subject to spells of deep discouragement and depression. Even in its simplest terms, Philadelphia presented a discouraging front to a man young, ambitious, and possessed of innovating tendencies. And Edsall was a complex and sensitive man.

A tempting line for Edsall in those years was pediatrics, which was just developing as a specialty. Edsall served for six or eight years as visiting physician in the St. Christopher's Hospital for Children and did a considerable amount of investigative work in pediatrics during that time. He followed this line, he says, "partly because I was interested in pediatrics and enjoyed it, and also because I thought it was a good training in the particular type of conditions I was interested in, namely nutritional diseases, dietetics, etc. It was valuable in this way, but I feel it came close to the mistake mentioned above, namely that many men have a tendency to try early to make themselves skilful in several lines with the result that they end by not being very skilful in any line." [15] This was not true in this shift in Edsall's career when he focussed his attention on the metabolic abnormalities of children and studied intensively in the laboratory these dramatic chemical changes.

Another very important element for Edsall's total development came at this time: his appointment as a district physician in Philadelphia. These were part-time appointments made by the city and carried an almost negligible salary. Edsall wrote of this experience:

I saw the very poor of a section of the city, both at my office and in their homes, all persons unable to pay for the services of a physician. At times I saw as many as five hundred in a month. It proved to be an experience of very little directly clinical value, though I had hoped in the beginning it would be, because the work had to be done under conditions that made accuracy and careful study impossible. It was, however, enormously valuable in the training it gave in meeting people as well as the conditions that they lived under, and in comprehending the social conditions of life and of medical practice. I would without hesitation recommend any man preparing to practice to get some intensive experience in the homes of the sick. I am convinced that the medical training will not be satisfactory until something of this kind is a part of it, preferably as part of the interne experience or immediately after it, rather than during the routine medical course.[16]

Dr. Geoffrey Edsall recalls his father saying that "his years in this work did more to clarify his sense of the importance of the relationship between medicine and the community than any other experience that he had had."[17]

It was also this period which set him strongly against alcohol, when he saw the disastrous wreckage of homes that resulted from its abuse. It was perhaps the most important source of his feeling that the world would be better off if there were no such thing as alcohol. However, he loved to tell on himself a story of a visit he made to one of the poor families, when he found the husband dead drunk on the sofa and the wife sweating away over her ironing board to keep the family going. When the young doctor said something sympathetic to the wife, criticizing the husband for wasting his money and his health and neglecting the family, the woman took after him with her hot iron, saying that her husband had a right to do whatever he pleased with his life, and chased Edsall out of the house.[18]

His work in the poorer districts took Edsall into the kingdom of the Vares, the political bosses who were so powerful they were known as the "Dukes of South Philadelphia." During one of the reform waves Edsall was persuaded by some of his friends to run for city alderman on the reform ticket. After being absolutely assured that he didn't stand a Chinaman's chance of winning the election, Edsall consented to run. He got something like 250 votes to his opponent's 2500—more votes than there were voters in the ward. This was standard practice in Philadelphia elections, where the art of fraudulent voting about reached its U. S. peak. The machine used to vote not only the Register of Births and the tombstones in the churchyard, but dogs, cats, even horses, as long as they had names.[19]

Edsall felt a contempt and disgust for local politics which was probably well expressed by his contemporary Owen Wister, who said, "Not a Dickens, only a Zola, would have the face (and the stomach) to tell the whole truth about Philadelphia." [20] Following the Civil War, Philadelphia had fallen into the hands of the "Gas Ring" who ruled the city from its stronghold in the Gas Department. A new charter passed in the '80's forced the Gas Ring to change its methods and began the decades of fraudulent voting by which the Republican machine kept control.[21] The three Vare brothers waxed fat as contractor-bosses in South Philadelphia (always the base of their power), and the strongest, William Scott Vare, was a power on the national scene until the Democrats swept the country in 1933. Until 1933 even the Democratic Party in Philadelphia took orders from the Republican Vare machine, according to Salter's study of boss rule.[22]

The Vares at least knew that their power was based upon service (of a kind) to the poor of their city. They stood for lavish expenditure on public improvements in South Philadelphia, and William Vare opposed child labor and supported workman's compensation, mothers' assistance, and other social reforms.[23]

Edsall used to tell a story about William Vare which he had from a medical colleague. Vare had visited the doctor during a time when the reform movement was more than usually troublesome and said to his doctor, apparently quite sincerely, "I don't understand why people don't appreciate what we give them. We give them 80% of value which is more than they'll find they could get anywhere else." [24]

Edsall's early adherence to the Democratic Party was no doubt strengthened by such experiences.

When Edsall first moved to Philadelphia, he settled on 16th Street between Spruce and Pine, in the same neighborhood as his friends Alonzo E. Taylor, A. H. Jopson, W. R. Nicholson, Joseph Sailer, and David Riesman. He and Dr. Alfred Stengel took their meals at a boarding house around the corner on Spruce Street, and it was there that Edsall met Margaret Harding Tileston of Boston, a teacher of mathematics at Miss Irwin's School in Philadelphia. Margaret Tileston was a graduate of Radcliffe. One of seven children, she came from a well-known Boston family. Her father, John Boies Tileston, was at one time a member of the publishing firm of Brewer and Tileston, at another period a member of the Tileston and Hollingsworth Paper Co. Her mother, Mary Wilder Foote Tileston, was the daughter of Caleb Foote, the editor of the Salem *Gazette.* This lady was the editor of several anthologies, of which the most famous was *Daily Strength for Daily Needs*, a best seller of its period and not unknown today.

Margaret Tileston must have been a truly charming woman, young, ardent, courageous, intelligent — and beautiful. Her obituary in *The Radcliffe Magazine* mentions that "the number of her friends was great and ever-increasing; constantly making new ones, she never lost an old one." [25]

Her friend Harriet Mixter wrote of her:

> She stood out as a thoroughbred might from among mongrels. There may have been heirs to empires there, but not one carried head and shoulders more splendidly erect or made you feel more surely that something uncommon was coming your way. Her walk was, indeed, one of the characteristic things about her. If she came to meet you, with her quick, eager step . . . you felt that she was the bearer of some precious message . . .
>
> The salient thing about her, of which you were always conscious . . . was a sense of unquenchable happiness . . . Yet at the same time she carried about with her the sense that each moment of her life was consciously dedicated to some serious purpose.[26]

In 1897 Margaret Tileston's parents were planning to visit her, and she wrote to describe the various diners at the main table of the boarding house (dinner 40¢). About Dr. Edsall she wrote, "Has sat next me for months & I like him very much. He is 6 feet four, and big in proportion — bearded." [27]

David Edsall began to pay attention to Miss Tileston. They went for walks. He sent her flowers. He took her to *Götterdämmerung* when Nordica sang Brünhilde. Margaret began to rely on his friendship, particularly early in 1898 when her father died after an operation for cancer of the stomach. He had gone into the hospital in January saying that he would only be there six weeks and then it would be March and the bluebirds would have come, but he never lived to see them.

In the spring after one of their walks, she wrote in her diary, "Dr. Edsall asked me my future plans, urged me to write and encouraged me about the success of my work at school." [28]

In March she wrote to her mother:

After rain last night we had a very mild day to-day, so that I wore no coat this afternoon. There was a wind, however, which is always troublesome when one wears a veil. Immediately after luncheon I went to the bank, my stock of money being reduced to 74¢. Isn't it a fine thing to have a bank account, so that one can draw whenever one wishes money?

B for my Banker – it's not that I'm proud,
But to talk of one's Banker impresses the crowd.

Then I took a bunch of violets and a few white grapes to Mrs. Dempster, my washerwoman, who has been ill for some days. Poor soul! She seemed very miserable, and I persuaded her to let me ask Dr. Edsall to go to see her. You know he is the city physician of this district, and he is glad to look after any poor people. Mrs. Dempster said that she didn't like to call a physician when she couldn't pay his bill, but I told her that there wouldn't be any bill. I had already spoken to Dr. Edsall about her, & I know that he will be glad to attend her. When I came back to the house I found that he had called. He knew that I always stay in till three except on the rare occasions when I go to the bank, which closes at that hour. He is engaged Monday, Wednesday, & Friday afternoons, so he must think me a pious fraud to be out when he was likely to come to see if I would go to walk.

I thought he'd come too! [29]

In May Margaret had her appendix out, which was followed by phlebitis, so that she was more than a month in hospital. After that she went home to Milton, where Edsall paid the family a weekend visit in September. He was still "Dr. Edsall" in her diary then, but overnight the situation changed. They became engaged, and the ardent wooer now was free to express his feelings, often writing to her twice a day. The engagement followed the leisurely Victorian pattern and was considered no reason to change Mrs. Tileston's plans for taking her four daughters on the Grand Tour. In October Margaret sailed for Genoa with her mother and Amelia, Edith, and Eleanor. Edsall met them in New York and helped to put them on the boat. Margaret gave him selections from Browning and he gave her a stickpin that had belonged to his aunt.

The Tilestons were abroad a year, traveling through Italy,

France, Belgium, and England. The letters followed her from city to city, letters that were kept carefully and read and re-read.

Edsall was busy during that year trying to put his affairs in such shape that he could maintain his future wife in the style he wished. In May he wrote of his plans for taking over a practice:

> I have been very busy since getting out my "removal" cards and the cards from Dr. Coley to his patients announcing his retirement from practice and my succession thereto. I can not say anything about the possibilities of my acquiring much of his practice but I think the outlook is fairly favorable. He has a dozen or more families in especial that I should like very much to retain for they are both people who pay well and very nice people.
>
> At any rate I am not risking much money and the only thing that troubles me, my darlingest, is that if the venture is successful it may involve our living in rooms there and I still fear that may be very unpleasant for you . . . we could measure our stay there by months and should therefore not need to anticipate a long time out of a home of our own if we could marry next year.[30]

In September of 1899 the Tilestons reached home, and in October Edsall was talking to his betrothed about his new appointment as assistant to Dr. John H. Musser who wanted help in handling his very busy practice. Of this short-lived venture he wrote later, that it came at a time "when I was imperatively in need of money, and although he was a most considerate man, after three months I threw up the thing and borrowed money because I felt that my future career was entirely cooked unless I could get out of that and devote myself almost entirely to my own development."[31]

From talking about a wedding "next year," they soon settled on the date of December 22 of that same year, 1899. Boston did not offer enough scope for wedding preparations and the bride went to New York for her wedding gown, though her white satin slippers came from Tuttle's in Boston and the wedding ring from Shreve's (the bride's for the groom, that is). The wedding veil came from Stearns' and was worn with a wreath of lilies-of-the-valley.

The wedding took place in King's Chapel in Boston and the ushers were Margaret's cousin Henry Wilder Foote, a prominent Unitarian minister; Orville Waring, who was a classmate of David

Edsall's at Princeton; and Dr. Stengel, who had shared the boarding house table in Philadelphia and was Edsall's close friend. The well-known musician, Arthur Foote, the uncle of the bride, was best man. Her brother Roger Tileston gave her away.

"The bride is tall and stately, and looked very attractive in a gown of white satin en train," enthused the newspapers . . . "The bride carried herself finely." After the honeymoon with friends in Richmond, Virginia, they returned to Philadelphia, living first at 1339 Pine Street, and in April 1900 moving into a house at 346 South 16th Street.

"It was an exquisite day, mild and bright, Easter and our first morning in the new house," wrote the new Mrs. Edsall in her diary. "We were in childishly high spirits."[32]

A few days later she noted, "I went to work at D's back office and spent much of the am. arranging the papers in piles and the pm. in sorting the magazines in order. We had a scrubbing woman. . . A man came to finish putting in the telephone. The 'Dr. Edsall' sign was put up."[33]

Edsall at the same time was excusing his delay in answering his class secretary's letter by explaining that he had been "excessively busy with my usual work and in purchasing everything from pins to pianos and fitting up a house."[34]

Here they were to stay comfortably until their son John was born in 1902.

CHAPTER 4

Fruitful Years in Philadelphia, 1900–1909

Despite the pressure of his work in these years, Edsall managed to share a happy family and social life. In general he not only dined at home, but had lunch there as well and afterwards would relax a bit, perhaps listening to music or playing the piano for his wife. Much of his daily round could be covered on foot, or in "the cars." Rural Hamburg was the favorite place for visiting with the children, and many summers were spent at Cataumet on Cape Cod. Mother and children would be settled on the Cape, but Edsall never felt comfortable for long at the seashore with its flat reaches of sand and water, and its damp relaxing climate. He loved the mountains best, and usually found time for a good tramping holiday with a congenial friend or two.

He and Margaret went abroad in the summer of 1901, and for part of the time were content to be sight-seeing tourists journeying through Italy—Naples, Rome, Florence, Venice, Belluno, Cortina, Toblach, and so to Graz in Austria. From June 12 to August 22 they remained in Graz, with Edsall working hard at the hospital every day under Professor Friedrich Kraus, who was director of the Medical Clinic. Edsall had been using a good deal of chemistry in Philadelphia and he wanted to be sure of his techniques, so this was his chief field of study. There were other interesting young professors there, Essherich, who later went to Vienna as professor, and Eppinger in pathology, while Kraus himself later went to Berlin. Edsall warmly recommended Graz to his brother-in-law Dr. Wilder Tileston, and Tileston went there to study later that year.[1]

Not only did Edsall make many stimulating contacts, but he strengthened his approach to the chemical studies of the urine which came in the following years. A paper that covered the work he had done abroad was published in the *University of Pennsylvania Medical Bulletin* on his return, "A Contribution Concerning the

Clinical Significance of the Readily Eliminable Sulphur of the Urine."

While they were in Graz, Edsall heard from Alfred Stengel that he had been elected to membership in the Association of American Physicians, then as now a premier medical organization.

After leaving Graz, they had a briefer visit to Vienna which was also put to good professional use. Margaret was already proving herself an ideal doctor's wife, and cheerfully amused herself while he was working, sometimes buying bread and a half kilo of strawberries to soak in wine and sugar which she and Edsall could enjoy together later.

Their first son John was born in 1902, and it is revealing that the best medical care of the day dictated that he should be born at home. Margaret had the care of a nurse experienced in midwifery, as well as the attendance of Dr. Norris (probably Dr. Richard C. Norris, assistant professor of obstetrics). Edsall gave the ether in the later stages and her diary notes: "Took ether badly and did not breathe for about two minutes and frightened D. who was giving it."[2] Nevertheless, mother and child did well.

Richard was born in 1905 and Geoffrey in 1908. Throughout the Philadelphia years, the household included a lovely nurse named Libis.

Edsall's advance in the University of Pennsylvania was steady, though not without difficulties and anxieties. In 1901 a humble instructor in clinical medicine and assistant physician to the University Hospital, in 1903 he became an associate in medicine, and in 1905 assistant professor, being thought of as a possible successor to his own professor of materia medica, Dr. H. C. Wood. Simon Flexner, concerned over necessary improvements in the medical school, had written to William Welch in 1901, "When Wood retires, which must be soon, it is by no means a foregone conclusion that a trained man will be chosen to succeed him in Pharmacology."[3] (Flexner originally wrote "scientific man" but scratched it out.) Nevertheless, when Wood retired in 1907, Edsall was chosen to follow him in a position renamed the Professorship of Therapeutics and Pharmacology. His first ten years in medicine were, as Edsall said, "exceedingly difficult and of very doubtful outcome,"[4] but by 1907 it seemed as if his future was assured.

The field of pharmacology and therapeutics was not one where the new light had penetrated very far; in fact, in Edsall's view, it tended to consist of a mass of ill-digested facts. Speaking before the American Medical Association in 1910, he alluded to a suggestion that the *U. S. Pharmacopeia* should be the basis of teaching "pharmacal therapy," but protested that many valuable new drugs were not in the *Pharmacopeia*, and further, "the Pharmacopeia contains so much at present that is absolutely non-essential that any attempt to teach what is in it would lead to still worse conditions than now exist . . . The too frequent lack of critical sense among practitioners in regard to pharmacal therapy may be largely attributed to the excessive demands made on the student in this branch."[5]

Theories of education which were to have a determining influence at Harvard some years later were already very clear in Edsall's mind. He continued before his AMA colleagues:

"In the elementary stages of the acquirement of knowledge, if time and energy are too largely used in remembering manifold facts, the pressure is too great to permit of their being reflectively and thoughtfully received and they come to be received, more and more, in a mechanical way; and, finally, dogmatic statements which require little or no reasoning for their appreciation become more acceptable than those that require thoughtful consideration, because dogmatic statements can be grasped in less time, and there is need for hurry if a great deal of ground must be covered . . ."[6]

Because of these weaknesses in the teaching of therapeutics, Edsall felt that "the average young practitioner [is] more sound in his diagnosis than in his treatment . . . He has formed the habit of accepting dogmatic opinions in the domain of therapeutics, while in other domains he has formed the habit of critical consideration of statements before accepting them."[7]

Edsall never had the reputation of a brilliant lecturer. We have a view of him at this time from Dr. Thomas L. Pellett, who was his student after he became a full professor. Although young Pellett had had every intention of going to Cornell Medical College, he was sent to Philadelphia because his father decided he had better go "where Dave was."

"I cherish the memory of little kindnesses he showed me," said

Dr. Pellett. "First after calling at his house I was invited with another local boy to a Sunday dinner, and once rode with him in his little old open car from the Medical Laboratories out the drive past a walking group of envious classmates."[8]

He was not the entertaining lecturer that Dr. Alfred Stengel was [Dr. Pellett added]. He was scheduled to give a clinical lecture on internal medicine to the 3rd and 4th year students from about 3:50 P.M. to 5:50 P.M., hours when our quest for medical knowledge was at a low ebb. We soon appreciated that he told about all his stuff in the first 20 minutes before he showed the patients. There was a handy exit up in the corner where the boys could slide out at opportune times.

One day he gave a talk on gastroptosis and visceroptosis, a popular subject at the time. A little skinny woman was wheeled in to be presented. The time was 6:00 P.M. As he sat down side-saddle on the edge of the bed (a terrible breach for a student to make), leaning over to percuss out the lower border of the stomach, a howl of laughter went up, and he turned to the student standing by and sternly asked, "What is the joke?" The fellow said, "They are wondering if you are going to stay all night."[9]

It is interesting to see how Edsall arranged his work and his life in such a way that he was functioning as a full-time academic medical professor on the clinical side decades before this new type of position had been accepted by the profession in general. But such single-minded application had already been developing on the scientific side, and it grew up among practicing doctors along with the research point of view. The application of scientific knowledge to the actual treatment of patients, the development of chemical and other tests that would tell the doctor more than his senses could tell him, the harnessing of the pure research in physiology, pathology, chemistry to the service of clinical medicine—this was the achievement of those first decades of the twentieth century. The doctors who had this point of view, and the training and ability to back it up, were few, scattered around the country in the main teaching centers. They all knew each other's work and cheered each other on.

Between 1900 and 1910 the young men began to form organizations where they could publish and discuss and carry their work forward. One of the most important of these was the Interurban

Clinical Club, founded in 1905 a year after the Interurban Surgical Club was begun. It included twenty-four men, six from each of the four leading medical cities of the United States, and at each meeting one or the other city played host and showed what new work they were doing. Members met twice a year for two days. The Club has been an important factor in the development of American medicine from the time it started. It represented the people who changed medical teaching from a clinical part-time occupation into the modern professional full-time professorships of medicine.

The members were largely chosen by William Osler, and the first meeting was held in Baltimore in April 1905, just before he left to take up his new chair of Regius Professor of Medicine at Oxford. It is worth mentioning the names of these men, for they brought into being the new medicine to which Edsall was devoting his life.

From Philadelphia:	David Edsall A. O. J. Kelly Warfield T. Longcope David Riesman Joseph Sailer Alfred Stengel
From Boston:	Richard Cabot Elliott P. Joslin Frederick T. Lord Edwin A. Locke Joseph H. Pratt Wilder Tileston
From New York:	Samuel W. Lambert Walter B. James Charles N. B. Camac Lewis A. Connor Frank S. Meara Theodore C. Janeway
From Baltimore:	Lewellys F. Barker William S. Thayer Thomas B. Futcher

Rufus Cole
Charles P. Emerson
Thomas McCrae

Another new society of which Edsall was early a member was the Society for Experimental Biology and Medicine, founded in 1903 by Graham Lusk and S. J. Meltzer. Edsall was elected to membership in 1905, and found himself in a group which included Walter B. Cannon, Otto Folin, Reid Hunt, and Ernest Tyzzer from Harvard—and men who were to play an important role as Edsall's co-revolutionaries at Pennsylvania in a few years, Richard M. Pearce, A. N. Richards, and Alonzo Taylor—and men who would fill the same role at St. Louis with Edsall, Joseph Erlanger, Philip Shaffer, and Eugene Opie—and from Baltimore, Harvey Cushing.

But neither of these new groups, nor the older Association of American Physicians (founded in 1886), gave the young clinicians the forum they needed, and in 1907 at a meeting of the AAP Meltzer picked up W. T. Longcope, J. H. Pratt, Edsall, and his brother-in-law Wilder Tileston on the boardwalk of Atlantic City and talked to them about forming a society of their own. Edsall was made temporary chairman and H. A. Christian temporary secretary, and the twenty-two men chosen for charter membership were the obvious leaders in this type of work. Thus was born the society now called the American Society for Clinical Investigation, in those days promptly nicknamed the "Young Turks" after the young revolutionaries of Turkey.[10]

That same year, the 1907 meeting of the American Medical Association brought forth a demand that the Association publish "a journal to bridge the gulf between purely scientific research and its application to practical clinical medicine." This led to the founding of *The Archives of Internal Medicine*.[11] Here Edsall served on the founding committee with Drs. George Dock, W. S. Thayer, Joseph L. Miller, Theodore Janeway, and Richard Cabot. Edsall was to have more serious contact with Dock within a very few years.

Another concern of the American Medical Association which held Edsall for many years was the Committee for the Protection of Medical Research, a defense against the anti-vivisectionists, which was appointed in 1908 with Walter Cannon as chairman.

Members included Joseph Capps, professor of medicine at the University of Chicago, Reid Hunt, then pharmacologist of the U. S. Public Health Service, Edsall (listed by error as professor of pediatrics at the University of Pennsylvania in Cannon's account), Simon Flexner, director of the Rockefeller Institute for Medical Research (a position he reached in 1903), and Harvey Cushing, brain surgeon.

> This group, with few changes, worked together for seventeen years [Cannon wrote]. Our first task was to examine the conditions under which experimental medicine was being conducted in the United States. Inquiry revealed that in a number of laboratories there had been posted for many years regulations defining the humane treatment of animals used for experimental purposes. We collected these scattered regulations and generalized them . . . They provided for caution in using stray cats and dogs by arranging a delay before their use, at least as long as that which was customary in the local pound; they stipulated the kind of care in the housing and feeding of the animals; they demanded use of anesthesia when the operative procedure involved more discomfort than that of giving the anesthetic; and they provided for putting the animals to death before recovery from the anesthetic unless the director of the laboratory authorized recovery for the purposes of the experiment. These regulations . . . were adopted by corporate action of medical faculties and medical research institutes throughout the U. S. . . . The members of the Committee were convinced that these regulations were chiefly valuable in assuring the interested public that the procedures in animal experimentation are conducted in a humane manner." [12]

In 1908 Edsall was also appointed to the AMA Council on Pharmacy and Chemistry, set up in 1905, which had published its eminently useful handbook "New and Nonofficial Remedies" in 1907. Edsall's report of the work of this Council appeared in the *Journal of the American Medical Association* in November 1910, and gives a picture of the behavior of an unregulated drug industry which was a worthy mate to those money-grubbing medical schools exposed by Flexner in the same year.

Edsall said in opening his speech:

> I do not need to offer reasons for the existence of the Council on Pharmacy and Chemistry . . . We all of us know the deplorable state

of things that led to its establishment. When lay journals could show that in centers of learning approximately one-half of the prescriptions filled were for proprietary remedies; when it was likewise evident that in nearly all instances these proprietary remedies were supervised as to their actual composition, their activity or potency, and all other important facts, by no one who was disinterested, and that we were dependent on the manufacturer or agent for all knowledge regarding them; and when it had already been shown of many of those that had been examined that a part or all the statements made regarding them were more or less completely false, it was quite clear that we could not maintain the honor of the profession or fulfil our duty to the public unless a complete revolution were carried out, and authoritative information obtained regarding all preparations intended for the use of the profession, and unless all those that did not conform to certain just and honorable demands of the profession were definitely discarded.[13]

Originally the Council had planned to investigate proprietary drugs only, but soon had realized that it would have to analyze "all substances, proprietary or other, that are at all widely used medicinally or that have even any reasonable appearance of medicinal value and are not contained in the U. S. P."[14]—a daunting and tremendous undertaking indeed. The field was broadened to include medicinal foods, organ extracts, digitalis preparations, forms of organic iron, serums, vaccines, and mineral waters. As a result there was a major investigative campaign undertaken which spread country-wide and involved chemists, pharmacists, pharmacologists, bacteriologists and hygienists, to say nothing of the clinicians on the committee on therapeutics, and behind them, the "whole thinking body of the medical profession." At first it had appeared that there would be a "drawn battle between the power that comes from many millions on the one side and the force that is given by a righteous cause and the support of the American Medical Association on the other."[15] But the manufacturers' opposition soon went underground.

Edsall continued, "A handful of the profession have at times taken exception to the decisions, sometimes because they belonged to the class of people that always fight for the under dog even though he be at times a quarrelsome mongrel; sometimes from far less altruistic motives, such as financial interest in the preparation or some interest in a medical journal that was advertising the preparation."[16]

The Council went through the drugs and medicaments one by one, checked and often tested the manufacturer's claims in the AMA's own Chemical Laboratory, and produced the immensely valuable 250 pages of "New and Nonofficial Remedies." After that new drugs were given the same careful going-over and the results published in a column which still appears in the *Journal.** I remember when I was Edsall's resident in Boston, he used to turn over to me the preliminary work on his share of this which came in voluminously every week. The chief problem in those days, before the First World War, was that many of the drugs were obviously silly, being ineffective if not harmful.

Edsall continued: "Previously the good, the indifferent and the bad were all in one class — we knew nothing reliable about any of them . . . The spectacle astonishes one now that it can be viewed from the perspective of growing distance: a learned scientific profession doing about one half of what is its chief work — treating the sick — on the strength solely of information received from advertisements and detail men."[17]

Drugs which were approved were listed in *New and Nonofficial Remedies.* There was no list of those rejected, as a blacklist would be illegal as well as unnecessary. However, information as to frauds or controversy appeared in the reports of the Council, of the Chemical Laboratory, and in "Propaganda for Reform in Proprietary Medicines."

The general level of ethics in the industry came in for some excoriating comment:

* This burst of activity resulted in a number of other essential publications. In 1913 appeared the first edition of *A Handbook of Useful Drugs,* subtitled "A selected list of important drugs suggested for the use of teachers of materia medica & therapeutics and to serve as a basis for the examination in therapeutics by state medical examining & licensing boards." This handbook, revised every year or so, was based on the preliminary list of the Council of Pharmacy and Chemistry, published in 1911, and was continued in print until 1952. To fill the gap after 1952 an annual *Handbook of Drugs* was planned, but this has not yet appeared. However, 1965 saw the first edition of *New Drugs,* reprinted in '66. Also dating from 1913 is the first publication of *Useful Remedies: An epitome of the properties and uses of the articles included in the list of important medicaments,* and in 1916 appeared the first of many editions of the *Epitome of the Pharmacopeia of the U.S. and the National Formulary.*

The Council is asked why it accepts anything, even though all right in itself, from firms that have once deliberately tried by direct or indirect means to deal unfairly or dishonestly with the profession . . . It is undoubtedly very trying and at times seems degrading to have relations with people when they have once been shown to be willing to adopt falsehood or dishonorable subterfuges in order to accomplish their ends . . . Such practices would in ordinary life rapidly undermine the confidence of the people in a business house with which they dealt; but the medical profession is trained, I think, into a large tolerance of human frailties, and this sometimes goes much too far. Such action as this the profession has not yet shown itself ready to follow. Were such a rule made now and enforced, even with great moderation, New and Nonofficial Remedies would be a very thin volume . . .

At present it is the custom of nearly all manufacturers to present any important article and especially any new articles that need the support and interest of the best men in the profession, to the Council in conformity with its rules, and then under this virtuous cloak to offer anything else they wish in any way they wish to the riffraff that is a part of our profession as well as of any other . . . We often see a firm that has articles in New and Nonofficial Remedies using most outrageously unethical methods in exploiting other preparations. In other words, ethics has no real place in the business policy of any except a very few firms.[18]

Edsall's appointment to this fighting arm of the AMA, his membership in new medical societies and on the boards of new journals, showed the respect in which he was held not only in Philadelphia but nationally. Much of his national reputation was based on his published work—over seventy papers appeared between 1897 and his departure from Philadelphia in 1910. Many of these papers came from his work in the Pepper Laboratory. He really did a fantastic amount of laboratory work for that time; the methods one used for chemical determinations were new and required using systems which were not standardized but had to be carefully worked out in the laboratory and refined. Edsall made an extraordinarily large volume of observations when one thinks of the difficult techniques involved. Edsall's reputation as a pioneer in medicine was well earned, not that he made extraordinary discoveries but that he was opening up the field of laboratory investigation in clinical disease at a time when this was very new and relatively unexplored.

His chemical studies expanded into some nine further papers on the estimation of various factors in the urine, after his return from Graz. In a few years he had made such a name for himself in this particular line that in 1906 he appeared in Keen's book on surgery as author of the chapter on "Examination of the Urine in Relation to Surgical Measures." His main concern in this article was to moderate the enthusiasm of his colleagues for the power of chemical analysis to provide definite answers. "Since I cannot entirely share the enthusiasm of some writers . . . but approach their discussion with the feeling that modern methods of examining the urine have in but few ways furnished really satisfactory additions to our diagnostic powers, my task is not altogether attractive . . . expecting clinical-laboratory methods to furnish specific facts of conclusive value in diagnosis and prognosis . . . is, of course, entirely out of the question."[19] This statement is not too surprising when we read his description of the difficulties of testing even for sugar or albumin.

On the other hand, a good example of what Edsall could do with the chemical means at his disposal occurs in "A Brief Report of the Clinical, Physiological, and Chemical Study of Three Cases of Family Periodic Paralysis," by Drs. John K. Mitchell, Simon Flexner, and D. L. Edsall.[20] Edsall was responsible for the chemical studies in these interesting cases, and was able to show to a certain extent what was and was not happening before and during these transient attacks of paralysis. While not attempting a definite answer, the paper focused attention on the metabolism of potassium, the attacks having been definitely shortened by the taking in of large quantities of potassium. This was a most interesting method of benefitting this rare disease, and a method which has subsequently been shown to be of value. A remarkable paper, so early to relate the disease to potassium abnormalities.

In the following year he and Caspar Miller published "A Contribution to the Chemical Pathology of Acromegaly," reporting on some very careful metabolic studies made on two inmates of the Pennsylvania Training School for Feeble-Minded Children. Their figures in general demonstrated a striking retention of phosphorus and nitrogen, less marked retention of calcium, which led them to the view "that there is in acromegaly a growth of abnormal bone rather than a mere abnormal growth of bone, and that there are very marked abnormalities in the soft tissues, as well as in the bones . . . chemical study of the bones, as well as chemical studies of the disease in general, may . . . demonstrate definitely that the alterations in the bones are the result of metabolic abnormalities."[21] This is a rather astute conclusion for that early period of 1903, but no more than what one might expect from a man who was deeply interested in metabolism.

The greatest number of Edsall's papers dealt with questions of metabolism. It is easy to understand why Edsall's discussions deal with urinary analyses and neglect the analysis of the blood, because in those days satisfactory techniques for blood analysis were undeveloped and difficult. The handsome techniques for these studies developed later by Folin and Benedict did not then exist, so one must recognize with considerable enthusiasm Edsall's understanding of metabolic abnormalities in man. His interpretation of their importance is quite remarkable. It is also clear from the symposia in which he was asked to participate that his work had given him considerable reputation; clinical advances then were largely based on bedside observations, so that his laboratory contributions represented a considerable advance.

Among the metabolic papers were a number of precise studies on the value of rectal feeding, then much in vogue. Edsall reported on a number of cases with Alfred Stengel and Caspar W. Miller, doing very careful studies of urinary and fecal excretion with diets largely consisting of eggs and milk administered by rectum, and coming to the conclusion that one couldn't adequately feed patients on this regime, that it was largely water which was absorbed and accounted for the patients' apparent improvement, and that analyses of protein and fat in urine and feces showed the woeful inefficiency of this form of nourishment. These were well-documented studies and disagreed with most of the other investigations of the moment which suggested that rectal feeding had certain merits.

These studies bore fruit later in his reports on proprietary foods, particularly in June 1909 when he excoriated the shortcomings of widely popular special foods before the American Medical Association meeting. He also had some things to say about his profession's knowledge of dietetics in general.

Many practitioners do actually make skilful use of drugs [said the Professor of Therapeutics and Pharmacology] . . . the intelligent man . . . knows how to obtain logical reasons for making his choices in drugs . . . With hygiene and diet, however, there is very commonly an actual inability to reach a rational preliminary decision as to what general line should be followed.

I have had some interest in learning with what degree of freedom and ease a large number of practitioners, especially recent graduates who have modern training, make use of dietetic measures. I think I am right in saying that a large majority of them believe – and, I imagine, believe rightly – that . . . they can accomplish much more with drugs than with diet even in those cases in which they recognize the especial suitability of dietetics . . . In dietetics, aside from a few conditions such as diabetes, in which specific things can be stated, their knowledge usually consists, on the one hand, of some physiologic facts which have not been presented to them in such a way as to adjust them very clearly to clinical procedures, and, on the other hand, of a more or less considerable mass of practical details which have usually been presented to them almost purely as arbitrary measures and their trend not very well adjusted to the physiologic facts . . . Indeed, in dealing with advanced students or even with practitioners who have had no special training in dietetics, I find that they are likely to be unable to give any regulations that could not be equally well or better given by a wise housewife; and the reasons for the regulations that they do give are generally not at all clear, while on the other hand they do very frequently know why they order any particular drug.[22]

Edsall was sometimes embarrassingly modern.

One of his main points about diet was that one should treat a disturbance of function, not a disease – a keystone of his approach to disease throughout his life. If it is a truism today, this is due to the concepts developed by Edsall's generation.

The fame of his metabolism work led to his being asked to deliver the Harvey Lecture in November 1907, when he chose for his topic "The Bearing of Metabolism Studies on Clinical Medicine." It is very interesting that in a period of only ten years Edsall should have taken the chemistry of the urine and stool and developed in the field so far that he was asked to give a Harvey Lecture. This was done before good methods of analysis were devised. Each paper consisted of only a few cases, many of them rare, but by dint of terribly hard work and careful analyses, Edsall was able to arrive at many definite conclusions. It is remarkable that he could get so many observations when apparently methods in use were extremely poor, not nearly as good as those which came shortly thereafter.

Another very interesting group of papers concerned his growing awareness of the dangerous and unappreciated effects of X-rays. In the few years since its invention, the medical profession had not had time to evaluate this mysterious force; Holtzknecht had only invented his measure of dosage in 1902. By 1905, when Edsall's first paper on the subject (written with Dr. John H. Musser) appeared in the *University of Pennsylvania Medical Bulletin* the medical profession had become aware of the danger of skin burns, but more general and serious reactions were just beginning to be reported. Edsall had been making metabolism studies on a number of patients, among them two of Dr. Musser's patients who had leukemia, and was checking the intake and excretion of nitrogen and phosphorus, and the uric acid and purine in the urine. Therefore he was in a good position to detect the more profound effects of X-ray treatment. The doctors reported: "Our investigation of the changes in metabolism that follow x-ray treatment of leukaemia have demonstrated in a striking manner that the influence upon metabolic processes is extremely rapid and very profound, and our results emphasize most forcibly the possibilities in x-rays for the production of both good and bad results . . . The effect upon metabolism that we observed is probably unequalled by that known to be due to any other therapeutic agent." [23]

By June 1906 Edsall was reporting to the American Medical Association on a "somewhat extensive series of studies of the effect of the x-ray on metabolism in various medical conditions that I have made in the last eighteen months," his own experience and the work of others leading him to believe that "the influence on metabolism is, in some instances, most violent; that it occurs even in normal persons, and that apparently when a general influence is clinically evident it is always associated with decided changes in metabolism . . . due largely, if not entirely, to tissue destruction." [24] He had been greatly impressed to find that a single "mild" exposure could have a profound influence on metabolism for days after. In one case, a patient with pernicious anemia was given an X-ray exposure over both thighs, which apparently caused death in a short time.

Another group of papers which Edsall produced concerned the management of hospital wards, with particular reference to cross-infections. Edsall had risen to the headship of the medical service at the Episcopal Hospital in Philadelphia where he was in charge for several months in the year on the rotating basis by which services were then handled. He lamented the fact that whereas in surgical wards the chief's "system" was carefully followed and any deviations were quickly detected because of a rise in deaths or infections—in medical wards, on the other hand, cause and effect was less easy to see, and therefore the incentive to work out and adhere to a proper system was less impelling. But Edsall carried the turn of mind of the investigator with him wherever he went—as he said later to the Wisconsin State Medical Society:

> I do not feel the slightest sympathy with the frequently expressed fear that the tendency at present is to train up investigators, rather than clinicians, my reason being the very definite one that, to my mind, a sound clinician is essentially an investigator, whether he works in a laboratory with all possible facilities or in hard country practice; whether he conducts extensive studies of special questions or simply attempts to learn as precisely as possible what needs to be done for each separate patient. In the latter case, as much as in any other, he will be successful and useful in direct proportion to the degree to which he has been trained into the point of view and the methods of a careful investigator.[25]

One of the first of Edsall's observations concerned the relapses and perforations in typhoid fever, which he traced in large measure to an infected milk supply. Certainly when he saw to it that the milk was pure to begin with (relatively speaking), was kept clean and cold, and that the nurse who handled it and who handled the special diets did that work exclusively and had not just come from scrubbing the bedpans, he saw a remarkable drop in these complications of typhoid. Through the years he reported on "A Small Series of Cases of Peculiar Staphyloccic Infection of the Skin in Typhoid Fever Patients," "A Small Hospital Epidemic of Pneumococcus Infections" (with Ghriskey), "The Influence of Infected

Milk in the Diet of the Sick—particularly in acute infectious diseases," "The Hygiene of Medical Cases, Particularly in Hospital Wards," and "Prophylaxis against Infectious Diseases from the Standpoint of the Practitioner." In describing his system at the Episcopal Hospital, he paid particular tribute to the chief nurse, Miss Payne, and to Miss Marion Smith, who had adopted many of the same regulations at the University Hospital.

Once he had reduced infections due to contaminated food, he found that the next most likely route for germs to travel from bed to bed was through the nurses themselves and he made a number of changes in details of nursing—the manner in which the nurses cared for the patients' mouths, for example—and trained the nurses in methods which protected them as well as the patients. He adapted a bedpan sterilizer which could be used on the ward, considered the fly problem, and forbade infectious patients to help nurse each other.

He developed a system of sorting and placing the patients by which he set great store. Typhoid being the main infectious disease, he kept his typhoid patients together in the main ward, away from other patients. In particular, patients with acute gastroenteric attacks, dysentery, etc. were kept away from the typhoid patients and if possible isolated in a side room off the main ward. Tuberculosis patients he also tried to keep out of the general wards. Pneumonia he regarded as highly contagious, and of particular danger to the typhoid patients, so he isolated cases of pneumonia where possible, or screened them from the rest of the ward. The general trend at that time in U. S. hospitals was to isolate the typhoid patients, but with his Philadelphia experience Edsall did not find this practical, and remarked, "I prefer to use my isolation space, first of all, for other diseases, particularly pneumonia. I have also had so many typhoid cases that I could not have isolated them unless the main ward . . . had been used for them alone."[26]

Apropos of the great amount of typhoid in Philadelphia, where he sometimes saw 1500 cases in a year, Edsall used to tell me a story which seems almost apocryphal, but he told it to me in all seriousness. He said that at the time when the Philadelphia water supply was to be purified, some of the doctors in Philadelphia objected on the grounds that most of their practice came from typhoid patients and that purification of the water would eliminate typhoid and reduce their practice. That horrified Edsall and certainly horrified me. It was an example of the things that disillusioned him with Philadelphia.

Edsall added another wry footnote to this story:

> I made a simple calculation, for example, of the approximate economic effect of water filtration (through its influence upon typhoid fever) in Philadelphia when I lived there. An appalling number of cases of typhoid fever had been occurring in that city for a long time. Within a short period, water filtration so largely reduced this that, considering only the most obvious and definite economic results of the disease, such as the definite loss of wages, the cost of attention by physicians and nurses, the cost of hospital care, it became evident that on an expenditure for the filtration plant of between \$15,000,000 and \$20,000,000 (with all the undoubted graft included in that cost) there was nevertheless a return of roughly \$5,000,000 per year of economic betterment to the city. Clearly the original cost was rapidly absorbed, and after that it was all financial advantage, aside from the humane aspects of the matter.[27]

No wonder he appreciated the initial use of the Widal test for typhoid. "I remember keenly the excitement, with dreadful numbers of cases of typhoid fever always around us, when the Widal test for typhoid came in and seemed likely to prove able to aid us greatly, as it has done, in the accurate diagnosis of the presence of that disease. We were wisely taught at that time that if we studied typhoid fever and its diverse complications thoroughly we would know most features of acute disease in general."[28]

Specialization had not really gone very far in this first decade of the twentieth century, if Edsall, who had nearly become a surgeon in Pittsburgh, could rise to be president first of the Philadelphia Pediatric Society in 1908 and then in 1909 of the American Pediatric Society. Edsall was not a dedicated child specialist in the modern sense, but he had a vision of the importance of the study of pediatric disease against the larger background of general medicine. He used to say that in pediatrics you saw simple disease pictures without the complications resulting from the multiple illnesses of longer life, degenerative disease, and so forth. Therefore pediatrics provided an ideal way to study disease, one that would prove informative and clarifying to one's understanding of it.[29]

Such crossing of the lines was not so unusual in those days. Osler, Forchheimer, Musser, and Pepper were all members of the American Pediatric Society. Edsall published a number of interesting papers on clinical problems, chiefly studies made along his particular line of metabolism and diet: "Some Biological Differences Between the Natural and Artificial Feeding of Infants," "Concerning the Accuracy of Percentage Modification of Milk for Infants" (with Fife), "The Dietetic Use of Pre-digested Legume Flour, Particularly in Atrophic Infants: with a Study of Absorption and Metabolism" (with Miller), and a paper relating his observations on children to more general considerations—"Observations Relating to the Nature of Atrophy of Intestinal Origin."

Edsall's main contribution to pediatrics was his belief that it was a major clinical subject, a point of view which was unusual when he put it forward. In 1910 he helped to set up one of the first full-time pediatric chairs in the country at Washington University School of Medicine in St. Louis.

As Edsall said diffidently in his presidential address to the American Pediatric Society, "I am in a somewhat dangerous position . . . Since my own work has come to be more largely outside the field of pediatrics than in it I may easily find myself presenting much the same appearance to you, the masters of pediatrics, as does the occasional maiden lady of philanthropic tendencies who would instruct mothers in suitable methods of training children."

This did not prevent him from stating, "In spite of all that has been done, proper dignity and regard are not yet generally accorded in this country to a subject which, it appears to me, is unquestionably the broadest and most complex and the most important division of general medicine, and proper standing will not be generally given to pediatrics in this country until better things are demanded for it by concerted action, and until also better things are in certain ways demanded of it."[30] *

He lectured his Philadelphia colleagues more freely:

In the past year, and in preceding years, there have been, among

* From Annual address of the President, American Pediatric Society, May 3, 1910, J. Pediat. 47: 809–816, 1955.

many interesting and instructive contributions, very few that added new knowledge or threw additional light upon matters through serious studies at the bedside or investigations in the laboratory. The contributions have been in too large proportion casuistic; they have been in too small proportion definite studies; and pathology, which is the bulwark of all clinical progress, has figured very little in our proceedings . . .

It is . . . not a criticism of pediatrics in Philadelphia alone, but is, I think, a fair arraignment of pediatrics everywhere in this country . . .

It is difficult for me to understand why pediatrics has not been a more fruitfully cultivated field than it has in this country. I have given special consideration to this matter recently; and I am unable to see why, with increasing numbers of young men constantly being trained in a way that makes them capable of being useful investigators . . . so few go into pediatrics with any deeper purpose than to gain a working practical knowledge of the subject; while, comparatively speaking, so many go energetically, not only into general medicine and surgery, but into neurology and other specialties. And yet, pediatrics is almost as broad a subject as general medicine.[31]

This is also the period during which Edsall made his reputation as an authority on industrial diseases. In Pittsburgh, at the start of his career, he had seen workers in the steel mills suffering from the effects of heat. His work as district physician in Philadelphia also brought him into contact with industrial workers and their particular ailments, while his wide reading in the foreign literature alerted him as to what to look for. England, with its many government studies of health in various industries, and Germany, with its compulsory health insurance of workers and its careful statistics, were both far ahead of the United States, where laissez-faire was still the rule, and any causal connection between the work and the state of the worker's health was overlooked or denied in many dangerous trades.

Edsall's first paper in this field appeared in the *American Journal of the Medical Sciences* in 1904: "Two Cases of Violent but Transitory Myokymia and Myotonia Apparently Due to Excessive Hot Weather," what are now called "heat cramps." One patient was a locomotive fireman, the other had become much overheated while working in the sweltering weather of a Philadelphia July. At this point the cause of the muscle spasms was a mystery, and Edsall pondered a disturbance of the body chemistry caused by a possible digestive upset, or the effects of heat on the skin in some kind of reflex irritation, noted a parallel with tetany (then also a mystery), effects of loss of fluid, possible arteriosclerosis, and a number of more esoteric possibilities, citing German and other foreign sources.

By 1908, he was at least able to delineate a disorder new to the medical literature, in a report to the American Medical Association entitled "A Disorder Due to Exposure to Intense Heat, characterized clinically chiefly by violent muscular spasms and excessive irritability of the muscles. Preliminary Note." In 1907 some further cases had appeared in his wards in the hot weather, and following up the heat clue at an Association of American Physicians meeting, and among doctors with patients employed in occupations that exposed them to intense heat, he found "that these physicians had all repeatedly seen such cases, many of which had been very severe and some even fatal. I found, too, that the men themselves who are engaged in such occupations, as well as their employers, are extremely familiar with the disorder," making a situation common in industrial diseases, when a popular clinical entity was entirely unrecorded as far as the medical literature went.

Having followed up numerous cases, he was now able to state: "it is, of course, a different disorder from heatstroke, and clearly quite different from heat prostration. It appears also clinically to be different from any other recognized forms of muscular spasm."[32]

Edsall's finding of high quantities of creatine in the urine of patients with heat cramps was questioned by Harvard's great chemist Otto Folin, who wrote to Edsall about it. But when Folin checked it, he found Edsall was correct.[33]

In 1909 Edsall gave a further report on the subject before the Association of American Physicians. He was still looking for the source of the disease in some kind of nervous disorder, though he had found in studying the urine of two cases a remarkable absence of chlorides. In the discussion he stated that the Navy had reported good results in using enemas of salt solution, though he had not had a chance to try it himself, but he termed it "the most rational suggestion that has been made."[34] The clue was in his hand, but was some decades in unraveling. In the '20's in Great Britain J. S.

Haldane was working on the same problem with miners, and ascribed the cramps to salt depletion, and their prevention to giving miners salt water to drink. In the late '20's D. M. Glover was using salt in plants in Cleveland,[35] but as late as 1934 at least one large steel company was advising against the use of salt.[36]

Like pediatrics, the field of industrial diseases had an intellectual as well as practical and humanitarian appeal to Edsall.

> I have long ridden a hobby [he told an audience at Johns Hopkins in 1918] that has the fascination of not only being an amusement, but of possessing both clinical and research interests and also large human interests, relating both to individuals and to groups of people. Industrial disease combines, to an unusual degree, the charm of the study of disease itself, of its etiology, and of the wide human relations of medicine, and it brings one into touch, besides, with an unlimited display of constantly changing technical processes that exhibit the ingenuity and resourcefulness of man to the full . . .
>
> I have always felt grateful to Sir William Osler for asking me to write certain articles for the original edition of "Modern Medicine," a task for which I felt myself very ill-suited, but which gave me a stimulus without which I should never have learned that a factory is a genial garden to the lover of etiology, with much fruit on many a tree.[37]

Edsall's contribution appeared in 1907 in the first edition of Osler's many-volumed work: *Modern Medicine: Its Theory and Practice in Original Contributions by American and Foreign Authors, Vol. 1.*[38] Edsall provided four chapters:

Chronic Lead Poisoning
Chronic Arsenic Poisoning
Other Metallic Poisons, Mercury, Phosphorus, etc.
Carbon Monoxide Poisoning, Illuminating Gas Poisoning, Combustion Product Poisoning, Chronic Carbon Bisulphide Poisoning

By 1910 he had such a reputation in this field that Professor Henry W. Farnam of Yale, the president of the American Association for Labor Legislation, appointed Edsall to a five-man committee which was to acquaint the President of the United States with these problems, "emphasizing the urgent necessity and practical expediency of a national expert inquiry into the whole subject of industrial or occupational diseases."[39] This organization had been active since 1908, investigating diseases such as phosphorus poisoning in the match industry and stirring up legislative action on the use of phosphorus, lead, mercury, the reporting of industrial diseases, and other reforms.

When he was invited to address the annual meeting of the Wisconsin State Medical Society in 1909, Edsall picked "Some of the Relations of Occupations to Medicine" for his topic. This paper was a full outline of a field still very new to the U.S. He pointed out, for instance, that due to the careful regulation of the white lead industry in France, there were fewer cases of lead poisoning in two years in Paris, one of the main white lead centers, than he had had in the Episcopal Hospital in Philadelphia from just one factory.

Edsall knew there were no quick and easy answers to such problems. "Pass a law" was not his solution to a complicated social problem, which required education of both employers and employed. Workers were slow to change their ways for mere safety, though employers were in general "very quick to see the economic value of hygienic measures if they can be convinced that they actually accomplish results . . . Suitable laws will come much more quickly and more certainly when there has been accumulated general knowledge of the conditions that actually exist in this country and of the things that are needed in correcting them. This must originate in very large part from physicians, for they alone have frequent occasion to see clearly what the effects are."[40]

On the other hand, he was not one to slight the usefulness of a well-timed piece of legislation. He used to cite the case of a law passed in Massachusetts forbidding the use of arsenic compounds in the pigments used in coloring wallpaper, because it had been shown that these compounds were oxidized to form volatile compounds that passed into the air and caused arsenic poisoning. As soon as any one state passed such a law, arsenic compounds disappeared from wallpapers, because no company wanted to make a product that could not be sold in every state.[41]

CHAPTER 5

Early Relations with Harvard, 1904–1909

EDSALL was no stranger to the medical profession in Boston, nor to the workings of the Harvard Medical School. After all, Boston was his wife's home town, and they saw a good deal of her brother, Dr. Wilder Tileston, who was still in Boston in those days, and of Dr. Edwin A. Locke, who became a close friend.

Beginning in the fall of 1899, the Harvard Medical School reorganized its courses according to a new system known as the concentration method of teaching. The student devoted the whole of each day in his first half year to anatomy and histology, the second half year to physiology and physiological chemistry, and the first half of his second year to pathology and bacteriology, with classes in medicine and surgery following in the second half. This was a very experimental course, but in a few years was seen to have made some very definite improvements in medical education.

Early in 1904 Edsall was sent by his own medical school to report on the new Harvard system, his interest and sagacity in such questions being already apparent. His nineteen-page evaluation of the system highlights the strengths and weaknesses not only of Harvard but of Pennsylvania medical teaching.

In 1904 the Harvard Medical School was bulging the walls of the big mid-Victorian building on Boylston Street where it had been located since 1883. The addition of the Sears Laboratory in 1890 had not relieved the pressure for very long. Typical of the crowded conditions was the laboratory of experimental pharmacology and therapeutics, one "not very large room"[1] which had been divided into two laboratories, one for hygiene and one for pharmacology. (As it was described to John D. Rockefeller, Jr. by his right hand man, Starr Murphy, "Where one alone had no place, two were put in."[2])

The old breach in medicine between "Theory and Practice" on one hand and "Clinical Medicine" on the other was still in full force, and as a result Edsall found the course in medicine "much less systematic than that in surgery," and "somewhat confused and inco-ordinate."[3] The clinical instruction in medicine as part of physiology and pathological chemistry was described to him as being "very bad." In the third year there were ward classes held both in clinical medicine and in theory and practice, with little relation between the two.

By contrast, Edsall found the separate department of pediatrics giving a course he considered "extensive and systematic to a noteworthy degree." Summarizing what he had seen, he wrote:

> The most striking point noted concerning the practical work is that so few hours are given to small sections of the class, and so many — particularly in Medicine — are devoted to so-called clinical lectures . . . The student at Harvard . . . sees quite as many cases as does the student at the University of Pennsylvania, but he sees them in a different way and much less intimately. The chief fault in the teaching of practical work there seems to me to lie in the fact that in both clinical lectures and ward-classes, it is the general custom to show a number of cases in each hour. The impression that I gained is that the students are taught facility in diagnosis and treatment, rather than careful, detailed examination and thoughtful reflection upon the cases.

Nevertheless, the attempt at a new and coordinated system impressed him.

> The whole atmosphere seems to be that of a carefully planned attempt at a general system. This is, in part, still unsatisfactory; but it is already sufficiently successful to make it evident that it will become a very forceful element in the future well-being of the school. This system is also apparent in the school's product. I had a good many opportunities to see the men examine cases and report them before the class without previous preparation, and was greatly struck with the fact that they usually went through this performance in a direct and systematic manner, which brought them quickly and with good reason to their conclusions . . .
>
> The students are well-grounded in pathology; and at the time I saw them, chiefly when they were in the middle of their third year, they appeared to be remarkably proficient in surgery and in pediatrics. So far as I could judge, however, they were by no means well prepared in medicine — a fact that I am inclined to attribute to the much less

finished and less systematic methods of teaching in the Departments of "Theory and Practice" and "Clinical Medicine."

While I feel by no means prepared to offer a definite opinion concerning the matter, I think there is a strong possibility that the concentration-system may have much to do with the excellent methods that many of the third and fourth year students exhibit in their work. I approached this question firmly convinced that concentration is a bad system; but I must admit that I am now in doubt whether it does not permit the student to learn a subject in a more logical and reflective manner than he can when many subjects are studied at the same time and may produce confusion.

Another factor in the school's success which he found worthy of note was the "active spirit of loyalty." Distinguished men of high standing were willing to hold minor positions at a nominal salary in the school. "There is no other institution in America," he wrote, "where social conditions and traditions arouse so general and warm a spirit of universal loyalty as that which exists at Harvard."

As the years went by, Edsall became increasingly conscious that an active spirit of loyalty was not very characteristic of Penn. Pepper was dead. In many departments seniority ruled. Edsall was appointed a full professor, though it was not a position which was considered to be in line for the top post in medicine, tradition decreeing that the professor of clinical medicine ascended to the senior professorship. Wherever he looked, the old guard seemed firmly in the saddle and blocked what he wanted to do; wider horizons began to beckon.

One of the first possibilities related to the Harvard Medical School and the Massachusetts General Hospital. This came to him through his friendship with Dr. Locke and the other Boston members of the Interurban Clinical Club. It was in 1908 that Locke first had the idea that Edsall would be a suitable successor to Frederick Shattuck, the distinguished Bostonian who was professor of clinical medicine at the Harvard Medical School and head of a service at the MGH. Shattuck's retirement was beginning to be talked of.

Locke had been an intern at the Massachusetts General and was a close friend also of Richard Cabot. Cabot was an outstanding doctor, an unusual generous-minded individual. Cabot had said to Locke, "Let us be brothers," and they had shared home and office after Locke's internship, though Locke's hospital appointment was at the Boston City Hospital because of his association with Dr. Henry Jackson.

Cabot was considered to be the likely successor to Shattuck, but in spite of their close friendship Locke was thinking of a real break with tradition, a man from "outside." Locke knew that a new era for medicine was coming, and felt that the Mass. General and Harvard needed a different type of medical department chief, one trained in the fundamental sciences, a man whose purpose was learning the mechanisms involved in disease and their correction. The younger men in a department should be developing and teaching progressive medicine rather than the established clinical knowledge of the day. Nowhere was this spirit so strong as in the Interurban Clinical Club, and this was a place where Locke had a chance to measure Edsall's quality.

Here is Locke's own account of the developments, written some thirty years later.

I had known Edsall rather intimately for a few years prior to 1912 and had come to consider him as one of the very few brilliant young medical men in the United States. He was one of the original 24 young teachers of medicine chosen by Dr. Osler in 1905 to form the Interurban Medical Club and our associations at the meetings of this club greatly enhanced my admiration for his medical training and ability. Edsall was at that time Prof. of Therapeutics and Pharmacology at the U. of P. Medical School (1907–1910) but very unhappy because of the conditions existing in the medical school at that time. His statements on several occasions when these matters were discussed made me feel quite definitely that he would welcome an opportunity to join the faculty of the Harvard Medical School though the subject was never at that time directly mentioned by either of us nor was there ever to my knowledge any discussion of such a possibility by the profession in Boston.

The story of the events leading up to his appointment as Jackson Prof. of Clinical Medicine at the Harvard Medical School and to the Mass. General Hospital are as follows. Dr. Arthur T. Cabot and I were attending the International Congress on Tuberculosis in Washington in early October 1908 and one evening as we left the meeting

I walked down the corridor at the New Willard with him. I said to him, "Dr. Cabot, in a few years Dr. Shattuck will be resigning at the Medical School and hospital and it seems to me extremely important that his successor should be chosen with great care." He expressed very affectionate personal regard and great admiration for Dr. Shattuck as a physician and teacher and then said, "You know what the young men think, have you any ideas regarding the best man to succeed Dr. Shattuck?" I replied that I thought there was a man in Philadelphia who was the one outstanding medical teacher and investigator in the country who should be very seriously considered notwithstanding the fact that Dr. Shattuck might not vacate the chair of medicine for some three years. Dr. Cabot said, "I have given the matter much thought and am deeply interested in what you say. Who is the doctor you have in mind?" When I told him, he said, "I have never heard of him but if you feel so strongly in the matter I want to talk with you at length. Can't you come to my room at the Raleigh and tell me more of Dr. Edsall and your ideas of what the Hospital and School should do?" We talked until almost two A.M. and when we finished it was agreed that sometime during the fall or winter I should, if possible, arrange to have Dr. Edsall come to Boston to meet very informally representatives from the trustees of the M.G.H. and the administration of Harvard College.

I think it was some few months later when Dr. Edsall came to visit me for a few days and the luncheon to which you refer was given (at 311 Beacon St.). Those present were Drs. A. T. Cabot, Frederic Washburn, Henry P. Walcott, Edsall, Mrs. Locke and myself.

My recollection is that nothing was done immediately following this meeting but during the year 1909 some discussions were held and presumably in particular by members of the Trustees and the corporation.* Very early in the affair I went to Dr. Shattuck and had a most interesting conversation with him. He was most friendly and frank in discussing the whole situation, first expressing great surprise that there should be any question that anyone but Richard Cabot should follow him. He said he had never considered any other man and felt him to be the logical and best qualified one from all points of view. Dr. Shattuck expressed some general apprehension of the trend in medicine towards too much emphasis on the laboratory and socalled research. He was eager to hear all about Edsall and his training as he had only heard rather vaguely about him. When we had finished our talk he said with considerable feeling "I still can't see it Locke. Richard seems to me the best man in the field but I am probably wrong. Certainly the

* The governing body of Harvard University.

suggestion of Edsall must be looked into most carefully and with open minds." Subsequently I had several conferences with Dr. Shattuck over the same subject and always found him unbelievably fair minded and he finally became convinced of the desirability of calling Edsall . . .

Soon after the luncheon I went to Richard Cabot and told him frankly what was in the mind of some of us. I told him frankly that I knew of his expectation and desire to follow Dr. S. and that I could not work for the succession of another without telling him about it. His reply was the noblest thing I have ever listened to. "Yes, Locke, I have always expected to succeed Shattuck, my whole life has been planned with that in view. I have done all the post graduate teaching with this aim. That however is of no importance whatsoever. Edsall is a better man than I am and if you can tell me that there is any chance of getting him here to Boston I will do anything and everything in my power to help get him and if he comes I shall be only too happy to serve under him." As you know he lived up to this promise and Edsall told me many times during the next few years of how constantly Cabot helped him and backed him up.[4]

As part of his campaign, Locke asked Edsall for copies of his papers and received the following relaxed and friendly reply:

Dear Locke,

Wilder asked me to send you some reprints – *all* he said, I am obeying orders and sending *nearly all* but I am sure neither you nor any one else will be benefited by the older ones. I have excluded a half dozen or so older ones which were too ancient in their views to pass muster now. I send an old one on "Dissociation of Sensation in Potts Disease" because it is an evidence of the frequent custom of clinicians – they have paid more attention to a simple chance clinical observation that required no work than to many of the others that did require work, simply because it was new clinically.

There are some other reprints that I can't find because they were burned two years ago after scarlet fever since they were caught in the infected room & some metabolism things that appeared only in the Contributions from the Pepper Laboratory & of these latter I have no copies. The most interesting to me of these latter was a study of the "Influence of Excessive Water Drinking on Absorption and Metabolism" in Vol. I of the Contributions. Among the others I remember as having interested me, A Study of Some Cases of Bothriocephalus Infection, which was published in American Medicine about 4 years ago and a report of some cases of Myotomia and Myokymia published in the

American Journal of the Med. Sciences 3 or 4 years ago. I have been especially interested in the latter as I have been observing further cases like them & I am now pretty well convinced that it is a definite clinical syndrome due to a definite occupational cause.

I have a half dozen or so things – clinical and laboratory either finished or "in the mill" but not published, of which I will send you reprints when they come out if you wish. I am too lazy to send out reprints except to people who ask for them.

You may be interested in the "Hygiene of Medical Cases" etc. It touches tuberculosis, indirectly at least.

Sincerely yours

April 22nd [1908] D. L. EDSALL [5]

Choosing between Philadelphia and Boston as a field for his future life, Edsall wrote frankly to Locke in May of 1909 that his preferences were with Boston. "I do not like the general spirit of this community," he wrote of Philadelphia, "and whatever I might be able to do here myself I should feel that my boys would be better off growing up in the midst of a finer spirit toward public things." On the professional side, he felt, "certainly I should be better off than I am now as to accomplishing things in my department. For now I have neither clinical nor laboratory opportunities to do more than get some work done myself and help some few younger men who voluntarily appeal to me for guidance – no real chance that is to build up a staff & do things by team-work."

However, the Philadelphians had got wind of his Harvard possibilities. Several of Edsall's colleagues spoke to him about it, he told Locke, "amongst others Dr. Musser, who is looked upon as the strongest, probably, of the men who would be candidates for the chair of medicine here . . . He would advise me not to accept it, i.e. a Boston call, for, he said, he thought I should be Prof. of Medicine here, with a much freer swing than now exists, & he said he would urge my appointment & he told me I would be in a stronger position here, geographically, to influence the profession in general so far as my powers might permit me to do . . . It is a rather remarkable statement for the 'logical man' to make to his junior . . ." [6]

It might well have taken his breath away, especially in the atmosphere of rather bitter in-fighting in Philadelphia.

Top positions at either Boston or Philadelphia were still in the indefinite future when a more definite and equally tempting possibility emerged. Edsall received an unexpected offer from the Washington University in St. Louis to reorganize their school of medicine.

CHAPTER 6

Invitation to St. Louis, 1909–1910

THE growth of Washington University in St. Louis at the start of this century was the work, in large part, of one remarkable man, Robert S. Brookings, who was for many years president of the Corporation and one of the university's chief benefactors. As early as 1905 he became interested in building up the school of medicine, which in 1899 had been formed from the two leading proprietary schools in St. Louis.

The Washington University medical department, developing out of the loose organization of the St. Louis Medical College and the Missouri Medical College, taught a four-year course, and required a high school diploma for admission. The outline of examinations given in lieu of more formal entrance credits shows how far this went: English grammar and composition, arithmetic, algebra as far as quadratics, elementary physics, U.S. history, geography, and Latin equivalent to one year in high school (a Latin deficiency could be made up during the first year in the medical school).

The school was also handicapped by the fact that medicine had reached the point where it had to be supported by endowment, but endowments generally did not exist. In 1899–1900, the Board of Overseers reported:

> The school is seriously hampered in its endeavor to supply the best medical instruction in the city, by the present lack of funds. During the past year the very thorough training in the fundamental branches has been made possible only by great self sacrifice on the part of the instructors . . . and by the generous contribution of money to defray the cost of two microscopes of higher grade, which were absolutely required for the teaching of bacteriology to an increased number of students . . . It is humiliating, however, that the faculty should be compelled to resort to such methods to meet running expenses . . . The very favorable prospect that the number of students will steadily increase,

does not in itself promise that the conditions of things will be relieved in the near future, because as students increase, the cost of laboratory instruction will also increase.[1]

The financial bind that the school was in because of dependence on student fees was even more clearly spelled out in the next year's report:

Practically all our income is from tuition fees. The cost of giving the best medical education is no where met by the tuition fees, and we could only increase our income materially by so lowering our requirements for entrance to and exit from our school as to attract a poorer, though larger class of students.[2]

By 1905, after having remodeled their dispensary into a 125-bed hospital, the school was really in financial difficulties. At this point Robert Brookings entered the picture. He turned them into an integrated department of the university, took steps to pay off the accumulated debt of $51,000 with $25,000 from Adolphus Busch and the same from himself, and set about finding them an endowment.

The winds of reform blowing in the first decade of the twentieth century were felt more or less by every medical school in the country. The Association of American Medical Colleges, founded in 1891, was working quietly from within. The American Medical Association set up its Council on Medical Education in 1904, to evaluate the country's medical schools, but without using the weapon of public exposure to persuade the low grade schools to improve. This final necessary step was taken by the Carnegie Foundation for the Advancement of Teaching, headed by Dr. Henry S. Pritchett, when he published the results of Abraham Flexner's revolutionary survey. Flexner visited every medical school in the country. He checked on facilities, nosed into laboratories, compared the records of entrants with the official entrance requirements, and sent the facts back to Pritchett, who then communicated them to the school concerned. When *Medical Education in the United States and Canada* appeared in 1910, one quarter of U.S. medical schools quietly folded up and disappeared.

Pritchett had taught astronomy at Washington University and was aware of the new developments taking place in St. Louis under

Brookings' leadership. He shared the pride felt at what had been accomplished, and wrote to Flexner before his first visit, "Now you are going to see something better, for my friend Brookings has taken particular interest in developing the medical school of the University."[3]

Thus the preliminary report of his visit which Flexner sent to Pritchett and Pritchett forwarded to Brookings came as all the more of a shock, for Flexner found that the school was "a little better than the worst I had seen elsewhere, but absolutely inadequate in every essential respect."[4]

As he summarized it later in his autobiography, Flexner said in this preliminary report:

> The present department fails signally to answer the requirements. Its laboratory branches are fairly but unevenly provided for; they can, however, be extended and completed without any considerable difficulty.
>
> The clinical branches are, to state the facts candidly, in wretched condition. The hospital facilities are inadequate and, such as they are, are poorly used. They ought to yield an immensely better training than students now get from them. The dispensary, which ought to furnish the student the very foundations of his clinical education, now does more to demoralize than to train him: the methods employed in it, as in much of the hospital work, are decidedly slipshod. The work is, of course, not all equally poor, but the point is that there is no "team work," no training in method, no governing purpose. Worst of all, the laboratory training which occupies two years and should provide tools for the clinical years, is in no organic relation to the clinical teaching at all.
>
> In order, now, to make the medical department of a piece with the rest of the university, heroic measures are necessary. There is no evidence that any one now on the faculty is preeminently fitted to undertake the work. I suggest in the first place the appointment of a trained, eminent, and successful teacher of clinical medicine as dean. Such a man will require a free hand in dealing with the existing situation. Unquestionably he will get out of the existing plant and its immediate resources a far better result than is now obtained; but the opportunity will not attract a man of the right stamp unless the university intends to develop a comprehensive scheme for the reconstruction of the department and is ready to set about its realization in the spirit which has produced the results now visible on the college campus.[5]

Brookings was not going to accept this verdict without a struggle. He demanded to be shown. Flexner returned to St. Louis and took Mr. Brookings through his own medical school. In one morning he had so convinced him that he summoned the trustees to hear Flexner's verdict.

According to Flexner, it went like this. "When I had finished, Mr. Brookings said: 'What shall we do?'

"'Abolish the school,' I replied.

"'Then what?' asked Mr. Brookings.

"'Form a new faculty, reorganize your clinical facilities from top to bottom, and raise an endowment which will enable you to repeat in St. Louis what President Gilman accomplished in Baltimore.'

"It was then and there voted that this plan should be carried through."[6]

It was to take more than a vote to get such a scheme off the ground. The Committee on Reorganization made its first report on December 17, 1909. After dealing with shortcomings of organization and staff, hospital and laboratory work, they too stated that "the first step to take is to secure a professor of internal medicine whose training is thoroughly scientific, who has recently been in touch with well organized medical schools and teaching hospitals, whose eminence as a physician is recognized, who shall assume immediate responsibility, under the Board and the Chancellor, for the execution of the plans of the Board."[7]

At the end of December they invited David Edsall to St. Louis, with the idea that he was their man. Edsall at this time was just forty. These requirements just about necessitated a young man.

Returning to Philadelphia, Edsall promptly wrote a remarkable letter addressed to the chancellor, David F. Houston, in which he spelled out the design of their ideal for the "Johns Hopkins of the Southwest" in terms of men, money, and plant.

1432 Pine Street, Philadelphia, Pa.
Dec. 24th, 1909

My dear Dr. Houston:

I have been thinking out the whole medical school scheme since I left you, and what follows is what appears to me essential if it is to be acceptable to the men to whom positions are to be offered, and is to take a front rank. I shall have to write it out as my stenographer is ill and I can't secure another immediately.

1. As to the hospital question. I believe that the Barnes * would do for a beginning if of the projected size — i.e. if there were 160 medical and surgical ward beds which *could be used for teaching* and if the University would guarantee that these beds *be kept filled* with patients (so far as material offered itself) if the Barnes Hospital trustees find that they cannot afford to do so. It would be necessary to plan to enlarge the Barnes before very long and increase the number of medical and surgical beds by 50 or 60 each and give some beds to the specialties — eye, ear and probably nervous cases.

It would be necessary that the children's hospital † be planned for at least 100 beds, better 150. Both it and the Barnes should be affiliated by definite agreement. There should also be an obstetric hospital which, at first, should contain at least 50 beds. It would be best if this were larger from the beginning and made an obstetrical and gynaecological hospital and obstetrics and gynaecology were put in one person's charge. This hospital should, at any rate, be so built and situated that it might later be enlarged to 150 beds for obstetric and gynaecological cases together. One minor point I would note here. I believe the Barnes Hospital trustees will want to have the private rooms in that hospital open to outside physicians. I see no serious objection to this from the *medical* standpoint, providing that it was fully understood that the physician and surgeon to the hospital and the other junior members of the University teaching force who are to be connected with that hospital *had prior rights* in case there were at any time a scarcity of rooms. It would of course be unsuitable to have the physician- or surgeon-in-chief made to obtain quarters for a patient sent, say, 500 miles, to be under his observation, while some outside St. Louis physician was treating a local patient just for convenience. This is a minor matter but it is well that it be understood by the trustees of the Barnes Hospital in order to avoid future misunderstandings. From the surgical standpoint this admission of outside men's patients is, I find, looked upon as being often *very* undesirable. First of all, a surgeon establishes a very rigid system in his work, and surgeons often differ greatly in details. They say that the admission of outside surgeons greatly disorganizes the nursing staff and

* The Barnes Hospital was then in the planning stage and not even formally affiliated with the school.

† The St. Louis Children's Hospital at this time was located at Jefferson & Adams Streets, and had 60 beds. It had been open to students since 1885 but was not formally affiliated with the school.

the operating-room system owing to the multiplicity of methods that thereby become employed. Also they say ill-trained, careless or unscrupulous surgeons get in inevitably and lead to occasional deaths or lesser misfortunes that are recognized by the internes and nurses, etc., as being wholly wrong and are gossiped about a good deal and hurt the reputation of the hospital. I would therefore urge that if possible this admission of patients of outside doctors be not undertaken, or if the trustees are set upon that point that they be strongly urged not to permit outside surgeons *to operate at all* in the hospital, except with the personal consent of the professor of surgery – and it would be better to leave out this latter provision (i.e. consent from the professor of surgery) for it would probably lead to jealousy and ill-feeling. Even the medical cases, if admitted at all under outside doctors' care, should not be admitted until the people in authority have by some established system convinced themselves that the doctor in charge, in each case, is a first class man. Otherwise these cases are likely to bring disgrace upon the hospital. The surgical question is the more important however, for the very good reason that *good* surgeons nearly always have hospital services of their own (they *become* good surgeons through having such services). Hence it is nearly always the mediocre or bad surgeons who want to send cases to a hospital that receives outside men, the other surgeons preferring to use their own hospitals where they have established their own system.

Now in addition to these hospitals the Mullanphy * will be desirable as will others that can properly be used. It is difficult without knowing your political situation at all intimately to say what could be done with the City Hospital,† but it would appear to me that one very important thing could be done – i.e. that your professors of medicine and surgery, especially the former, could be given the entree there and could by using some tact get the present form of staff to act as their lieutenants in providing for them, weekly or twice-weekly, a group of cases illustrating what they wished to discuss; so that with that wealth of material they could give their systematic lectures right there, illustrating their main points with the actual cases. This would make an

* The St. Louis Mullanphy Hospital was the oldest general hospital in the West. The faculty of Washington University had endowed ten teaching beds there, but otherwise the teaching facilities were used by the Catholic St. Louis University.

† Students had had visiting privileges at the City Hospital since 1899, but it was not until 1913 that an agreement was worked out with the Commission on Public Health to share in the management of the hospital, with the University responsible for one third of the work done there.

ideal method of systematic lecturing, illustrating the didactic by the objective. Such an opportunity exists nowhere else that I know of. Even abroad the opportunity does not usually exist, if ever, because the cases are divided among various services. It appears to me very desirable that the method of running the City Hospital be not disturbed at all from the present system until a very careful plan that is feasible be worked out.

Besides the Barnes and Children's and Obstetric Hospitals there should be a building which we have discussed before in which post-mortems could be conducted as part of the teaching, clinical lectures could be held, and also the routine clinical laboratory work for the hospitals, the clinical laboratory teaching and the research by clinical staffs could be carried out. This, you see, will need to be a building of very considerable proportions, so considerable that the Barnes Hospital trustees would probably consider it scarcely their province to provide all of it. There would need to be the post-mortem quarters, which need not be very large but should be large enough to accommodate a considerable group of students. If it could be arranged that this room could be made big enough to hold the whole class, upon occasion, it would be advantageous. The best method of using such material is to have most of it gone over with small groups but if cases that come to autopsy have been formally lectured upon before the whole class it completes the subject in a most illuminating way to have the professor of medicine, or other teacher as the case may be, take his class as a body right to the autopsy room and show them how far right or wrong he has been and why – a method that I have seen splendidly used in Gratz, in Austria, by one of the best teachers I ever saw. This room whatever its size could be in a commodious, high ceilinged basement, with tiers of standing room for the students closely arranged. There should then be a large lecture room for general clinical lectures in medicine and pediatrics (in surgery this would be chiefly done in the larger operating room). The slanting space on this floor could be used for one or two smaller class rooms, which I think are not provided for in the hospital proper. The clinical laboratory for teaching would need to be large enough to provide a small desk space and locker for each student, as Terry now does in his anatomical laboratory. This room and the large lecture room would need to be planned for the full-sized *clinical* class that you expect to have in future as the building would have to be so planned that these rooms could be enlarged if necessary – this latter would, I should think, be difficult. Then the building should contain two or three fair-sized rooms for routine hospital clinical laboratory work and a set of rooms for clinical research. It is apparent that this building, for economy of labor and accessibility, should be reached by basement corridor or tunnel, from the Barnes and Children's hospitals, and it would save a great deal of labor in transporting beds (in showing patients in clinics, etc.), if it were also reached from both hospitals by elevated corridors running to the lecture room floors – i.e. this would save taking patients down to the hospital basement and up again from the basement of the lecture building and vice versa when they were returned to the wards, which would amount to a good deal. This could be avoided if clinic rooms could be provided in the hospitals, which would be very convenient but would take considerable space as there would necessarily then be a large lecture room in each hospital. It is quite apparent also that it would be an economy to have the obstetric hospital so placed that it could use the routine laboratory for its examinations of urine, blood, etc., and for its autopsies. This could easily be done if it were just across the street, especially if it could be connected by tunnel, but the tunnel would not be essential as autopsies would be few in that hospital. Indeed they could, if necessary, be carried out in a room set aside for that purpose in the obstetric hospital.

It would be entirely feasible if space permits to have the dispensaries for the hospitals in the building for the clinical laboratories, etc., and this would economize in building. But that would make the building one of very considerable size and height. The dispensaries when built should be planned for teaching as well as care of patients, as some of the most valuable teaching is done right in the out-patient department.*

The buildings for the teaching of the medical sciences I need not go into now, as that is, I suppose, pretty well comprehended as to scope and is too big an affair to make it possible to say anything now that would be profitable.

As to the chairs which it would be wise to have: – I am deeply convinced as I think further of the matter that it is an unparalleled opportunity to take a stand that will not only render wonderful service to the whole of that great region of the country but will also act as an example and a standard for the rest of the country. It must be remembered that Hopkins started a medical school not only with a "big four" but with several other men who were very carefully chosen and who have since become the biggest men *now* there. To start with four big chairs well filled would give your school enough impetus to put it close to the leaders. I question whether it would be enough to make it get

* This came into being as the Clinic Building in 1914 in the new medical center.

recognition as of the *first* rank, and I question whether the very best men could be secured to go to a school that has not been of the first rank unless more than this was done. If on the other hand all the chairs that are at present recognized as being of the first importance were to be filled at once with first class men and suitable provision made for them as to laboratories, equipment and budgets, I feel the greatest confidence that, except for a possible man or two who might be tied to his present place by special reasons, the best men available could be secured; and once that were done I am sure every one acquainted with the medical situation would say that, with a complete faculty free of "deadwood," you would have without question the commanding situation in the country. To accomplish this I should consider it necessary to fill the chairs of medicine, surgery, pediatrics, pathology, physiology and pharmacology, and preventive medicine. I should recommend that Terry be retained in anatomy. As to chemistry I am reasonably clear in my belief that that ought to be filled by another man. I have looked up Dr. Warren's writings in the last seven years and found only one paper of any moment at all and that was a pretty thorough piece of work but a study that was essentially technical pharmaceutical chemistry with no evidence of broad physiological relations. Dr. Warren also is not medically trained and I know of no man without medical training who is suited to the teaching of physiological and pathological chemistry. Even Dr. Folin, of Harvard, who is extremely prominent as an investigator and a remarkable man, has been criticized a good deal by many of his colleagues because he teaches the technical relations of his subject rather than what the student needs to know in medicine. This and what I observed while in St. Louis make me feel that Dr. Warren would not be satisfactory to the other group of men and if any changes were to be made it had best be made when others are made in order to avoid embarrassment, especially since you said he had generously suggested that it might be thought advisable. This subject is of such extreme importance in medicine at present and is so very poorly cared for in most places that I look upon it as a very important one to arrange with great care. A strong man in that place, together with the other places well filled, would make your position impregnable. Half way measures would put you in a situation somewhat superior to Tulane, for instance, but not a certain position and you might readily have the experience that Tulane did. I declined Tulane myself because, among other reasons I felt that they were planning only a half way reform and reconstruction, and I know others did the same thing. There were of course other reasons, such as the fact that Tulane provides a much more limited field of influence and this would be in your favor in securing a limited number of men, but how far this would help I do not know.

Now as to the budgets for the various departments. Salaries for the major positions are, in the best medical schools, usually $5,000, except that some of the best schools pay their clinicians less. But no clinician of importance would go to another city and leave thereby the income he had established for less than that and there is a tendency to raise the clinicians' salaries above that point if they do only consulting work. I believe that Dr. Osler and Dr. Halsted were given $5,000 by the medical school and a further salary by the hospital, in Baltimore, but I have never known definitely that this was true. There are a few non-clinical chairs that pay more than $5,000. For instance Dr. Howell, of Baltimore, was I am told just offered $7,500 to go to Columbia in Physiology but declined. They probably raised his salary in Baltimore from $5,000 to at least $6,000 to hold him. I think, however, that all the men I should recommend to you for chairs would go to St. Louis for $5,000, if it were made attractive by being made a new fine school to build up, and one or two men might be secured for $4,000. On the other hand one or two especially firmly fixed and very desirable men might be secured with $6,000 when they could not by $5,000. The extra duties as Dean, etc., would demand some extra pay. (I would say parenthetically that whatever my relations with the school might be I could not undertake to act as Dean and carry out the details of that office but I should be willing to act as a sort of Director, in the early days of the new arrangement at any rate). In addition there should, of course, be provided the usual heat, light and janitor service, and the janitor for each department needs a little more than ordinary pay for such persons as he needs to be trained into special work and kept when once trained. Then for assistants' salaries and for the expenses of the departments (material for teaching and research) the budgets should have added to them about $6,000 for anatomy, pathology, chemistry and preventive medicine, about $2,000 or $2,500 more than that for physiology and pharmacology as that department would have extra work to do in teaching two subjects, and $5,000 each for medicine and surgery, possibly more for medicine as it would have a good deal of clinical laboratory teaching, etc., to conduct, and $3,000 for pediatrics. These budgets are a little more than in some of our best medical schools, but less than in at least one school, in some departments at least. In order to get men to go from old established fine schools, it would, however, be necessary to pay as much as the best, or at times more, for the conduct of the departments and assistants. Otherwise they would look askance at the offer. I think they are at most very little more than some others offer, and I know they are rather less than one fine school offers, and I know Pittsburgh has offered at least one man more than this but did not get him because he was uncertain as to his colleagues. The further cost of running the school I cannot tell you. It would be increased somewhat over what it has been with you owing to the increased size of buildings, etc. Anything that I have omitted from this I should be glad to be reminded of.

I have attempted to outline only a scheme for the best type of school. It certainly would not profit any of the best men to become part of the school unless they were given better opportunities for work than they now have. I know all the men I have in mind have been offered as large salaries as these I suggest, already, with one or two exceptions that I do not know about definitely, and large sums have been offered them for running their departments but they have declined the offers – one man because he did not want to go to New York, the others chiefly because the schools from which the offers came were not, *as a whole*, satisfactory to them. It is in the entire reconstruction that is possible that your tremendous opportunity lies.

I have been thinking over the Catholic University problem * and it seems more and more desirable to enter into relations with them if they will adopt the same entrance standards and do their best otherwise. It would add wonderfully to the power for good to wipe out all important opposition. I have gone over the writings of the men in their medical science departments and they have two or three good men and some others that are pretty fair. Their greatest weakness lies in their clinical departments and these would be wiped out if any affiliation were undertaken. This matter would appear to me probably capable of being suitably arranged.

I would say that I find my own affairs in a somewhat complicated situation upon my return owing to one fairly direct offer on an important position here – i.e. the directorship of the research hospital, etc., that Mr. Phipps has given to the University for the study of tuberculosis, (and of other medical diseases if thought desirable). He has given a $500,000 hospital and laboratory and $50,000 a year to run it. Dr. Flick, the present Director, has been asked to retain his position but will probably decline it as it means really taking more out of his hands than it has been, and I have been asked if I would take it in addition

* The Carnegie Foundation was urging some form of cooperation among the St. Louis University Medical School (Catholic), the University of Missouri, and Washington University, perhaps a clinical school to serve all three for their last two years of medical school.

to my present work in that case. It would mean a largely increased salary and broader opportunities but a somewhat complex situation and I am not sure what I should do about it. It is, however, important that I should be able to reach a decision within four or five weeks. I am also told that there is some probability of a much more important change here within two months as our Professor of Medicine has suggested resigning. There would be a good deal of a fight for that place — into which I should not enter — but some of my friends think they could have the chair offered to me without my applying for it. That is an uncertain matter but I would have to act upon that at once if it came, and if I accepted it it would be unsuitable for me to go away from here for two years at least even if I desired to and of course with a fair prospect that that will occur very soon it increases the feeling that I have had all along that I do not care to leave here unless I am sure I should be going to an exceptionally good opportunity. Aside from the bearing of this upon myself it is another concrete illustration of the fact that the supply of clinical men who are interested in medical education as their chief work is small so that positions that would tie fast elsewhere some of the other men you would want are likely to turn up at any minute. I find in fact that another of the men I should recommend to you, besides the one I spoke of, is being considered for a very important position though I do not think it has yet been offered to him. The Carnegie Foundation people are stirring the leading schools to activity, as well as the others, and there is an air of "something about to happen" in most of them.

I mention these latter things simply to indicate to you that I am not trying to plan extravagant and unusual conditions but simply to meet what I think is necessary in order to rank at the top or at any rate fully alongside the leaders.

Very sincerely yours,
DAVID L. EDSALL [8]

Edsall's remarkable letter took the committee's breath away. In January they reported on Edsall's proposals, and money was the major problem — an additional $60,000 each year for salaries and $10,000 more on hospitals, with $30,000 needed for the improvement of laboratories and half a million needed for hospital construction — this in a school spending $80,500 a year on medicine and overstrained at that. It would nearly double the operating budget.

"It is a great pity," said the committee, "that the large plan endorsed by Dr. Edsall, and by such men as Doctors Welch and Flexner, cannot be immediately put into execution. They [the committee] believe that it would be the greatest thing that could happen to St. Louis . . . But they do not at present see where the money is coming from. They can only suggest at this time that the plan be kept in mind as an ideal towards which all our efforts should be bent, and that every step we take should lead in this direction."[9]

Brookings wrote sadly to Pritchett that money made Edsall's plan "utterly impractical . . . unless we can secure the great bulk of the money from some outside source." Nevertheless, he was not going to give up his dream. The same month he wrote, "This larger scheme appeals to me very strongly, and is so much the best worth doing of anything that I know educationally in this or in any other section of the country that I would be willing to make sacrifices which I can ill afford to do at present to bring it about, and in fact, would have to personally stand behind this increase of endowment from $30,000. to $60,000. a year . . . If I had the money I would not permit any other man in the country to have a hand in it."

Brookings sailed for Europe on January 22, 1910. Before he left, he set his great scheme in motion. He led off by promising the university $400 to $500 thousand for the new medical center, together with $30,000 for improving laboratory buildings, and $12,500 which was to be half the sum needed to improve hospital operation meanwhile. Further, he guaranteed to pay half the expected annual deficit of $45,000 a year. One of his stipulations was "that a sufficiently strong group of men may be secured for the heads of the various departments both in the first two years comprising the medical science department of the university and in the last two years comprising the clinical school; and I herewith make Chancellor Houston and Dr. Edsall of Philadelphia the sole judges as to the personnel of the teaching staff . . ."[10]

He saw both Edsall and Dr. Harvey Cushing in New York, and both were "seriously considering the move to St. Louis." His increase of the annual budget for both the science and clinical departments followed every suggestion Edsall had made. This was only the beginning, he told the two, and as he wrote to Pritchett, "I am sure

I can get all the money from year to year that they can make good use of."[11] He sailed for Europe confident that both doctors would join the St. Louis faculty.

If David Edsall was regarded as one of the country's leading lights in medicine in the early 1900's, Harvey Cushing was obviously outstanding in surgery. Because of the remarkable and creative work he had been doing in brain surgery at the Hopkins, Cushing had many offers from such places as New York Hospital, Yale University, and Bellevue. Harvard had been interested in him from the time he visited Boston in 1903 and met President Eliot. In 1907 the planning of the new Peter Bent Brigham Hospital had begun in Boston, and Dr. Cushing had been consulted from the start, his friend Dr. W. T. Councilman writing him in March of that year, "I have tried to have you appointed so that the Surgical Department of the hospital could be formed by you. But they will not do this. I presume the trustees will be ready to make an appointment in two or possibly three years."[12] * Others in Boston, including Dr. J. C. Warren, were working towards the same end.[13]

Nevertheless, Harvard had made Cushing no definite offer when Edsall and the Flexners were planning the new faculty of the Washington University medical school, and they made every effort to induce Cushing to go to St. Louis. Abraham Flexner brought Brookings and Cushing together when Brookings was passing through New York on his way to Egypt. As Cushing wrote in his notes, "I had no inkling what the meeting was about, thinking it might have something to do with the Research Defense League in view of the recent attack on the Rockefeller Institute and Flexner by the antivivisectionists." * What was his surprise to have Brookings unfold his plan "to Johns Hopkinize the Medical Department of the Washington University."[14]

As for the Edsalls, they were all set to go to St. Louis. As Mrs. Edsall described it in a letter to her friend Edith Johnson, Feb. 23, 1910:

> At first [the St. Louis people] were staggered by the size of the sum needed but they became fired with zeal for raising it. David was to be Professor of medicine, the head of the school, with the power to choose

* From Fulton, John F., *Harvey Cushing*, 1946. Courtesy of Charles C. Thomas, Publisher, Springfield, Illinois.

all the heads of departments, all the men in his special department (all the present faculty have resigned), also. Like dictatorship, really, as to spending of money, deciding on the buildings and their equipment, his own salary, etc.

Their plan was to make their medical school the Johns Hopkins of the Southwest and it seemed like the biggest opportunity in a generation to do a great work in medicine – an irresistible offer. David's best friends, with whom he talked over the matter, thought he couldn't do anything but go – nothing so big could be offered him elsewhere. It was very exciting, and in some ways made my heart stand still to think of going west and being so far from everyone I have known, so distant from my family, bringing up my children in St. Louis and probably ending my days there, but I saw many good things, even from a personal point of view, and I tried to think solely of what was best for David. He became quite high-spirited in the prospect of doing such a big work, so unfettered – and the Chancellor of the University and its chief trustee (and benefactor) were the most delightful men to deal with.[15]

Here at last was the great opportunity for the man who wanted to "influence the profession in general so far as my powers might permit me."[16]

The next word from St. Louis (and a prophetic one, as it turned out) was that they were having difficulty in securing the expected funds. Not long after, they announced further prospects, and asked Edsall to wait. "This sounded as if the whole thing would fall through and we felt quite downhearted," wrote Mrs. Edsall to Edith Johnson. On the 20th of January St. Louis was on again, and Edsall went to New York to see Cushing. As Mrs. Edsall tells it:

Of course I thought our whole future would be settled (for St. Louis) before he returned and I waited at home in the most intense suspense – That (Wednesday) evening one of David's best friends called me up to ask if I knew what was going on – some of David's friends had got together, first collected a meeting of 40 men, then got up a petition setting forth David's importance to the University and begging the Provost to make arrangements that would keep him here. (This petition was signed in 6 hours by nearly 500 doctors!)

He wanted to know where David was staying in New York and he called him up by telephone and made him promise not to commit himself to St. Louis until after his return to Philadelphia to see what his friends were doing to keep him here. Thursday Mrs. Frazier came to see me and worked on my feelings by telling me how much bigger an influence in medicine David would have at this old University if he were put at the head of the medical school, etc . . . Friday night David returned and from this time till Wednesday the suspense was most acute . . .

David arrived all agog with excitement and gratification – the St. Louis men were ideal, everything was what he wished – he had postponed saying Yes only because he thought it more decent to his friends here. He was profoundly touched when he learned how hard they had been working for him, how eager they all were to second any move that benefitted the U. of Pa. That Sunday we went to Baltimore to see one of the big men that David wanted for his St. Louis Faculty * (four or five had promised to go *if* David went and all the arrangement was carried out.) and also to take supper. Dr. Welch was there and his opinion was what David wanted (largely). He cross-examined David as to all the particulars of the St. Louis scheme and when David told him who would be the heads of departments, he exclaimed, "with such men you would have the second great opportunity in medicine! We had ours and took it a generation ago at Johns Hopkins." But when David spoke of the U. of Pa. he said, "If they give you the chair of medicine – with all the traditions of that school, I regard it as the premiere place in American medicine."

So he didn't make the problem easier to settle!

We reached Broad St. Station a little after midnight, still expecting to go to St. Louis, and were met by an emissary of the U. of Pa. who came to the house † and unfolded the detailed plan which was to be laid before the Provost for his approval – but if David refused to consider it, and was going to St. Louis anyway, they would not lay it before the Provost. David listened and said that he *would* consider such a proposition – he couldn't help it. It is this plan which I am just going to unfold to you that I will ask you not to speak of till I let you know that it is public (all I say to my friends here is that the authorities at the U. of Pa. made David such an attractive offer that he gave up St. Louis).

That Sunday night (1:30 a.m.) I went to bed feeling as if it was more than I could bear, wondering which plan would be best for David's career and his happiness. The University plan (which David accepted

* Her diary says they saw Dr. Cushing, Dr. Welch, and Dr. Reid Hunt; the man was probably Cushing.

† This was Dr. Pearce, according to her dairy.

the following Wednesday when he saw the Provost and was unanimously passed by the Board of Trustees on Feb. 1st) was to make David *Professor of Medicine* (that chair is the leading one of the school, and the present occupant, who is 68, resigns at the end of this year), to let him reorganize his whole department (30 men), to get rid of the heads of two or three other departments who have been *deadwood* and replace them by men whom they practically asked David to name — to set an age limit, which would soon leave two other chairs free for better men — in short, to shake up the whole staff and begin a new era at this old University, which has been going backward for 10 or 15 years past — and all these vital changes were made to depend on David's staying — so he decided that if his own Alma Mater would do so much for him, he couldn't help accepting the offer here, and it would probably be a wider opportunity for influencing the future of medicine through this old Eastern University than through a new medical school in the west. All the men who most urgently advised him to go to St. Louis have told him that his bigger opening was now here . . .

P.S. I forgot to say that in the new arrangements at the U. of Pa. David (and some other heads of departments) promise to give half their time to the University so we shall never make a *very large* income! [17]

While it seemed as if the choice must settle on St. Louis, Margaret was all prepared to make a major change, though it was doubly difficult for one who had such close ties to family and friends. She wrote to Mary Kirkbride, "We thought at once of you all as the family we should miss most—but at least our summers will be in New England and you *must* visit us!" [18]

Once the decision was made to stay in Philadelphia, she told Mary, "I do feel that Fate has cheated us out of a wonderful chapter—one forever unwritten!" [19]

By this move, the University of Pennsylvania broke the succession to the chair of medicine and gave the crown to their professor of therapeutics and pharmacology. Tyson retired and Edsall stayed at Pennsylvania.

His decision had considerable influence with Cushing, who wrote, "Things moved along pretty actively for a while, but I wobbled a little owing to Edsall's final withdrawal . . . father's death upset me greatly, and St. Louis seemed very far away." [20] In 1911 Cushing wrote to his future colleague, Dr. Henry Christian, then dean at Harvard, "If Edsall's going to St. Louis had been assured and if I had felt confident that the Washington University Hospital would be ready before the Brigham I would unquestionably have accepted the offer."[21]

Osler had written him, "It is too bad that Edsall is not to go to St. Louis. Look over the ground carefully. You have such a grip on the East that I can understand how loath you must be to go beyond the Mississippi. But there are plenty of good fellows there."[22] *

As it was, however, Cushing received the joint offer of professor of surgery at the Harvard Medical School and surgeon-in-chief of the Brigham Hospital in April 1910, and accepted. Angry feelings were left in St. Louis where people felt that Cushing had just played them along until he could get the offer from Harvard, which he had intended to accept all the time. They were to say the same when Edsall, after some extraordinary changes of fortune, left them to go to Harvard.

* From Fulton, John F., *Harvey Cushing*, 1946. Courtesy of Charles C. Thomas, Publisher, Springfield, Illinois.

CHAPTER 7

Explosion in Philadelphia, 1910–1911

EDSALL was the man chosen to head the new developments in the University of Pennsylvania medical school, but university politics played a large part in carrying him to his high post quite apart from his own efforts and ambitions. The reform party had been gathering power for some ten years; its most important figures were the provost, Charles C. Harrison, and his nephew Charles H. Frazier, a leading neurological surgeon and a firm friend of Edsall. Powerful on the other side were Dr. Allen J. Smith, who became dean when Frazier resigned the post in 1909, and his brother Edgar Fahs Smith, the vice provost who later succeeded Harrison. The most important body was the board of trustees, and here Dr. Weir Mitchell for all his eighty years was a tower of strength for the reformers.

Good appointments, a raising of entrance requirements, or the building of a new laboratory were not enough to swing the medical school forward into the age of scientific medicine. But Provost Harrison made good use of the uneasiness caused by Abraham Flexner's survey of the school in March 1909.[1] Apparently Flexner was no more favorably impressed with it than he had been with the Washington University medical school in St. Louis, for we have an account of his chief, Dr. Pritchett, telling Provost Harrison that "the only worthwhile change would involve eliminating the entire teaching force."[2] Harrison's ambition was nothing less than to make a clean sweep of the deadwood and revitalize the school by putting top-ranking scientists in the leading chairs. The petition of five hundred doctors asking him to keep Edsall at Pennsylvania gave him his opportunity.

> More must be done [stated the petition] toward fulfilling its second function as an institution of research, in order to keep abreast with the modern tendencies in the development and growth of medical schools.
>
> In order to accomplish this, other schools are expending huge sums

in the erection of new buildings and in the acquisition of clinical facilities. At the University of Pennsylvania this expenditure has already been made, and we are now abundantly provided with facilities for clinical and laboratory investigations.

It remains, therefore, for the University to provide men with the proper qualifications and a comparatively small additional outlay of money.

In this country, at least, there is a dearth of clinicians who have an intimate knowledge and the proper appreciation of the value of laboratory methods for the advancement of clinical medicine. The University now has on its teaching staff one who has already gained preeminence as a clinician and scientific investigator, and it has recently been brought to our attention that he has received an attractive offer from a sister institution. Should Dr. Edsall accept this call, it is our firm conviction that the medical department and the University at large would lose one who is regarded, not only here, but throughout the country, as one of the strongest and most influential members of our faculty. We sincerely believe his loss would be irreparable, not only because we would be deprived of his services to the University in his present office, but because we would lose the tremendous influence and the stimulating effect he exerts upon the younger men in medicine . . .

For the future development and needed reforms of the medical school . . . we appeal to you . . . to take such steps as will insure Dr. Edsall's continuance in the University.[3]

Harrison went into action at once. He bypassed the faculty medical council and went to the board of trustees, put pressure on old Dr. Tyson to retire from the chair of medicine, set up a new department of physiological chemistry which would supersede Dr. John Marshall's department of chemistry, offered a new chair of comparative pathology to Dr. Allen J. Smith who was to resign the pathology professorship in favor of a new man, moved toward the university system with a proposal of half time for top men, and with an eye to the future, proposed an age limit.

WOULD MAKE PENN HOME OF RESEARCH WITH STRONG STAFF . . . DR. EDSALL TO TAKE CHAIR OF MEDICINE . . . MAY PRUNE MEDICAL FACULTY AT PENN BY FIXING AGE LIMIT . . . TEACHERS WORRY . . . FAMOUS MEN WILL STRENGTHEN U. OF P. MEDICAL FACULTY . . . UPHEAVAL IN U. OF P. MEDICAL TEACHING . . . said the papers.

At first Edsall had thought that the most his Alma Mater could offer him would be Tyson's chair and this would not have been enough to hold him. As Margaret Edsall described it, "with all the old hindrances, dead wood in various departments, &c, &c — to contend with, he thought he could carry out much more work in St. Louis."[4] When he saw that the whole scheme of reform hinged on his staying, and with "the new spirit now shown, with 500 loyal alumni ready to back him up in everything, [he thought] that in common decency and loyalty to his Alma Mater, he *must* stay here."[5]

The word promptly got around, of course, and the exaggerated rumor was that Dr. Tyson was to be turned out of his chair that very week, while Edsall would deliver his lectures for the rest of the year. A counter petition was circulated, but for the time being it was ineffective.[6]

In the full flush of optimism about new developments in Philadelphia, Edsall wrote Simon Flexner, who had shaken the dust of Penn. off his feet some years earlier: "I have no doubt you were as much astonished as I was at the plan that has been adopted here and that has kept me here. You knew the situation well and you would not recognize the atmosphere any more . . . I could not ask a broader and more progressive and generous spirit than they have shown in everything they have done and unless those of us who will have charge of things fail we ought to get back into the fine old traditions in full force."[7]

On February 25, 1910 Edsall wrote to Locke a final rejection, as he thought, to the Harvard possibilities.

> After having had so much done for me here, and having offered me so fine an opportunity, it will of course be out of the question for me to consider leaving here, and I should not, so far as I can see, have any desire to leave here for any other place in the country. Hence this puts the quietus, it would appear to me, upon the charming possibility that we have talked over in relation to Harvard, a fact which in many ways I regret, particularly because of the possibility that it meant of bringing me into close relation with a group of men of whom I am very fond in Boston and because, of course, of the domestic reasons why it would have been a very attractive life.
>
> As I understand it, your Medical affairs will be settled next year.

It is within the bounds of possibility that my mind might change before that time, but it is so entirely improbable that I feel that it is due to you to say quite definitely that you should make your plans to work for some other person, who seems to you satisfactory, and entirely drop all consideration of myself.[8]

By March of 1910 enough details of the Philadelphia reorganization were settled for Edsall to write to Dr. H. A. Christian (then dean of the Harvard Medical School) as follows:

In the first place Dr. Tyson will retire and I shall take charge of the Department of Medicine and reorganize it as much as seems desirable. Drs. Musser and Stengel will remain, I am glad to say. My present department will be turned into a department of pure Pharmacology, and will be put on a purely scientific basis, the therapeutics falling into the department of Medicine. The Physiological and Pathological Chemistry will be wholly reorganized under the new department and Alonzo Taylor will have charge of that. The pathological department will be entirely reorganized and some man, as yet undetermined, will have charge of that. Pearce will be in the department of Research Medicine. It is not definitely established what we shall do in regard to the Phipps' Institute; that will be largely for research purposes but will also be used for teaching to a considerable extent. In addition to these facts the trustees are about to establish an age limit which will probably, from rumors, be 62 for surgeons and 65 for medical men. This will mean an entire re-organization in the Surgical department in a year or two. There will probably be some minor reorganizations in the other departments, whose heads will remain the same, though that is not yet quite clear. Then it is now clearly recognized and established in regard to myself and to others who are already in such positions or who will take them in future that the heads of the great clinical departments will be expected either to receive very small salaries or more usually, will be expected to devote at least half their time to the work in the University and to do only consulting work outside, and to do that in such form that it will not interfere with their University work and influence. This is a definitely established policy and agreement between those working for the University and the University authorities.[9]

This news was announced to the world at an alumni banquet on March 30, 1910. It was a gala occasion. The provost announced that the new chair of physiological chemistry had been endowed with $100,000 and would go to Alonzo E. Taylor, then professor of pathology at the University of California. Dr. Howard Taylor Ricketts of Chicago was named to the chair of pathology. Dr. Richard M. Pearce, of the University and Bellevue Hospital Medical College in New York, would fill the chair of research and experimental medicine. Edsall was named the new professor of medicine. A new policy whereby professors were expected to devote half of their time to their university work was announced, but the question of the age limit, which had caused charges of "oslerizing" * in the papers, was quietly passed by.

In the general storm of good feeling, the fact that Drs. Marshall and Smith had been demoted was glossed over. The provost "paid high tribute" to Dr. Tyson in announcing his resignation, and "the banqueters sprang to their feet and cheered repeatedly. Dr. S. Weir Mitchell led in this demonstration. The author-physician waved his handkerchief and gave a lusty 'hooray,' and a group in the corner took up the refrain, 'He's a Jolly Good Fellow.' Then scores of handkerchiefs fluttered from pockets, while cries of 'Tyson!' 'Tyson!' came from all parts of the hall. The outburst ended in a stentorian declaration ensemble, 'He's all right.'

"Fully as cordial a reception was given to the announcement that Dr. Edsall was to succeed Dr. Tyson. The class of '93 led in this demonstration, the physicians waving napkins, candelabra and clinking glasses in attesting their satisfaction." [10]

Tyson certainly was one of the best-loved of the medical school professors and a fine physician of the old school. As a teacher, he stuck to the book, reading his lectures from the printed page while the students followed him in their own copies. At the end of the lesson, a mark in the book would tell him where to begin his next lecture. In 1909, at the opening of the new Medical Amphitheatre

* In his farewell address at Johns Hopkins, Osler had stated his belief that almost every "great and far-reaching conquest of the mind" had been made by a man under forty, while many of the world's evils could be ascribed to men over sixty. He had referred in a humorous aside to Anthony Trollope's ironic novel, *The Fixed Period*, in which old men were retired and then chloroformed, but newspapers across the country had a field day with the story, running such headlines as "Osler Recommends Chloroform at Sixty," and the true facts never caught up with the picture of the ruthless and demonic scientist. "To oslerize" entered the American language. (Harvey Cushing, *The Life of Sir William Osler*. New York: Oxford U.P., 1925, Vol. 1, pp. 666–69.)

and Clinical Laboratory at the Pennsylvania Hospital, Dr. George Dock (then professor of medicine at Tulane) was congratulating all and sundry, and added, "Nor should we forget at this time the thanks due to Dr. Tyson, who in 1874 brought the only methods of laboratory work then used in internal medicine, those of pathological histology, from Europe, applied them in the education of medical students in America for the first time generally."[11] To the "new men" he was hopelessly old fashioned. The only research he was known to have carried out was a fermentation method of determining the urea in urine.[12] But his following was strong among the alumni, and, as the months passed, the story of his retirement under pressure began to have its effect.

Although Edsall himself was a thorough Pennsylvanian, a graduate of the home medical school, member of the Pepper Laboratory, and teacher at the school, only two of the new men had any ties with Penn. Taylor, heading the new department of physiological chemistry, was an 1894 graduate of the Pennsylvania medical school and had got his start in the Pepper Laboratory. Pearce had taught at the school when he worked with Simon Flexner in pathology, leaving when Flexner did. He was the happy solution to the trustees' perplexity in handling the new department of research medicine. This has been endowed by Miss Harriet C. Prevost in honor of Dr. Musser with the stipulation that he was to be permanent director of the department. Medical dynasties had every bit as much importance in Philadelphia as they did in Boston, and just as it did Stengel no harm to be married to Dr. Pepper's daughter, it may have helped Pearce in this odd situation that he was Dr. Musser's son-in-law.[13] There is no doubt that he was the man for the post in any case; after finding their seven applicants inadequate the committee, headed by Edsall, looked over the whole field and chose to invite Pearce.[14]

A. N. Richards, however, was a complete outsider. He had his Ph.D. in physiological chemistry from Columbia (1901), and had been a research fellow of the Rockefeller Institute. Edsall's friend Howland had worked with him for four years on experimental disease of the liver, which may have been one way Edsall learned of his stature and promise. Ricketts was a Chicago man with his M.D. from Northwestern. He was a pioneer immunologist, and already noted for his work in tracing the etiology of Rocky Mountain spotted fever. He never lived to take up his duties at Philadelphia, though he visited Philadelphia in March and had dinner with the Edsalls. He then went to Mexico to work on typhus fever and died of it in May 1910, at the age of 29.

Edsall wrote to the provost's assistant who had asked for details:

> You and Mr. Harrison may be interested to know as a further evidence of Dr. Ricketts' heroic way of looking at things that Dr. Miller told me in St. Louis the other day that Dr. Ricketts found he was not very well and simply quietly stayed at home instead of going to his laboratory, without saying anything about his reason, and then told Dr. Wilder that he had found that he was developing typhus fever and that he knew he was not in good condition physically and he did not know what the outcome would be, so he had been very carefully arranging all his notes and mapping out a further plan of campaign so that Dr. Wilder and others might go on and make use of what he had done in case the necessity arose, wishing so far as it was in his power to be useful then, even sick as he was, in continuing the work against that scourge. He continued his work of planning out further investigation to be carried out by others up to the time that he became delirious.[15]

(This whole group of disease-producing organisms, including those responsible for typhus and Rocky Mountain spotted fever, were later named the Rickettsia in his honor.)

The issue of half time for professors, soon to become the much argued "full time" under the strong advocacy of the Rockefeller boards, caused a great deal of feeling. In general, the top men had a busy practice, and did not devote half their time to teaching and research. To compensate for their loss of income, they were to be on salary, but for many of the men used to the older scheme of things, this seemed an impossible reduction in income. Resignations followed.

Dr. William G. Schleif, who had been in the pharmacology department, was quoted in the papers as saying: "I could not afford to give up my practice and devote all my time to instruction in the medical department at the salary offered."

The interviewer asked how much time a clinical lecturer devoted to the institution, and Schleif replied, "He may go to his classes once or twice a week and usually leaves after the lecture.

Sometimes he will go into the hospital with the boys. I grant that the man who devotes his whole time to instruction will prepare with greater care, yet he does not have the opportunity of acquiring the first-hand knowledge that falls under the notice of the doctor who sits by the bedside of the sufferer and who has a personal interest in the patient."[16]

Edsall's adherence to the "university plan" for his own work had unexpected repercussions, since Philadelphia doctors took it as an announcement that he wanted *no* consulting practice, almost (in his phrase) as if he had sent out cards saying so.[17]

A year later Edsall wrote to Simon Flexner about this vexed question:

> Last year I felt that half time was enough. My brief experience however has shown me that it is not, even if more than adhered to. I am much more thoroughly convinced than ever that the most important thing in Medical education is to put the clinical chairs on a more unselfish and less commercial basis, but I cannot feel quite sure yet what the wisest basis is . . .
>
> I do not care at all for the views of most practitioners, but I think the most unselfish and wisest of them do have to be considered a little . . .
>
> I am still uncertain where the line can best be drawn, though I am sure it ought to be made tighter than I made it last year.[18]

In replying, Simon Flexner said:

> What it has meant to medicine in this country to have put scientific medicine upon a university basis, the experience of the past twenty years shows. The amount and quality of scientific production now compare favorably with that of the older European countries. What it has meant also not to put clinical medicine upon a university basis in the country, the past twenty years equally show. Clinical medicine is infinitely in the rear of scientific medicine and, in my opinion, will stay there until the reformation is effected. May I add, parenthetically, that I do not precisely feel the force of the argument that puts a group of able men into scientific chairs on university salaries and demands their entire time; and permits another group of men, who are not initially more able, to carry university titles and to give a fraction of their time to university work, and then to use a university position and standing to run a large and profitable business.[19]

Over the summer of 1910 everything seemed to be going Edsall's way. With the new reforms in mind, he sailed on the *Bremen* in June, visiting Cologne, Maintz, Paris, Rouen, and London before his return on August 11. The British Medical Association had invited him to be the first speaker on acidosis in the section devoted to medicine, at their annual meeting in July.

Back in the United States, he went walking in Vermont, and took time out to write to Harrison about who was to be dropped from his department. But as the academic year opened the hurricane signals went up. The conservative reaction was gaining strength. On September 3 he was writing to the provost:

I am obliged to refer a matter to you since I have no other means of reaching any conclusion in it, and it is of some importance. I came back in order to be on hand on September first to take up my duties in preparation for the coming teaching year. The chief thing necessary was to get the hospital work in good order, Dr. Musser and Dr. Tyson both being absent, and it being my understanding that I was to take charge of Dr. Tyson's service in the wards on the first of September, and to act for Dr. Musser at that time. I spoke to the gentlemen who were substituting for them and found that they understood as I did, that I was to take charge at that time. Before I visited the hospital on September first however, I was called up by Dr. Dulles,* who, with considerable violence told me that he heard that I was talking of taking charge of some of the cases in the wards and he assured me that I had no rights with any of the cases until the Board of Managers had informed me to that effect, that I could not assume charge of them; that whatever had been done in regard to my University position, the Board of Managers had never assigned any beds to me and he would request me to keep out of the wards until the Managers could act upon the matter. I look upon this communication in itself, as being a very marked indignity and discourtesy which is exaggerated by the fact that it was given over the telephone in the presence of one or more assistant physicians, resident physicians and nurses from the University Hospital, and has caused a good deal of gossip. This however is the minor side of the question. Dr. Dulles said that he had little doubt that when the Board had had an opportunity to act it would arrange that as Professor of Medicine I should take charge of the wards. It is however, desirable that this should occur soon, as there is very little time remaining now

* Dr. Charles Winslow Dulles, on the Board of Managers of the University Hospital.

before the actual teaching begins, and I shall be unable to take up any teaching, and be unable to arrange the schedule, until I am at liberty to assume the usual and normal charge of the wards.

I should be very much obliged if I can learn through you how this matter can be concluded as speedily as possible. The personal element in it is of no consequence at all and I mention that only to show you that I am not in a position to take any more steps in the matter now myself. The matter of importance about it is simply that the activities of the Medical Department should not be interfered with, and the teaching should be ready to go on at the proper time. I am sorry to annoy you with such a matter but as you see I am unable to meet it in any other way. I would say for your information that so far as I could see at the time there was not the slightest reason for my getting any authority from the Board of Managers, and I am still unable to see that there was any oversight or discourtesy on my part, because I had already been made physician to the hospital last year, but without a service and I had not the slightest intimation that there was any other official action on the part of the Board of Managers necessary, as so far as I know the Managers have not assigned cases to those gentlemen who have had services previously, but have simply elected them to the position of physician to the Hospital, which as I said had been done with me previous to the changes at the University.[20]

This difficulty was overcome, but a more serious blow soon followed it: Harrison resigned. His new plans, particularly the proposal for half-time for professors, were meeting formidable opposition from Dr. J. William White, the professor of surgery, and whether it was the prospect of such a bitter battle, or the weight of sixteen years as provost, or the discouragement of a lost cause, Harrison left the administration. Shortly afterwards Weir Mitchell resigned from the board of trustees.[21] This left Edsall and the other new men with no university backing for the medical revolution they were supposed to be making.

Typical of Edsall's frustrations was the incident over the Pepper Laboratory, headed by his former good friend, the professor of clinical medicine, Dr. Alfred Stengel. In developing his new department of medicine, Edsall wanted to have additional space in the Pepper Laboratory for his work. Dr. Stengel agreed, provided that Edsall could get an extra $5,000 from the trustees to cover the expense of the changes. This the trustees declined to provide, and

Edsall felt his plans had received a severe setback.[22] He had other trouble with Stengel as well, and Mrs. Edsall's diary had noted on May 8, "D. had fruitless talk with Dr. Stengel at Dr. Musser's."[23]

In a hasty postscript Edsall told Locke in the middle of November, "The new Provost is not exactly gratifying to me & there are some very important things that must be acted upon soon that may determine me in the desire to fight through here or on the other hand to go away."[24]

As early as November 21, Edsall had made up his mind against Philadelphia, as appears from his wife's diary. The Interurban Club met in Philadelphia December 2 and 3, and Wilder Tileston and Howland stayed with the Edsalls. Howland, a close friend of Edsall, and newly appointed professor of pediatrics at the Washington University medical school in St. Louis, was quick to pursue any chance that Edsall might be brought out there after all. On December 6 he wrote to Eugene L. Opie, the new pathology professor in St. Louis:

> For your very private ear . . . Edsall is almost sure to leave Philadelphia. I don't suppose there is the faintest possibility of our getting him [i.e. in St. Louis] but I'm going to see what can be done. If he goes they will likely come to a big bust up in Pennsylvania & I doubt if Richards would stay there or rather I think he could be moved . . . Edsall, Pearce, Taylor & Richards are alone against fourteen. Pennsylvania is surely doomed & it seems so strange when a year ago everything seemed more promising there than any where else.[25]

Edsall had also been to New York on November 25 to see Simon Flexner, and Flexner may have taken a hand in the renewing of an offer from St. Louis, but this is only speculation. Howland's enthusiastic partisanship, as well as Edsall's remarkable ideas for a brand new type of department, are revealed in a letter from Howland to Opie, Dec. 29, 1910.

> I have seen Edsall and just telegraphed you. We can get him if certain of the facilities he wants are provided. His feeling always has been that preventive medicine has been gone about in the wrong way, that it is really a subject that touches very close to practice and that when it is taught by a laboratory man, such as Rosenau, it hardly gets any farther than the serum and vaccine treatment and prevention of

disease. He believes that preventive medicine includes these and much more, that as I have told you before it included eugenics, occupational diseases and their prevention, infant mortality, prevention of infectious diseases and a study of the etiology of many obscure conditions and also to a certain extent the investigation of new methods of treatment such as meningococcus serum, Ehrlich's "606," etc. etc. Many of these subjects which touch most intimately the body medical are almost entirely disregarded in our schools. In order to properly study disease and treatment certain clinical facilities would be requisite. He by no means desires a large, active hospital service but he wishes beds for investigation of problems in which he is interested and also for the demonstration to students of the diseases capable of prevention and the methods for doing it.

This is a new conception and yet I feel that such a department headed by such a man would be of the most enormous value to the community and to the school. It may be argued that were Edsall to vacate the department we would have a position on our hands which no one could fill. It came out, however, in conversation that Edsall would want to take Longcope * as one of his assistants and it seems to me not improbable that Longcope could grow up into such a position. Edsall could head this department and head it wonderfully, he would also give strength where we are very weak and should it become necessary could fill, as I think no one else could, a position in medicine . . .

There is no denying the fact that we have been marking time during the last six months so far as the most important matters are concerned. He would enable us to hurry things along in absolutely the right way and we could rest assured that there was no doubt about the future of the school. I should not have a qualm and would feel that we were at once of national importance and calibre. With Dock's † poor health another dean is inevitable and nowhere could there be found a better one than Edsall. I went over with him very carefully the position that a dean should occupy; his opinions are exactly our own. In the original arrangement he was willing to hold the sceptre merely until the men [who] were chosen for the various departments so as to be sure that only good men were obtained but he was then voluntarily to abrogate his authority and occupy the same position as the rest of us, an eminently sensible stand it seems to me.

I realize that the chief hitch is likely to come in regard to wards

* Warfield T. Longcope, then assistant professor of applied medicine at Penn., went on to a distinguished career at Columbia and Johns Hopkins.

† The position Edsall would have had in St. Louis had gone to Dr. George Dock, then professor of medicine at Tulane.

for Edsall and that the chief opposition will come from Dock, but to my mind Edsall we must have . . . There is also in the plans of the Barnes Hospital a wing to the East provided for future extension; this might be built at once for that department . . .

Mr. Pritchett . . . knows Edsall and the difficulties we have had with Dock. I asked him this afternoon, after going over the matter with him, if it were sound policy from the University's standpoint and from our own to make every possible endeavor to obtain a man (Edsall) by creating a position especially for him which without him could not be filled. He replied that it was the soundest policy to do everything honorable to get a man that was needed. This with my own unbounded admiration for Edsall and my feeling that with him we are firmly entrenched made me all the more anxious to get him if possible. Mr. Pritchett will write Mr. Brookings . . . urging him as strongly as possible to secure him.

I have had my say.

Yours,
J. H.[26]

If many of the new men in St. Louis not only expected but hoped that Edsall would replace Dock, a strong personal motive is added to the basic incompatibilities of point of view between the two men. These emerged later, as will be seen.

St. Louis shortly made Edsall a new offer, and we have that story as told by Mrs. Edsall to her mother:

Jan. 3

DEAREST M . . .

I don't know whether all our vague accounts of the troubles at the U. of P. have prepared your mind for any possibility of a change. Things have simply grown worse week by week & some time ago David became convinced that in the conditions and general atmosphere here, & most of the arrangements he was supposed to have done for him *not* being carried out, he could never do his best work or make the big thing of his course that he had planned — he would simply be beating his head against a rock and fighting against odds all the time, & he would waste half his powers that way and never be happy — so in a few days he will hand in his resignation as Professor of Medicine!! After our high hopes and great exultation of last year, it seems a very sad thing, but let me hasten to say that Fate, which usually knocks but once at a man's door has offered him a chance which he thinks equal to that of last year — only different. He is to be Professor of Preventive

Medicine at St. Louis!!! The thing has been seriously planned only two or three weeks (Dr. Howland wanted it long ago) and Sunday afternoon Dr. Howland & Chancellor Houston (of the University at St. Louis) came here & spent the afternoon & evening talking it over. Chancellor Houston took the 1 PM. yesterday back to St. Louis, promising to telegraph this PM. as soon as he could find out whether all David's conditions could be granted. It came this evening, saying, "All settled. Satisfactory. Will wire details" – David had been sitting reading a Book of Limericks, in great suspense lest the thing should not go through. He will write to you himself of the scope of his work & the splendid conditions. He feels he can be very happy in St. Louis & do more for medicine than anywhere else. He went to Baltimore Friday to consult Dr. Welch, who was extremely enthusiastic over it – thought it the biggest opening in medicine for a long time.

Chancellor Houston thinks it the biggest field in Medicine or indeed in any line – David is to have $8000, & freedom to do some consulting work – It has practically all come into acute possibility within the last week & was settled only two days ago so I am still dazed. Don't pass on any of this except carefully to those I spoke of – in a few days David will think it his duty to resign & it will all be public.

For a while we thought that possibly Harvard might be our refuge – always the goal of my dreams – but Dr. Locke wrote within a short time that it now looked more like Richard Cabot & wouldn't be settled for another year anyway.

This is surely the bigger chance, David thinks.

One of the brightest features is that David is to take a year of preparation & he thinks he can spend it to most advantage working in Boston – you see the possibilities in that! Haven't I given you something to talk about?! The new year has come in with a startling bang!

Lovingly

MARGARET [27]

When the second offer came from Washington University, one of the first people Edsall told was A. N. Richards. "Richards," he said, "they say that lightning never strikes twice in the same place, but in my own case it seems to have struck a second time. I have had a second offer from Washington University and this time I think I shall accept it." [28]

Edsall had finally decided that his choice of Philadelphia as against St. Louis had been a mistake of the largest kind. Not trusting his own judgment, he had allowed himself to be persuaded to remain when he had really wanted to accept the challenge of the St. Louis position. Now, instead of leading a new school, he was left to make the best of a new type of professorship in preventive medicine. This time, he wrote Flexner, he made the decision to leave in spite of the fact that he could ill afford to lose the income from consultations which he had built up.

> [I] preferred decidedly to run that risk rather than to risk my whole career here since I can see nothing else in General Medicine that is at all likely to appear and to offer an equal opportunity for a career if I had not accepted the St. Louis offer in Preventive Medicine. My mistake in staying here last year was chiefly due to my feeling that the demonstration the Alumni and authorities made, indicated that they wanted what I wanted and were willing to do so much to accomplish it that I could hardly fairly turn them down. I got this viewpoint from the men with whom I was in touch in that harried period, and in whose opinion I had confidence. I was at first strongly against staying here but feared to follow my own feelings because of my long standing bitterness toward conditions here.[29]

And as he wrote to Dr. Christian, "It is all very well to be a professor of Medicine and have a department that is reasonably satisfactory in itself but that is not what I want. I want to be in a school that is willing to adopt a policy that is likely to benefit medicine in general as well as to put the school upon a high plane that is enough to make one proud of it." [30]

This time he did not intend to let anybody work on his feelings or try to make him change his mind again. "You speak of their having allowed me to leave," he wrote Christian. "I was very careful indeed not to give them any chance to do anything that might tie me here and they knew nothing about my leaving until my resignation was handed to them."

1432 Pine Street, Philadelphia
January 6, 1911

Provost Edgar F. Smith
University of Pennsylvania
Philadelphia

DEAR MR. SMITH: —

I have for some time been very carefully considering whether I could in justice to myself continue longer than this year in the service of the University since circumstances have become so largely altered from those that I supposed existed when I arranged to remain here last year. A decision has unexpectedly been forced upon me now.

Lest we have any misunderstanding regarding the entire suitability of my withdrawing so soon after the extensive changes that have been made in the Medical School within the past year, permit me to point out my main reasons for declining at that time a very handsome offer from elsewhere in order to remain. The chief reasons were as follows: —

1. I had such assurances from Provost Harrison as to convince me that I could work in entire accord with him in the Medical School.
2. I felt assured that all problems would be considered by the governing bodies of the School with a large disregard of personal relations and with customary methods of academic procedure.
3. It seemed ungracious to the Alumni to go against what appeared to be a gratifying display of regard for my usefulness if I could arrange to stay.
4. I was assured that my immediate colleagues desired me to accept the offer made me. This I made an unequivocal condition of my staying.
5. I was offered the Chair of Medicine and the opportunity to reorganize completely the Medical division.

The fifth was an essential reason, but the others were still more important in leading me to a decision. Upon these grounds I made a large personal sacrifice because I thought that I could accomplish even more here than in St. Louis.

Since then Mr. Harrison's resignation has removed my first reason for remaining. My relations with you, cordial as they are, have nevertheless made me see in past conferences, as you no doubt recognize fully, that we have in some important ways differing views regarding the needs of the Medical School and I cannot properly either oppose your views or alter my own.

As to the second reason given, recent occurrences have clearly demonstrated that I differ widely from a large majority of my colleagues as to proper and wise methods of academic procedure, in that I believe that matters of grave importance should not be settled by the Medical Council until they come before a meeting of the whole Council nor until the whole Council has been offered opportunity to consider any reasons for whatever action is determined upon. I have been unable to alter my feeling that this is so important a matter as to give sufficient reason in itself for severing my associations, not because of any particular decision that the Council arrived at, but because of the manner in which decisions were reached and the fact that there is every reason to expect that they may be reached in the same way in future.

The third reason has been wiped out in ways that I need not detail.

The matters thus far mentioned are not remediable to the extent that I could feel here the contentment that is essential to productive work.

As to the fourth reason I find that I was incorrectly informed on this point and that an important colleague cannot work harmoniously under present conditions, if these conditions were to continue. His loss would be serious to the University. It would for personal reasons be a matter of the deepest regret to me to feel that I had been instrumental in this particular person's being disaffected and since everything pointed under any circumstance to my probably remaining here but a short time, it appeared to me also to be entirely unfair to the University to permit anything to happen that would further his possible resignation.

I have within a few days been confronted by another offer from St. Louis which is even more gratifying to me than that made last year. I could not decline it at any time without great sacrifice and under the present conditions I am of course under no obligations to remain; and gratifying as that course would have been I felt that to discuss it with you or with Mr. Harrison could lead to no difference in the result, and would only increase regrets. As soon therefore, as the scope of the work and the facilities offered me was made evident to me I stated that I would accept it as it was evidently more attractive to me than anything else that I could undertake. I have within a few hours learned that all the minor conditions that I suggested have been met and I have therefore notified the Washington University authorities that I definitely accept.

I am glad to feel that accepting the offer at this juncture will remove any possible chance that the University might lose a valuable man through me. I ask you therefore to present to the Trustees my resignation as Professor of Medicine to take effect as is customary at the end of this academic year, that is on September first 1911.

I need scarcely say that however enticing my prospective work elsewhere may be, I regret that my hope that I might be an active and useful factor in the progress of my Alma Mater is not to be realized, and I regret equally that I am obliged to sever my connection with the University of Pennsylvania at the beginning of your administration.

You will note that I have made this letter explanatory to an unusual degree. This as you are likely to surmise was with two purposes in view, namely first of all in order that I might have no misunderstandings with you or the Trustees; secondly there has been so strong a tendency on the part of many persons to attribute unworthy and even dishonorable motives to acts the purposes of which they made no real effort to understand that it appears to me possible, since one owes something to one's standing in the community, that I may find it necessary to permit the text of this letter to become public, though I should of course regret very much being obliged to add that evidence of discord at the University and see no occasion at present for doing this.

With very warm regards, I am, sincerely yours

DAVID L. EDSALL

I have told the Washington University people that they are at liberty to announce it after Saturday, January 7th.

Since writing the above I have just had an urgent message to go to St. Louis to give advice. Since this is the only time for a week to come that I can do this without interfering with stated hours at the University here, I am leaving on the next train, so that I can not find opportunity for three days to come to offer you any further information if you should desire it.[31]

Once Edsall's resignation was announced, the public looked for the resignation of his colleagues, Taylor, Richards, and Pearce. The *North American* quoted Dr. Henry W. Cattell as saying, "There will be no peace at the University of Pennsylvania until Dr. John Marshall is returned to his full professorship of chemistry, and Dr. Allen J. Smith to the chair of pathology."[32]

But there was no immediate breakup. Richards remained to become one of the most famous of Pennsylvania professors. Pearce stayed until 1920, when he became head of the Division of Medical Sciences of the Rockefeller Foundation, where (in Edsall's words) "by his wisdom and judgment [he] had more influence in developing Medicine throughout the world than any other one man."[33] Taylor resigned in 1921 and became director of the Food Research Institute of Stanford University. Since the professorship of pathology had not been filled after Ricketts' death, Pearce taking care of that department as well as his own for a year, Dr. Smith was easily restored to his old chair, and Dr. Marshall (who according to some newspaper accounts had been confined to teaching dental and veterinary students) continued as professor of chemistry and gave lectures in toxicology.[34] Dr. White was elected to the board of trustees and Dr. Stengel's accession to the position of professor of medicine from that of clinical professor restored the normal succession in that department.

In 1912 Edsall described the situation to Simon Flexner:

> Pennsylvania is doing badly enough. They have legislated Pearce and Taylor out of the Medical Council by making a new Executive Faculty, thus eliminating all the "Young Turk" element except Richards who is not aggressive; and they have split Dr. Musser's chair * between Riesman and Sailer, after vainly trying to get Thayer. The latter act they have kept quiet but I am directly informed that they made him what they felt was a very generous offer. That act of Edgar Smith's was, however, the only suggestion of a recognition of the need of regeneration that I have heard of. They all seem outwardly deeply satisfied with themselves. White seems to be in almost absolute control of the Medical School.[35]

* Dr. Musser had just died of the heart ailment which had kept him from seeking the top chair in medicine.

CHAPTER 8

"A Second Time Ambushed," 1911–1912

EDSALL finished out an uncomfortable year at the University of Pennsylvania. By the close of the semester, in the spring of 1911, feelings were so bitter that his friends Hamill in Philadelphia and Locke in Boston decided they would not even attempt to give him a traditional farewell dinner; they held a small one in New York City instead. Edsall had a year's leave of absence from St. Louis to use for study and research, and the research in metabolism going on at the Carnegie Nutrition Laboratory in Boston drew him. On leaving Philadelphia, he took his family to Cataumet on Cape Cod for the summer and then moved to his wife's family home in Milton, close to Boston. There he was right in the medical swim, dining with Locke and Shattuck, and seeing much of the men who were to be his colleagues after another surprising turn of Fortune's wheel. At the Carnegie Laboratory he started some promising experiments on problems of respiration.

Edsall was invited to give the principal address at the winter dinner of the Aesculapian Club, one of the most influential medical groups in Boston. In comparison with the ribald spring meeting, which was likely to be entertained by a play staged by new members from the graduating class of the medical school, the winter meeting was an excellent forum for serious ideas, and Edsall treated it to a forward-looking address on "The Clinician, the Hospital and the Medical School."[1] This speech shows how relationships between these three elements were in a state of flux at that time, with no clear policies for educating the new full-time professor in clinical medicine, or for arranging his work in the best way, or for establishing workable arrangements between medical school and hospital. Academic medicine was coming into being as a new field with a new balance of responsibilities. Said Edsall:

The life of the clinical teacher is and always will be more difficult than that of his academic colleagues because necessarily more complex, for under any circumstances he must have equal academic responsibilities, the personnel of his department must always be more elaborate, and also he must carry the very serious responsibility of the lives and health of the hospital patients under his care. Further than this, entirely aside from the financial aspects of the matter, it is certainly questionable whether he can ever, without greatly endangering his usefulness, relinquish, as is being suggested, the added responsibility and complexity that comes with outside consulting work.

This last point was to become a major battle in medicine for the next decade.

Edsall always had a wide idea of "research"—what it was and what it meant in clinical medicine. He pushed for wide fundamental training to be focused on clinical problems, to clarify the causes of the manifestations of diseases and direct logical treatment. It proved to be a logical approach which tended to eliminate many traditions of the past and substitute the more modern physiologic knowledge. Clinical research began the modern surge of physiologic understanding of human disease.

Continuing before the Aesculapian Club, Edsall said:

Aside from the fact that in the broad sense of the word research alone advances any subject, it is especially important that clinicians be trained into the judicial and inquiring spirit, and especially into the critical spirit that a little experience in research gives, for no one is more constantly an investigator than any thoughtful practicing physician and no one has more important or more complex problems to solve, and his mind must be trained to meet these. By training in research I do not mean simply training in discovering new things. Valuable as these results of research are, there is a more essential influence upon the clinician as an individual. His work is such that he must frequently act empirically and I think all thoughtful clinicians will agree with me that his greatest struggle is in avoiding the danger of falling into the habit of deciding all things upon an empirical basis. While a hypercritical attitude makes a man ineffective, there is nothing more dangerous perhaps in practical medicine than an uncritical attitude.

After considering the best hospital systems operating in Germany, France and England, he offered his own idea of the best system, not by any means a purely Germanic one.

With a suitable combination of the German academic system and something approaching the English ward clerk system, the best results now possible for both students and medicine would seem to be likely . . . It is quite apparent how essential it is that in carrying out this plan there be entire freedom in the academic use of large hospital facilities; that appointments to these hospitals be made by the medical school and hospital authorities together, but on a purely academic basis, and that medical school and hospital work in entire unison . . . But is it possible to have any such extensive change as this involves in the usual relations with hospitals?

He answered this question in the affirmative, pointing out the positive benefits to hospitals from such a scheme:

It is not simply the important fact that these changes will provide men of higher training for the hospitals from the internes to the chiefs, and will imply that the attendants give more time and thought to their hospital work that recommends them to hospitals. It is not merely that they will increase the usefulness of the individual medical school affected in each instance. It is chiefly that they will change the work of the hospitals from simple ministration to the sick of one period to ministration to the sick of the future as well as the present, through providing advanced training as well as new information for the future.

While he was working in Boston, Edsall kept in touch with things in St. Louis. And as 1911 drew to a close the news he received was more and more disquieting.

Despite the wholehearted backing of Brookings and the general approval of the administration and faculty, the new men in St. Louis found that they had embarked on a formidable task. The Johns Hopkins Medical School had started fresh, on an ideal model. The St. Louis men were faced not only with a medical school having two bodies of alumni jealous of each other, and a series of tenuously related hospitals with a variety of trustees; as Marjorie Fox describes it in her history of the St. Louis School, the city had been "a hotbed of professional rivalry . . . since the origin of the second medical college."[2] The new heads were younger than many of the distinguished clinicians in their departments, and they lacked a well-organized tradition. When the faculty accepted the idea of reorganization from top to bottom in May 1909, one faculty member pointed out that it should really be called *organization*, the fact being, as Fox describes it, that "there were 150 teachers for 185 students and each teacher worked independently, without knowing what the others were doing. In some departments work was duplicated, other work completely omitted. There were no heads to the departments in the third and fourth years except for the department of obstetrics and gynecology. The faculty had attempted organization without success and were ready to place the matter in the hands of the Board of Directors." [3]

In a city where many of the profession thought well of what had already been accomplished and considered that things had gone far enough, the newcomers were trying to drag the medical school forward into the twentieth century in one grand swing. Even the machinery for handling routine affairs needed to be created afresh, in a school where costs and endowment were to increase out of all expectation within a few years. In 1909 the school was spending $80,500 for medical education; by 1917 it was the second in the country to put its major departments on a full-time basis with the help of a million-dollar grant from the Rockefeller General Education Board.[4]

The new group were jeeringly called "The Wise Men from the East," and Fox quotes one of them as saying, "We were bumptious, aggressive, perhaps conceited," but goes on to point out, "They knew where they wanted to go, the opportunity had been given them, and their chance at a fresh start and their refusal to compromise their own high standards was the real reason that the reorganization was complete." [5]

Counting Edsall himself, there were seven new department heads in the medical school, so it is no wonder that the shakedown period was stormy. Edsall had had much to do with selecting four of these men: Eugene L. Opie, professor of pathology, Joseph Erlanger, professor of physiology, Philip A. Shaffer, professor of biological chemistry, and John Howland, professor of pediatrics. He had also strongly recommended St. Louis' professor of anatomy, Robert J. Terry, the one St. Louis professor to hold top rank in the reorganization.

Erlanger and Opie were both Hopkins graduates. Opie was a member of the first class in the Hopkins School of Medicine and had worked under Dr. Welch in the pathology laboratory. After ten years at the Hopkins, Opie had moved to the Rockefeller Institute early in the new century, becoming pathologist at the Presbyterian Hospital in New York in 1907. He had been editing *The Journal of Experimental Medicine* for five years when he went to St. Louis. After thirteen years in St. Louis, he taught as professor of pathology at the University of Pennsylvania and at Cornell, and was for three years scientific director of the International Health Division of the Rockefeller Foundation. Since 1941 he has been engaged in research at the Rockefeller University. In 1945 he received the Weber-Parkes Medal and Award of the Royal College of Physicians, in 1946 the Banting Medal, in 1959 the Jessie Stevenson Kovalenko medal of the National Academy of Sciences, to mention only a few.[6] In 1969 he is still working effectively at the Rockefeller University. He expresses great respect for Edsall and his viewpoint and ability.

Erlanger had also taught at the Hopkins, and in 1906 had become professor of physiology at the University of Wisconsin. He stayed at Washington University from 1910 until he retired in 1946. He was the author, with Herbert Gasser, of *Electrical Signs of Nervous Activity* published in 1937, and in 1939 was one of five contributors to *Symposium on the Synapse.* In 1944 he and Dr. Gasser received the Nobel Prize for "discoveries concerning the highly differentiated properties of single nerve fibers."[7] Dr. Gasser became director of the Rockefeller Institute and Dr. Erlanger spent his life in St. Louis. On his retirement the St. Louis Medical Society gave him their certificate of merit and gold medal for scientific accomplishment.

When the St. Louis committee approached Otto Folin of Harvard to head their department of biological chemistry, Folin refused but recommended a former pupil, Philip A. Shaffer. Dr. Shaffer had trained with him at the McLean mental hospital of the Massachusetts General Hospital. Shaffer had received a Harvard Ph.D. degree and then, as assistant and instructor in pathological chemistry at the Cornell University Medical College, had written several papers with Eugene DuBois on the metabolic abnormalities of endocrine disease. When chosen to go to St. Louis he was not only working for Cornell as physiological chemist for the Huntington Fund Cancer Research, but was also head of the chemistry section laboratory at Bellevue Hospital in New York under Eugene DuBois and Graham Lusk. He stayed at Washington University Medical School as professor of biological chemistry until 1946, serving as dean for fourteen years, the first time from 1913 to 1919 and again from 1937 to 1946.

The fourth of this brilliant group was John Howland, a New York University Medical School graduate who had worked for years with Emmett Holt and had risen to become the head of the Children's Clinic at Bellevue in 1908. He was universally loved and respected. After leaving St. Louis in 1912, he became the first full-time professor of pediatrics at Johns Hopkins, where he remained until his death in 1926.[8] Under his leadership, the full-time idea in pediatrics was so successful that Wilburt C. Davison in his Profile said that it "established alone and by itself the full-time movement in American medical education."[9] * Davison went on to quote one of Howland's students: "Howland was a combination of Winston Churchill, Esculapius, and the Czar of Russia. I don't expect to live to hear a man who was even in his class at giving a clinic."

The only member of the old staff whom Edsall recommended to head a department was Robert J. Terry, professor of anatomy, a graduate in 1895 of the old Missouri Medical College which later was made part of Washington University. Dr. Terry had the kind of background they were looking for. After he received an A.B. degree in Washington University in 1901, he worked abroad in Edinburgh and Freiburg and was an Austin Teaching Fellow at Harvard in 1906–07. He taught at the Washington University School of Medicine as professor of anatomy from 1901 on, and was head of the department after 1928. In 1941 he became emeritus professor.[10] (When we visited him in 1963 his memory of these years of teaching was mellow and uncritical, but in the memory of his former students he had been a very able, successful instructor.)

The group was headed by the man who filled the position that Edsall refused when he elected to stay at Pennsylvania in 1910—George Dock, professor of medicine and dean of the medical school.

* From Davison, W. C.: John Howland (1873–1926), J. Pediat. 46: 473–486, 1955.

George Dock came to St. Louis from the top medical position in New Orleans (another position which Edsall had turned down when it was offered to him). He had been professor of pathology at the University of Texas and for many years professor of medicine at Michigan. He had a reputation as one of the most scholarly and best-known clinicians of the time. Dr. Dock received his M.D. from the University of Pennsylvania in 1884, and had been made an honorary doctor of science in 1904. He also had an A.M. from Harvard University. He started out as assistant clinical pathologist at Pennsylvania, where he was one of William Osler's assistants. Fox recounts that "he was a personal friend of Osler and shared with him his delight in old and rare books. From this interest the Washington University Medical Library was to benefit greatly."[11] He was president of the Association of American Physicians in 1916–17. His stay in St. Louis was a series of battles and he resigned in 1922 and went to California without a teaching position.[12]

Erlanger, Opie, Dock, Terry, Shaffer and Howland met together for the first time at the AMA meeting in St. Louis in June 1910. The world of medicine was watching. Pritchett wrote Brookings, "You have certainly got a stunning group of men together. I do not think the like is to be found in America." Dr. Welch said that the new developments "marked a second epoch in medical education as the Johns Hopkins had the first."[13]

It is worth noting that it was a young group. Dr. Dock was the oldest at fifty, while all the others were in their thirties except for Philip Shaffer who was twenty-nine.

For administrative purposes this new group were soon organized into an "Executive Faculty," an idea which Erlanger had brought from Hopkins.[14] To begin with, Howland was not one of them but went abroad to study with Czerny before taking up the leadership of the department of pediatrics and the St. Louis Children's Hospital. As Fox describes it, "Dr. Howland felt that pediatrics, at the time, was the weakest subject in American medicine, and in organizing a large children's hospital on a modern basis the University was doing something that had not yet been done in this country. There was no model to follow. The organization and equipment must be different from anything in existence."[15]

Dr. Terry remembers that the new men were suspicious of

everyone connected with the medical school, and that the atmosphere was one of constant bickering.[16]

One of the first changes to appear was the enforcement of a one-year college requirement for entering students. As a result, enrollment in the fall of 1910 dropped to 122 from 185 the previous year; in 1911 the enrollment was 107. By 1912 the requirement had been raised to two years of college, and the entering class had begun to grow again, but with the requirement of three college years, the entering class had only 5 members.[17] (Many of the missing students were taking the University's premedical course.)

An early problem that the Executive Faculty tried to solve was that of revamping the hospital scheme and providing "relatively adequate laboratories" until the new buildings were ready. The 125-bed Washington University Hospital, housed in the old Missouri Medical College building on Jefferson Avenue, was, they considered, in shocking condition, and they conceived the idea of altering it "so that it might be used as a model and the new men could live in a scheme that could be moved into the new hospitals." [18]

Next they made two appointments to take effect February 1911: Fred T. Murphy as the professor of surgery, and David Edsall as professor of preventive medicine.

Fred Murphy had his M.D. from Harvard Medical School in 1901, and after being assistant in anatomy became an Austin teaching fellow in surgery and later assistant in surgery at Harvard. From 1904 to 1908 he was assistant surgeon of the Infants Hospital and surgeon to outpatients at M.G.H. from 1907 to 1911. Later, he was to spend two years in France in the First World War, winning a Distinguished Service Medal, and in 1918 he became director of the medical and surgical department of the Red Cross, representing the chief surgeon of the American Expeditionary Force. After the war he moved to Detroit.[19]

At the time of Edsall's appointment, Brookings renewed his efforts to get financial backing from Carnegie. He wrote to Pritchett about Edsall:

> Although we have recently classed him as an internal medicine man, he occupied for a long time the chair of pharmacology at the University of Pennsylvania and his training generally has been so broad and thorough that it seems to be no surprise to our people here

> that he has concluded that preventive medicine offers the widest possible field for usefulness. I will say to you frankly that I have been compelled to personally assume the growing added expense to this movement, which, unless I am able to secure considerable money either here or out of the city, will greatly cripple my ambition to develop other departments of the University — more especially our graduate school . . . It does seem to me that the unequalled group of men we have secured for our medical department should meet with some encouragement both at home and abroad, and I have great faith in being able to secure one or two million dollars within the next two or three years. I have never given up hopes of your ability to interest Mr. Carnegie after we have shown our faith as we have done.[20]

This particular hope never did bear fruit. Next Brookings tried to interest Mrs. Russell Sage in endowing the department of preventive medicine, but with the same lack of success.

Edsall wrote to Locke about his St. Louis appointment: "I am going to St. Louis after all. They have made me a wonderful offer with the broadest possible lines of teaching, research and public work in Preventive Medicine with generous and fine clinical facilities for research and teaching and with general consulting practice. It does not take me away from medicine or clinical work — it rather skims the cream off. I have the widest opportunity yet offered me — and now it's up to me to make good."[21]

His salary was to be $8,000, his department having a budget of $15,500, and he was promised approximately 30 beds in the hospital.*[22]

Edsall indeed had a sweeping view of the scope of the new department and his vision made the front page of the St. Louis *Republic* when his appointment was announced.

> The purpose of this department will be to conduct instruction and research in the various phases of preventive medicine. It will probably consider such problems as heredity and environment; influence

* It seems likely that the 30 beds were not in the existing University Hospital but in the plans for the Barnes Hospital. As remodeled in the summer of 1911, the University Hospital had only 31 medical beds in all. (Fox, *History of the Wash. U. School of Medicine.* Ch. 4.) When Edsall was leaving Washington University he wrote Locke to give as the reason "the slowness in building the hospital & other buildings made it inevitable that I should have at least a year & perhaps more time from now on without any facilities for my work." (Edsall to Locke, April 20, 1912. [Aub])

of occupation on disease, especially of prolonged strain in producing disease of the vascular system and nervous system; influence of occupation on infectious disease, on children, and on women directly and in relation to progeny; the principles of diet and nutrition; foods, their values and preparation; choice of economical diets and determination of diet values in types of disease; exercise – physiological effects of it and of the lack of it; infant mortality; institutional and school hygiene; infectious diseases, the manner of their propagation and prevention; public and private sanitation, the responsibility of the State and of the community, and methods of State and community control in regulation at home and abroad.

Undergraduate and graduate instruction and research will be conducted in these various fields. Opportunities will be furnished especially for practicing physicians to do special work. Certain courses will be open to those working in health departments and to social workers and teachers . . .

Public lectures will be provided on topics of importance to the community, such as water, milk, foods, diet, drug and other habits, infections and epidemics.

What has been done in Cuba, in Panama, and in sections of the United States in stamping out diseases such as yellow fever, malaria, tuberculosis, smallpox, typhoid fever and, more recently, hookworm disease, is matter of public knowledge.

It is not too much to hope that very great progress will be made in other directions . . .[23]

As the year 1911 advanced, the struggles to get the new system working became fiercer. Hoped-for outside money had not been gained. The agreements with the Barnes and Children's Hospitals had not yet been committed to paper. Dock's shortcomings as an administrator had become all too obvious. He seemed unable even to get out a catalogue for the coming year. As for Chancellor Houston, he appeared to believe that the "reorganization" had been accomplished with the arrival of the new men, and seemed oblivious to the problems of creating a whole new school.[24]

Dock and Houston, in a pacifying gesture to the old guard, had put two former heads of departments, Dr. Washington Fischel and Dr. Norman B. Carson, on the Executive Faculty. To men who were on fire with enthusiasm for building a truly outstanding new school, this was a nullifying act. Opie wrote later, "The Executive

Faculty so constituted, we believed, gave little promise that a school of the best type could be established under this direction."[25]

The old rivalries between the original medical schools continued to plague their councils; the active alumni of each original school far outnumbered those graduated from the combined school, and when in the summer of 1910 the old St. Louis Medical College group gained a "victory" over the Missouri Medical College men in the reorganization of departments, it made for renewed bitterness.

By the end of 1911, the news reaching Edsall on his year's leave in Boston was so alarming that, as he described it later, "I decided to break up my research &c here to go to St. Louis because some other general matters in the school had been giving me very grave concern and I felt that if on the ground I might help."[26]

He was planning to start teaching as soon as he got out there and had written to Terry about his course. He wrote to Opie in January, "I judge from Terry's reply to my letter outlining my course that I can smell war from Dock. I have expected it & indeed wanted to draw his fire early."[27]

He went into more detail about his problems with Dock in writing to Simon Flexner:

Last year when I arranged to go to St. Louis it was only after I had told Chancellor Houston that the arrangement as to hospital wards, method of teaching &c. was such that Dock might easily object decidedly & I had to have an official statement of his entire approval before I would accept. Mr. Houston sent me word that everything was completely satisfactory. Everything seemed to be until one day in May in New York when, to my astonishment, Dock stormed at me. Said the St. Louis people and I deceived him absolutely, that he had had no notion that I was to have wards or any clinical relations in teaching or otherwise, that he had legal evidence of control of all the medical clinical facilities and would fight it all to the end. I went to St. Louis and saw the Chancellor & he went over it all in detail & certainly Dock *seemed* to approve at the time but he made certain reservations which had seemed to the Chancellor of no moment, but which he (Dock) now says meant that *all details* were to be arranged in the future . . .

It is perfectly apparent that he (Dock) will fight absolutely everything about my department that makes it worth while to me, and even if the authorities suppress him entirely, as I think they would try to do, it would be a wrangle over everything and constant hampering in every way, which he could well do in ways that could not be controlled, because he has entire control of the dispensaries which feed the wards & for other evident reasons. I can not yield anything, for the department is a little exotic anyway, for a man of my interests & training, and I need all I was promised to make it a success.[28]

Later Edsall learned that Dr. Dock had opposed him from the beginning "and agreed to it only on condition that I go out of general medical relations entirely, and work in connection with patients simply by courtesy in his wards, & in a purely investigative way. Confining myself to *preventive medicine purely*."[29]

Apparently each man felt the authorities had promised him what he wanted, but the promises proved to be contradictory.

To make the situation all the more bitter, Edsall learned in January of 1912 that the Harvard place was again within his grasp. He wrote to Flexner:

Within the last few days I have been approached in such a way here that there is no doubt, I think, that I could have Shattuck's chair. One of the high authorities here told me all the details of what has been done & asked me *officially* to let him have some word from me within three weeks. He said they were considering everyone in the field & making inquiries of some other men as to what they would demand *whether they thought they could get them or not*, but I know that I head the list. I thought they were not considering me after what I had previously said. This brings the whole question into very concrete form. I am quite aware that owing to the strange situation that has come up in St. Louis I am a second time ambushed.[30]

Much as he might have wished things to work out differently, it never occurred to him not to honor his agreement. He left for St. Louis January 30, leaving his wife and family at home in Milton, and moved in with his friend John Howland at 5227 Westminster Place. He had arrived in the midst of a very complicated situation. The Executive Faculty were trying to build Rome in a day, and their difficulties were compounded by the fact that the university's chancellor had well earned his nickname of "The Sphinx," and used silence and procrastination freely to help him ride out the difficulties. The situation made the worst impression on Edsall. It was probably in early February that he wrote to Locke, "If things don't come out right here now, and show definite evidences of doing so in another year I will go down on Cape Cod and grow oysters."[31]

On the 11th of February he wired his wife that he was staying on in St. Louis, and on the 12th she noted: "Three letters from D. all gloomy of St. L."[32]

Howland had arrived in January and his situation seemed typical of the way things were going. As Edsall described it to Christian later:

He was told when he arranged to come out that they had the money to build and maintain a children's hospital of 150 beds and that it would be finished by this Fall. The money has dwindled to $125,000 and the managers (women) say they have no fixed funds to run it on & are frightened at the prospects of the expense of a new big hospital. Last January the University authorities told him they would proceed at once with the plans and construction of the hospital (& would get the money needed). And with the formulation & signing of a contract with its managers. Outline plans were drawn weeks ago by Goldwater & Howland, but although Howland has inquired repeatedly he can not learn that anything else has been done. The little hospital that he has to work in now & will at best have to use for eighteen months or longer is a wretched place — I have seen few less attractive.[33]

To Flexner, Edsall wrote: "Poor Howland is in a pickle, with a wretched hospital, no *definite* prospect of a new one and not adequate funds to run it if built (no endowment at all) and the plans they made with the local men, that he knew nothing of and was promised the contrary of, have put him in a most embarassing & annoying position."[34]

Edsall's own situation seemed impossible. In Philadelphia he had been kept out of the laboratory; now it seemed as if he was going to be kept out of the hospital. However, he had an interview with the chancellor on February 10 which promised better things. He wrote cheerfully to Flexner:

I had a long talk with him this morning, and suspicious things all disappeared. He wiped away all possibilities of my not being able to do my work properly and suggested himself that he had better put it all into writing so that there could be no possible thought of error or charge that I had been underhanded in any way. He also explained all the delays in a way that is satisfactory. Howland's hospital is pledged and he is authorized to go ahead at once and see that it is made what it should be . . .

Suspicious as I have grown of games of any sort, I feel satisfied that I am doing the right thing here and I especially feel that it was necessary as while the chancellor and the other men were very fair about everything I know they would have thought it unfair of me to have left here now, and under the circumstances I think they would be right.[35]

For a while, it seemed that things might be going to work out. Dr. Terry said of Edsall at this time, "His conduct in our meetings was so sound and trustworthy, that I was very much attracted to him."[36] But various things happened in the following weeks which Edsall termed to Christian "active and passive rank bad faith on the part of the authorities."[37]

As early as February 22 he was writing to Locke, "the situation here is, to my mind, much less satisfactory than it seemed to me I had any right to write officially to the committees at Harvard . . . everything may turn out all right . . . I do not feel the deep confidence in the fairness of the authorities that I did but I got into such a position that the question was whether I would take some risk—I think, myself, still a large risk—of things going wrong for the sake of the moral support it would give or run the very imminent risk of damaging the enterprise deeply."[38]

Between "Dock's utter incompetency" and the chancellor whom he regarded as "the most dull and lethargic administrator possible"[39] he at last gave up hope of a better outcome. One of the last straws seemed to be: "The Faculty even heard at afternoon teas from women and by chance meetings of faculty wives with the Chancellor's wife that things they had been anxiously asking about—such as the hospital agreement—had been settled for weeks."[40]

That spring in Boston the committee charged with finding Dr. Shattuck's successor wanted Edsall if they could get him, both Harvard Medical School and Mass. General Hospital members being in agreement. (One of the committee members, Dr. James J. Minot, has left an interesting sidelight on his selection: "It was said to me 'of course you will select a man from the M.G.H.' I said I did not care where he came from, provided he could speak

English and was able. I wanted the best. There was quite an objection to a stranger . . . I believed we wanted a young man who had vision and could do constructive work and not one who was too old to start and invent new things and do research, &c. Some of the committee flattered me by asking me to be a chief, I said no, not under any conditions. I am too old. I had served in the O.P.D. some 20 years and finally reached the house. 60 years was close to my age then."[41])

In February Edsall had given Harvard a definite refusal. But that spring Edsall spent a good deal of time traveling between St. Louis and Boston. In February his wife's youngest sister Eleanor came down with "quinsy" which was soon clearly scarlet fever and ten days later she was desperately ill. Her brother Dr. Wilder Tileston and Edsall both fought to save her, staying at her side night and day. Several times she rallied, but her heart had been affected and she died March 7th.

Returning to St. Louis March 13, Edsall came home again on March 25th to preside at a two-day School Hygiene Association meeting in Boston. This was the week before Easter, and when Dean Christian heard he was in town, he called up Locke to find out if Edsall had had a "change of heart." Locke replied, "I knew the situation in St. L. was awful and David very unhappy." Christian empowered him to find out the exact terms on which Edsall might be willing to come to Harvard, and Locke and Edsall walked up and down the train platform at South Station before Edsall returned to St. Louis, while Edsall outlined his terms.[42]

On his return to St. Louis, Edsall wrote to Locke, "I find things much more upset than ever out here, and all the men are on the verge of an explosion and this time it is clearly the fault of the authorities and the men are all pretty well convinced that they are only puppets in the game that is being played. This is of course largely confidential. If they make me an offer from Harvard in a form that I can afford to accept, I shall beyond question accept it, and now I could accept it with a feeling that my colleagues would not only not think it unfair, but any of them would probably do the same thing themselves if the chance offered."[43]

Within a week came the big explosion, which nearly brought an end to all their endeavors. The Executive Faculty were ready for desperate measures, and sent a manifesto to Chancellor Houston, which Edsall described as "the stiffest letter I ever saw."[44] Howland, Opie, Erlanger, Edsall, Terry, Murphy, and Shaffer all signed this formal letter, which expressed their multiple dissatisfactions in one grand blast.

They pointed out that the Executive Faculty was supposed to have "exclusive medical control not only of the medical school but also of the hospital," but against their wishes, the former professors of medicine and surgery had been added to their number and "in consequence for one year the Executive Faculty became confessedly a useless body." The position of superintendent of nurses had been filled without consulting them, as had the chief position in the new department of social service. Their recommendations to the university had not been acted on, including changes in the obstetrics department, the affiliation with the Skin and Cancer Hospital, the agreement with the Children's Hospital, and improvement in communication with president and trustees. They complained of endless administrative delays. Plans that they had made for the Barnes Hospital had been changed. They ended:

> A group of men has been brought together in a school without efficient organization. Almost every item of the older organization of the Medical School and Hospital has proved so defective that it has been necessary to devise new methods and put them into effect. Until this time-consuming work is complete, the School cannot even begin to render the service expected of it, or start on a progressive career and in the meanwhile all time lost endangers its reputation. It has already to our knowledge, suffered sharp criticism because of certain inefficient executive actions.
>
> We feel that the most serious responsibility of the School is to the lives and careers of the large group of senior and junior men who have been brought here. Lack of confidence in the future offers a serious danger to our organization.[45] *

* This incriminating letter vanished from the archives of the school and from the hospital files and a copy was only found in the personal files of Philip A. Shaffer. I asked him for it when I was in St. Louis a few months before Dr. Shaffer died. This letter details the causes of such dissatisfaction among the new appointments that it resulted in Edsall feeling justified in accepting his latest offer at Harvard, while Howland left before he had another post, to do private practice in New York City. Dr. Veeder and Dr. Shaffer both told me that Howland resigned because Negroes were not to be admitted to the Children's Hospital, which he thought was unfair and limited the source of patients. Apparently Dr.

Edsall felt that signing this letter had forced his hand. He telegraphed to Dr. Christian, who went at once to President Lowell. Lowell, confident of the Corporation's backing, acted at once. Edsall told Locke, "I have a telegram from President Lowell saying I will be appointed next Monday by the Corporation. It was in response to a telegram from me to Christian saying things were at such a point here that my hand might be forced at any time by the action of my colleagues." [46]

Edsall resigned on April 18. In reply to his letter of resignation, Houston said he did not wish to discuss the resignation, nor the executive faculty's letter, except to comment: "As to the essence of the communication, I shall only say in passing here that much more has already been accomplished than it was thought could be accomplished in so short a time, and that all the essential recommendations made by the Faculty, so far as the Board of Directors of the University had power, have been accepted." [47] At that, it had taken him two weeks to reply.

Once the decision had been made, Edsall wrote to Locke:

> Things are in such shape here that I cannot see how it could have been anything but bitter unhappiness for me, and Mrs. Edsall would have shared that . . . I have never felt so indignant in my life as I have over the treatment I have received here since they put things in such a way in February that I turned down what I wanted so much. Mr. Houston made to me & to Opie, Howland and Terry statements as to what he would do at once and he has done nothing at all about the important things and I have not *seen* him, or heard from him about them, nor have the others, in spite of the fact that three weeks ago and more all the faculty except Dock met him by demand, talked absolutely plainly to him about everything, and he said that he would act within a week. I believe (though I don't know) there are difficulties in collecting the money they said they had, or that local pressure has gained the upper hand — or else they have just plain decided that they will not do what they promised from the beginning. However I can put that all aside now and start among people I trust . . . I can not tell you how much indebted to you I am for all that you have done throughout several years past and which has had so much effect. I

Howland thought the Negro children at Johns Hopkins were among the most interesting cases of the clinic. *J. C. Aub.*

feel that matters never could have materialized but for the influence you have exerted. It means even more to me than you realize.[48]

"Worse than Philadelphia," Edsall felt about what he had been through, and wrote to Simon Flexner:

"The action of the authorities is still simply absolute silence and inaction. It is almost certain they are in some plight—probably money has a good deal to do with it, but they won't say anything. Many details are turning up to show that they have bungled things far worse than was done in Philadelphia."

Then he added, "Personally I have grown to feel that the way I was treated in Philadelphia is almost agreeable to look back upon in contrast with the experience I have had here since January but it involves all the men here and I am dreadfully sorry for them."

And scrupulous as ever, he added, "One thing I am very glad of. If I had left in February my colleagues would have felt aggrieved & I should have been in some doubt myself. Now I know those to whom I have talked of it feel that I am perfectly right and I think the others will. Indeed most of them would do the same thing if opportunity offered."[49]

Apparently the only bright point in the whole thing was his association with his colleagues. "The fine teamwork among the men in St. Louis (Dock excluded) is the only thing that has saved that place from wreckage of the most complete character," he wrote to Christian.[50]

One thing that particularly exasperated Edsall after his resignation was the fact that he was supposed to have been playing politics.

The Chancellor also tells the Trustees & the Faculty that I promised him to stay there in February when I had no real call from Harvard but fully intended to accept it if it came & did so as soon as it came,—although I told him that I definitely put myself out of the running and only took it the second time around after Thayer had declined & his whole Faculty had meanwhile got up in arms in their alarm over the situation. The Faculty all told him they knew this to be true but he just waives that aside and says he knows I meant bad faith. It is curiously near to the same experience I had in Philadelphia when I made a decision (to stay there) against my desires and better judgment because I felt I should damage a general scheme if I did otherwise and also because my friends urged me to do so as the right thing to do, and then I was accused by the enemy of having played a game to increase my own power and returns, and had no other thought than to stay there.[51]

As a final gesture, the university authorities stopped Edsall's salary while they still owed him $1200 for services rendered and asked to be reimbursed the $4000 he had received during his leave of absence. This he promptly and indignantly paid them, though he had to borrow money to do it. He then asked for the $1200 due him which was paid.

Simon Flexner wrote to Edsall in July that the situation in St. Louis

makes an altogether extraordinary and depressing impression . . . I take it that their heart is in the right place, but when the crucial moment came they found they could not command all the resources that they had expected to command, and then, probably, still hoping to accomplish these things in the future they adopted a policy that was not, strictly speaking, sincere. I fear that they have had a terrible awakening, and at the same time have felt aggrieved toward yourself and probably toward Howland. On the other hand, I do not see that your conduct is in any way open to criticism, but in a contest of that sort one rarely sees justice done in the heat of the contest. I congratulate you, however, upon your escape.[52]

And a little later in the month, still trying to see the best of both sides, Flexner wrote, "I take it that you will put the whole matter now as much as you can behind you. I am, I must confess, loath to believe that your views on general university administration are quite justified by existing facts."[53]

I quite freely admit that I am overstrained in the matter [Edsall replied] and that what I chiefly need is to put it all behind me. But I have had two experiences so far removed from what can justly be called common honesty on the part of University authorities that I am a little misanthropic, I imagine, and I have heard so many things of similar kind of other University authorities, especially in the middle & far West, that I can't help thinking the whole system is pretty rotten. But it has got on my nerves a bit — they have been rasped a good deal for fifteen years past — and the thing for me to do now is to take all the advantage I can of the happy surroundings I am in now, where I continue to find that promises are never *more* than what is done and usually less.

I fear I have bothered you a good deal and wholly unnecessarily but conditions proved to be such this spring that if the Harvard call had not come around to me again I would have resigned nevertheless and gone into practice – and the crises I have been going through, which have really been rather severe, rather shook up nerves that have been pretty tense for years.

I have finally got to a place where I fully understand the *moral* attitude of the authorities & faculty. In Philadelphia & much more in St. Louis (as to the authorities) I felt when it came to any "show-down" that we spoke different languages.[54]

Fortunately for him, he had reached the place where he belonged at last and was welcomed by both colleagues and administration.

After Edsall came to Harvard, he did his best to ease the way for Howland to succeed the Harvard professor of pediatrics, Dr. Thomas M. Rotch, but that did not work out. Howland resigned anyway. After the explosion in April, Edsall noted, "all the other men in the Faculty (except Dock) have said to me that they would most seriously consider any offer from elsewhere."[55] Shaffer was inquiring about the chair of chemistry at Cornell, but felt he should probably try to stay in St. Louis and did so. The storms in St. Louis blew themselves out. Brookings, William Bixby, and Edward Mallinckrodt had moved fast to save the situation. Opie became dean, and the deanship was put on a yearly basis thereafter. Drs. Fischel and Carson were removed from the Executive Faculty and placed on an advisory committee. Howland went first to New York and then to the professorship of pediatrics at Johns Hopkins. The rest of the team carried on.

The agreement affiliating the Children's Hospital with the medical school was signed in July 1912. Construction of the new medical center at Euclid and Kings Highway began the same summer. In the spring of 1913 Chancellor Houston resigned to become Secretary of Agriculture in Wilson's cabinet.* April 1915 saw a

* Close-mouthed to the last, he journeyed to Washington on the same train with several Missourians, including a member of the university corporation, without ever telling them that he was going to take the government post. (Opie, "The Reorganization of the Wash. U. School of Medicine." MS.)

big celebration to mark the completion of the medical center, with distinguished delegates and speakers from all over the country. In his address Dr. Pritchett said that the Washington University School of Medicine "perhaps more completely realized the ideals of our day than any other institution in the world."[56]

Endowments had come slowly, and in the end it was largely Brookings' money as well as vision which carried the school forward. (In 1914 he estimated that of the three and a half to four million dollars he had given Washington University, about one and a quarter million had gone to the medical department.) He was seconded generously by other St. Louis people, including Adolphus Busch, Edward Mallinckrodt, William E. Bixby, John T. Milliken, and Mrs. Mary Culver.[57]

From the early days of the reorganization, Brookings had been interested in the question of full-time appointments, and had hoped to see them established in the school. After Johns Hopkins accepted the Rockefeller grant in 1913 and became the first medical school in the country to have its major departments on full time, Brookings appealed to the General Education Board to give the same assistance to Washington University, and the Board eventually granted them one million dollars if they would provide half a million to complete the fund. In 1916 and 1917 the three major departments were established on a full time basis, with Dr. George Dock and Dr. G. Canby Robinson heading the department of medicine, Dr. Fred Murphy and Dr. Ernest Sachs the department of surgery, and Dr. William McKim Marriott and Dr. B. S. Veeder the department of pediatrics. This was strictly the Rockefeller system: by contract with the General Education Board, "the members of the full-time staff were free to 'render any service required by humanity or science,' although they would derive no pecuniary benefit from it, and the fees charged by the hospitals for professional services rendered to private patients, within or without the hospitals, must be used to promote the objects for which the fund was created."[58]

Edsall was to see a different system worked out at Harvard.

Brookings' services to medicine were recognized in 1929 when Washington University gave him not only an honorary LL.D. but an M.D. as well.

CHAPTER 9

Edsall's Third Revolution, 1912

"A Minor Medical Revolution: Dr. Edsall's Coming" — "Dr. D. L. Edsall, New Professor at Harvard Medical School, Has Gained World-Wide Fame" — said the Boston papers when Edsall's appointment came out. Dr. Richard Cabot wrote a long article for the *Boston Transcript* announcing the major new developments in Boston medicine. Introducing Dr. Edsall, Cabot wrote:

> His special technical equipment represents something rather new in clinical medicine. He does not represent a clearly defined new order of things, with a division so sharp as that which Lister brought into the practice of surgery; yet in a very real way he does stand for that new combination of chemical and pharmacological research which is the most striking and most promising movement in the medical world of today. This new movement has developed gradually, Ehrlich being one of its earliest and still most famous figures. Dr. Edsall's technical equipment is therefore something rather new, and no Boston medical man has as yet really matched it.

As to what this would mean to the Massachusetts General Hospital, he wrote:

> Boston has until very lately had no adequate inducements to offer to a man like Dr. Edsall. The archaic organization of our hospitals gave no adequate opportunity for such work as Dr. Edsall's until the staff of the Massachusetts General generously sacrificed their own personal interests to a reorganization which made it possible to offer Dr. Edsall an adequate place.

Dr. Cabot then went on to outline the changes at the hospital, which had instituted continuous service in its wards a few years earlier, and at the Harvard Medical School, where a new system of appointments made jointly by school and hospital was beginning to bear fruit.

Hitherto it has been necessary [Dr. Cabot continued] for the school to appoint as a professor of medicine some man who already had a hospital service which provided him with the cases to serve as the basis for his instruction of students. It was impossible to call from the outside the man whom the faculty of the school might most desire, because the school had no way of providing him with the hospital material necessary for purposes of instruction. In Dr. Edsall's case, the generous cooperation of the staff and the trustees of the Massachusetts General made it possible for the school to call a strong man from outside, and assure him of adequate hospital material. A similar situation is already assured in the Peter Bent Brigham Hospital, where the school nominates men for the positions of surgeon and physician to the hospital. Drs. Christian and Cushing, as professors in the Medical School are therefore assured of ample teaching material in the Brigham hospital. In addition, the Brigham hospital, as a departure from the usual practice in this country pays salaries to Drs. Christian and Cushing. Still another teaching asset for the school will be obtained by the appointment as a professor of medicine of Dr. George Sears of the Boston City Hospital staff.[1]

In the *Harvard Alumni Bulletin*, Dr. Cabot wrote of Edsall at this time:

His achievements within the vast field of medicine have revealed both breadth and depth. They have been characteristic of the best tendencies in modern medicine. He combines as no other man in this country does today the scientific and humanitarian interests . . . Dr. Edsall's researches have put him at the head of the practising physicians of America in the application of chemistry to medical problems.

But he is also in the front rank of those physicians who recognize public duties of the physician for the prevention of disease by education and through public authorities . . . He has done valuable work in the field of industrial hygiene.[2]

This was really very handsome of Cabot, who had wanted Shattuck's place very badly, and had expected that the choice would fall on him. But he settled down under Edsall, and, when I was a house officer, was not only very active on the ward, but behaved extremely well towards Edsall all the time. After a while Cabot gave up his position as MGH chief of staff and went to Harvard as Professor of Social Ethics, although he continued active teaching of medicine for years. However, I do not think that he moved because Edsall was at the General, but to keep his enthusiasm about his work high. He once told me that everyone should change his job at fifty, and I think that's what he did when he transferred to Harvard College. Cabot always remained a good friend of Edsall's, always thought extremely well of him and his work.

Edsall, who had left St. Louis May 9, went to a medical meeting in Atlantic City before reaching home May 15. Two days later he went to a dinner given by Dr. Cabot and Dr. Shattuck for him to meet the faculty, and the following night to an MGH dinner. Then Edsall and Cushing were introduced as new professors at the triennial dinner of the Harvard Medical Alumni Association on May 22nd. Dr. John Collins Warren presided, and speakers included Francis G. Benedict who spoke on the work of the Carnegie Nutrition Laboratory, Dr. Milton J. Rosenau who spoke on "The Department of Preventive Medicine and the new degree of Doctor of Public Health," and Dr. Ernest Tyzzer, director in charge of research at the new Collis P. Huntington Memorial Hospital. Both Cushing and Edsall spoke briefly, and the meeting ended with three long cheers for Dr. Warren and the singing of "Fair Harvard." [3]

So Edsall arrived at the Massachusetts General Hospital as head of the East Medical Service, to succeed Dr. Frederick Shattuck. In June he delivered the Shattuck lecture before the Massachusetts Medical Society, largely a report of his respiration research at the Carnegie Nutrition Laboratory.[4] On July 1 he began his service at the MGH, taking an occasional week off to climb Mt. Monadnock with his family, to go walking in the mountains with Carl Binger, and to visit Cataumet in August.

The old hospital as it appeared to Edsall can still be seen behind the imposing facade of the White Building, the Phillips House, and the Warren Building which now dominate the skyline along the Charles River. In the sunny quadrangle at the back,* with its green grass and flowering horse-chestnuts, its honeysuckle and old roses blooming in the spring and magnolias flowering by the sweep of stairs up to the Bulfinch Building, we can see the hospital as it used to appear in the days when the Bulfinch was the main building. Designed by Charles Bulfinch (the noted Bostonian who created

* Obscured momentarily by the present round of new building.

Boston's State House and completed Latrobe's plan for the National Capitol in Washington), the Bulfinch Building is not only a beautiful example of early nineteenth century style but is still a practical and useful hospital more than a hundred and fifty years later. When it was built it even had water closets installed, which makes it the first American institution of this type to include such a daring innovation.[5]

Edsall had his office at the Blossom Street end of the Bulfinch Building. Francis Rackemann, who was a house pupil that year, has given us a description:

> The main entrance to the hospital was at the east end of the Bulfinch Building. Here were the offices of the Superintendent and his assistants, as well as the offices of the Nurses' Training School. There was a small information desk with a telephone switchboard behind it. Patients were brought to the hospital by horse and carriage, including an M.G.H. horse ambulance driven by Tom Lee, the hackdriver. On arrival they were carried on a stretcher to the wards upstairs, or else wheeled on a stretcher-truck to the Accident Room, which was in the north end of the old Bigelow Building . . .
>
> In our day, the four medical wards were in the east end of the Bulfinch Building and the four surgical wards in its west end. Each ward held about twenty-eight beds, and each was a large, almost square, room, with a big chimney in the center and fireplaces in the front and back of it. On cold bleak days in winter, the open fire was cheerful and comfortable, but it was not often lit. When the Bulfinch was made fireproof in 1934, the chimney was removed; comfort gave way to efficiency.[6]

Between the Bulfinch Building and the river was a group of old buildings, most of which have been torn down since: the old outpatient building, the operating building, and the domestic building. Then there was a series of one-story wards which had been built when Listerism was just beginning. It was expected that they would become "infected" and would then be torn down, but they stayed in use until the hospital's needs for more economical use of the space dictated their removal. At one time the Charles River came almost to the foot of the small rise on which the Bulfinch Building stands, but in 1912 the bank had long been filled in and patients no longer arrived at the hospital by boat.

What did Edsall find at the MGH? One of the things he found was a first-class clinic. For years the MGH served as the clinic of the Harvard Medical School, although not always to the complete happiness of the MGH. There is a notation in the hospital records that in the middle of the nineteenth century, when the medical school wanted to move away from its Fruit Street house next door to the hospital and did in fact move to Dartmouth and Boylston Streets, several of the clinicians at the hospital held that it was a stroke of luck since the anatomy, the "chemistry," and the pathology staff of the medical school made life for the MGH clinicians much more difficult.

The staff of the MGH at the time of Edsall's arrival was a very interesting one. I was a fourth-year student assigned to the medical wards and I became a house officer about a year and half after Edsall arrived. I was very lucky, for I remember a good many of the old staff that were still around. Most of the old stalwarts at the Harvard Medical School retired during my student years, but it was exciting to see them and learn what medicine of that era was like at the hospital.

An important clinician of this period at the MGH was William Smith, who was called "Big Bill" to distinguish him from the younger William D. Smith. "Big Bill" was a bachelor and had a tremendous practice among the Boston elite. He had an ulcer and when he made ward rounds he had to be served crackers and milk. Since his ward rounds were usually quite long, it only increased our yearning for food. "Big Bill" was only interested in organic disease, and if a patient had what he called "ptosis and neurosis," he would try to make him leave his ward. "Big Bill" didn't do much laboratory work, but he was an excellent diagnostician. He did, however, have a way of annoying us particularly when he spoke of his wealthy patients, boasting of prescribing something that was beyond the means of ordinary people, such as installing an elevator for a cardiac patient.

Another distinguished clinician of that day was Dr. Frederick Lord. Dr. Lord began studying pneumonia in the late nineties and was one of the first doctors to have a bench in pathology at the hospital. His mind was completely fixed on lung conditions, and he was particularly concerned with the bacteriology of pneumonia.

Like many others in that early period, he worked hard and well trying to find an antiserum for pneumonia, but with little success. The different classes of pneumococci were only beginning to be differentiated, so that specific antisera became possible.

In all of this discussion I should not leave out Dr. Richard Cabot, because he made more than one notable contribution to the development of medicine in the United States. One of his important contributions, which unfortunately is frequently overlooked, was the development of social service in the hospital. "Peggy" Reilly, the nurse devoted to this activity all of her life (she deserves a chapter herself for her great spirit and her gift of laughter), told me that Cabot personally supported thirteen social service workers and so made the early activities successful. To my mind it was a handsome contribution, and Cabot deserves great credit for it. More important than the foregoing is the fact that Cabot was a great teacher and wrote several books on the physical examination of patients in relation to internal medicine. Blood very early intrigued him, and he helped introduce Boston medicine to the value of blood counts and smears as procedures in diagnosis. Cabot, however, was not a great clinical investigator, although he was a good clinician. The little laboratory work he did in relation to blood, of course, did not make him a real research worker despite his pioneering in hematology.

His great and lasting contribution to medicine was his introduction of the CPC or Clinico-pathological Conference, which is now used in hospitals throughout the world. On the basis of a patient's history, and the accumulated laboratory work, Cabot attempted to arrive at the diagnosis before an audience of students and doctors; his diagnosis would then be confirmed or rejected by the court of last resort, the pathologist. It must be remembered that this was the day of Osler's medicine, and Oslerian medicine was statistical medicine. Cabot's medicine was also statistical medicine. He would frequently say, "According to the statistics this is 75% likely to be so and so, therefore I will diagnose it as so." I don't want to take anything away from Cabot's use of the CPC as a teaching device; I think that it is well established. I would like, however, to point out that teaching medicine by case histories was not entirely new. Just before the turn of the century Walter Cannon, when still a student, introduced Harvard medical students to a similar system of teaching by case history, and the method can be traced back into history in both law and medicine.

Dean of Boston medicine was Fred Shattuck. Some of the younger men who were imbued with laboratory work would make so-called foot-of-the-bed ward rounds where they would look at the laboratory work and make a diagnosis on what were very inadequate laboratory data. As clinicians, however, they just didn't compare with the old man. Locke speaks of Shattuck as a good clinician who did not compare with Osler. To my mind, however, Shattuck stands out as one of the really great clinicians of that long past era. His diagnoses were brilliant although sometimes made on what we would now call inadequate evidence. He was, for example, a remarkable examiner of the chest and abdomen. I remember once when I was a house officer seeing one of the younger visiting men, sitting at the foot of the bed of a very fat woman patient, completely incapable of deciding the diagnosis. Then Shattuck came in, listened to her history, tapped her abdomen several times and said, "It's obvious, she's got stones in her gall bladder." I need not add that the diagnosis was correct.

It seemed to me then that he could make a diagnosis merely by walking up to the patient's bed. He didn't do any research work, though changing the diet in typhoid fever from starvation to a soft solid diet was due to him — a very important improvement. He had a fantastically handsome personality. He was full of jokes, teasing patients and making them feel well the moment he came in.

Dr. James Howard Means, a house pupil in those days, used to call the dramatic rounds Shattuck conducted "glamor rounds," partly because of Shattuck's remarkable taste in dress.

> He wore a garment which was morphologically like a morning coat [said Dr. Means] but it was made out of tweed so it wasn't black like a morning coat. It was quite a fancy looking garment and with it he almost invariably wore a red necktie and a fancy waistcoat and spats etc. so he was quite a dressy looking figure in those days . . . His coachman drove him down every day in his private turnout and left him at the east end of the Bulfinch Building . . . He came equipped with his dachshund and they all went into the entry at the east end of Bulfinch and his house officers were all stationed there waiting for

him . . . The senior, of course, greeted him first and shook him by the hand and wished him good morning. Then the junior promptly took his hat and coat and took care of them . . . The subpup's * job was to take care of Hans, the dachshund.[7]

After Dr. Shattuck retired he would get homesick for the hospital and would occasionally come down in the afternoon to see how things were going. There were many stories of his brilliant diagnoses under these conditions. A story that may be apocryphal even though characteristic concerned Dr. Al Jennings, a young house officer. One day Dr. Shattuck appeared on the ward when Jennings was the only doctor there; he told Jennings he would like to see a difficult case. There was one patient no one could diagnose; Dr. Shattuck looked him over. He then went to Jennings and said, "Doesn't anyone know that the patient has a large pericardial effusion?" The diagnosis had never been suggested, and shortly thereafter the staff met on Grand Rounds and could not make a diagnosis. Dr. Jennings, who was only a pup, said he thought the patient had a pericardial effusion; no one agreed with him. But he asked for permission to tap the patient and to everyone's chagrin he was right.

Dr. Shattuck was *the* finished clinician: his diagnoses were brilliant, his manners charming, his stories often a little off-color and very amusing. The house officers followed behind him and kept as close as possible to him for they did not want to miss a word—and so it was with the nurses. When the party approached a door the pup would rush forward to open it and everybody swept through. To a young man like myself, Shattuck always seemed to be right, the perfect doctor.

Edsall was not the kind of man to put on this elegant performance. He did not even appear on the wards well dressed, but got into a white surgical gown, very loose and shapeless. His heavy, enormous figure appeared even stranger in the white gown, and it was always too small for him, so his long arms would stick out from the sleeves, while the ribbons which tied the gown in the back fluttered in the breeze. The Boston doctors looked on rather puzzled.

In the second place, he had none of the formalities of the regular Boston hospital visit, where everything was extraordinarily mili-

* MGH interns were called house pupils, a hierarchy which ran down through senior, junior, pup, sub-pup, and sub sub-pup.

tarized. Edsall didn't mind if the house staff walked ahead of him or if they waited for him. While he didn't mind, I think a great many of the house staff did, because they rather liked the formality of a visit.

When Edsall went on ward rounds, he was always at your command. He was exceedingly modest and never laid down the law. Playing the all-knowing "Herr Geheimrat" was never part of his makeup. Discussion of a case meant for him give-and-take, no matter who the principals were. He had a way of listening to everyone: no one was too young or unsophisticated for him – he would listen. It would make me cross when a particularly dull student would tell him what was wrong with the patient. Even when the rest of us knew a student didn't know what he was talking about, Edsall listened with great interest and patience, and got what he could out of the conversation. In return, his part of the discussion was very illuminating and instructive, tending to stress the physiological mechanisms and so make the problem into a unit which simplified diagnosis and treatment. With his vast knowledge of clinical medicine, this approach made sense. He simplified problems and showed the progression of the abnormalities.

On the other hand, he was always aware of possible difficulties and complications. He usually took his cases in hand by mulling over them and ruminating about the simple explanations and the rare possibilities that might combine to give a patient a particular disease. His experience and his broad reading of the literature gave him a most profound insight into the complexities of disease. It was a very different approach from the relatively simple clinical work that Bostonians at that period usually engaged in.

Because Edsall was never opinionated, he was a particularly good teacher for house officers. Knowing the complexity of disease, he would never say that "one, two, three, four, five spells so and so." Instead he was diffident and tentative. These very qualities which made him such a brilliant teacher for house officers did not make him the best of teachers for second- and third-year students. I suppose that in the latter teaching situation more certainty and simplicity were needed, like the French clinic of the previous century.

In the little world of the hospital, Edsall was indeed something new, and it took time for people to get used to him. J. H. Means told me how his coming appeared to the staff:

I can remember that a certain amount of unrest was building up and a great deal of speculation as to what was going to happen . . . When I came on here in the autumn of 1911, both services were working on the old style of volunteer visiting physicians . . . This worked very well from the point of view of the house staff. It was a wonderful experience. You got a good deal of personal instruction from these visiting people, and we thought it was just about perfect – and that it might be changed in some subtle fashion disturbed us no end . . .

Well, we began to hear rumors that a fellow from Philadelphia named Edsall might be chosen. We didn't know anything about Edsall but we tried to find out what we could about him. We found that he was a man who had more sides to him than the people we had been accustomed to – Shattuck, etc. – because he had done a considerable amount of clinical investigation and was very much interested in investigation and it was he who brought the idea that a modern teaching clinic must become engaged in research . . .

This didn't sound too bad and we got reports about his ability as a clinician so that there wasn't any hostility against him being built up on the house staff anyway . . . Edsall was a totally new breed of cat. Indeed you might say Edsall was the *anlage* of the modern academic visiting physician or the modern full time professor of medicine in the teaching clinic.[8]

When it came to comparing Edsall's style with that of his predecessor, Dr. Shattuck, Dr. Means said:

Edsall was very informal. He came into the wards and somebody helped him into his white coat. He started right off with a group of interns to go all the way round the ward seeing all the new cases and the old ones he thought needed to be seen. It's hard to compare these visits. I think Shattuck's were awfully well done and very informative . . . He was also giving clinical lectures in the amphitheatre and he was a past master at that kind of teaching. Edsall didn't do that so well. I don't think he liked it as well, but in the wards he was superb, and the things that I remember about him as a clinician were his very genuine concern for the patient's welfare, and the primary importance of making the patient as comfortable as possible.[9]

When I was a house officer, I remember Edsall once coming down to the emergency ward to look at a patient with a large spleen. We had struggled with the patient but could make no diagnosis; we just didn't know what was wrong with him. Edsall examined and talked with him and came up with a diagnosis of malaria which was later confirmed. He also made several diagnoses of rare worm diseases which still stick in my memory. I mention these cases to illustrate that Edsall's vast experience with queer and unusual diseases, plus the broad inclusive way in which he looked at disease, allowed him to make diagnoses which one who merely followed the textbook could never make.

Dr. Means told me some of the things he admired in Edsall's methods:

> He had all sorts of little tricks that he would tell us about and put at the patient's disposal . . . He made a very thorough and complete physical and he had new stunts in physical examination . . . For example at the time he was very fond of feeling patients' abdomens under ideal circumstances. To that end he would have one of those big tubs drawn full to the brim of warm water and put the patient in it lying in a hammock and the rest of us would get around the tub and Edsall would sit on a stool and he would feel the patient's abdomen in this bath. The New England proprieties were satisfied because they always poured a little sulphur naphthol in the bath, or some milk, and that made the water opaque so there wasn't any indecent exposure . . . He'd let the patient lie in this warm tub for fifteen minutes before he would palpate the abdomen and this way the patients were beautifully relaxed.[10]

Edsall's first steps in Boston were anxiously watched by many. He himself felt it was rather a risky experiment. "I had just gone through two attempts, in two different cities, at rather radical efforts at progress," he wrote later. "Both were largely failures. It was, probably, in part due to mistakes but partly bad luck. I felt, however, very conscious of Napoleon's statement that any one who has been repeatedly unlucky is a poor person to put in charge of things because his previous experience destroys the morale of the organization he has charge of. When I came here, I thought people might be suspicious of my personal capacity in carrying anything out when I had had such luck before. I also had in my thoughts the feeling I had always heard, as an outsider, that this was a cold and critical community and very unfriendly to outsiders. Indeed, one very distinguished person, when I was called here, advised me not to 'put my head in the lion's mouth.'" But things went so well that a friend a year later said to Edsall, "With your previous experience you should keep pinching yourself to see if you are dreaming."[11]

Even in his first few weeks he was reviving in a different atmosphere from the troubled one he had left in St. Louis. He wrote to Simon Flexner in June:

> Things are extremely pleasant here. Trained in the school of suspicion and politics as I have been, in recent years especially, I can scarcely comprehend yet the pleasant spirit of general affection for Harvard and of desire to further the wishes of any one brought here to work at Harvard, that I meet constantly, and also the evidence that grows daily that what they promise is the least they intend to do. The only "outs" here are limited laboratory space and limited funds for salaries for men who will devote themselves to clinical research. The need of both is recognized, however and things are directly under consideration that will probably remedy the first very soon and steps are being taken to overcome the latter . . . I feel really at home and happy for the first time, professionally.[12]

And to Dr. Christian he wrote a month later: "As to affairs here—they have been very pleasant indeed. Everyone has been as nice as possible and at the hospital, where thus far my sole activities have, of course, occurred, I have been agreeably surprised in almost everything and disappointed in none. I find constantly that Dr. Minot's * description of conditions there (I got nearly all my official information as to conditions there from him) in nearly every way put the worst aspect of things in the foreground and never painted the pleasant things in too high a light."[13]

He frequently mentioned with warm gratitude the welcome that he received. "I think I should have failed completely had it not been that from Dr. Cabot down I have had the warmest and most helpful cooperation," he wrote a few years later.[14] And the feeling of satisfaction was the same on both sides, apparently. Dr. Locke tells how something like a year after Edsall joined the MGH

* Dr. James J. Minot, who was on the MGH committee for choosing Shattuck's successor.

staff, Locke met Dr. Shattuck, who told him, "I was not altogether confident when Edsall came here that he was from all points of view the best man but I have seen a great deal of him for a year and I want to tell you emphatically that we have made no mistake." [15]

CHAPTER 10

The Harvard Medical School as Edsall Found It, 1912

IN 1912 the Harvard Medical School had been settled for six years in its present buildings on Longwood Avenue, a noble Grecian quadrangle designed by architect Charles A. Coolidge. To build it, J. P. Morgan had given the university $1,350,000 (in a conference of not more than three minutes, as Dr. J. C. Warren delightedly reported[1]), John D. Rockefeller had given a million, Mrs. Collis P. Huntington had given for cancer research the building named for her husband (now replaced). Across Van Dyke Street (its name was changed to honor Dr. Frederick C. Shattuck after his death) opposite the administration building of the medical school rose the skeleton of the Peter Bent Brigham Hospital, which was to open the following year. The Infants' Hospital and the Carnegie Nutrition Laboratory lay a stone's throw to the northwest. The Harvard Dental School and the Angell Memorial Animal Hospital also stood on nearby Longwood Avenue.

Scattered through the rest of Boston were other major and minor hospitals affiliated with Harvard: the Massachusetts General, the Boston City Hospital in the South End, the Boston Lying-in, the Boston Dispensary, Children's Hospital, McLean Hospital and Boston State for the mentally ill, the Mass. Charitable Eye & Ear Infirmary, the Long Island Hospital in the harbor, Carney Hospital, and the Free Hospital for Women. The preceding year the MGH alone treated about 6500 ward patients and 6000 in the accident ward, plus all the outpatient work; the Boston City Hospital had 15,631 ward patients, and 184,000 in its various outpatient clinics.[2]

Despite this wealth of clinical opportunities, outsiders were puzzled by the absence of a teaching hospital fully controlled by Harvard. It was one of the points Abraham Flexner complained of when he reported on the Harvard Medical School in his survey of U.S. medical schools in 1910:

"While the university is free to secure laboratory men wherever it chooses," Flexner wrote, "it is practically bound to make clinical appointments by seniority, in accordance with the custom prevailing in the hospital which it uses, or to leave its professor without a hospital clinic. In general it follows that the heir to the hospital service is heir to the university chair. In consequence there is a noticeable lack of sympathy between the laboratory and the clinical men. They do not represent the same ideals." [3]

Such a condition he considered "fatal to freedom and continuity of pedagogic policy." [4]

This lack was to some extent filled by the arrangement between the university and the trustees of the Peter Bent Brigham, providing that the hospital's chiefs of staff should be appointed by the medical school. Nevertheless, there remained a strict separation of powers. As President Lowell stated in his first annual report in 1910:

"The University has no desire to manage the Hospital, nor have the Trustees of the latter an ambition to manage the School. But it is essential to the efficiency of a Medical School that its clinical instructors should have positions in hospitals, and hence an eminent surgeon or physician cannot be called from a distance to a chair in the School unless he can be offered at the same time a clinic in a hospital. This is impossible unless the appointments are made jointly." [5]

Dr. Harvey Cushing and Dr. Henry Christian were appointed chiefs of the surgical and medical services at the Brigham under this new arrangement, Dr. Cushing coming from Johns Hopkins, and Dr. Christian also in a sense an outsider, having his M.D. from Hopkins, though he had been dean of the Harvard Medical School since 1908.

Other working agreements gradually accomplished the same thing with the other major hospitals, and it was under such an arrangement that the next outside appointment was made: that of Edsall to the joint position of chief of service at the MGH and Jackson Professor of Clinical Medicine. The selection was made by a committee of three from the school and three from the hospital.

As Jackson Professor, Edsall was to hold one of the oldest and most esteemed chairs in the school. It had been set up in the 1850's to honor Dr. James Jackson, one of the moving spirits in the founding of the Mass. General Hospital. Oliver Wendell Holmes wrote of Jackson that both as instructor and practitioner, he "was as nearly a model one in both capacities as I can find anywhere recorded."[6] Dr. Holmes said further, "He would have it that to *cure* a patient was simply to *care* for him . . ." A doctor's duty was "to stand guard at every avenue that disease might enter, to leave nothing to chance; not merely to throw a few pills and powders into one pan of the scales of Fate, while Death the skeleton was seated in the other, but to lean with his whole weight on the side of life, and shift the balance in its favor if it lay in human power to do it."[7]

Dr. Jackson may have continued to wear knee breeches when they had long gone out of fashion, and to carry two watches to ensure punctuality, but his mind was not old-fashioned.[8] Late in his life he became one of the chief exponents in this country of the methods of the great French doctor, Pierre Charles Alexandre Louis. Louis is famous for the so-called numerical method, whereby he could state in terms of percentages the likely course of a disease, or the significance of certain symptoms. Louis' figures were based on his own extraordinarily minute observations of patients in the great hospitals of Paris, where for six years at the start of his career he followed cases without ever treating a patient, retiring to the country for months at a time to compile his statistics. He and his students and colleagues formed a society for medical observation and pushed back the borders of what had been thought possible to observe, assisted as they were by the recently invented stethoscope and methods of auscultation which brought the unknown territory of the chest within the doctor's purview. The results at autopsy, the patient's past history, all were gone into with a new degree of thoroughness. Once the natural history of a disease became clear, Louis proceeded to test out various methods of therapeutics with the same logical care. Louis was one of the strongest influences on American doctors of the mid-19th century, and his method flowered years later in the statistical medicine of Osler and Richard Cabot. Indeed, Dr. J. Collins Warren called Osler "the modern Louis."[9]

A generation of young American students walked the wards of the Paris hospitals with Louis – among them the Bostonians Henry I. Bowditch, Oliver Wendell Holmes, J. Mason Warren, and James Jackson's son, James Jackson, Jr.[10] A warm friendship sprang up between Louis and the young Jackson, who wrote home frequently, fully, and enthusiastically.[11] When the son tragically died just as he was starting his practice in Boston in 1834, his father became one of the chief advocates of the Louis methods. In his appendix to the American edition of Louis' book, *Researches on the Effects of Bloodletting*, Dr. Jackson published an analysis of ten years of pneumonia cases at the MGH according to the numerical system. He wrote in his preface to the same book:

> Let each mode of treatment have its fair trial; and let the results be compared with each other, and with similar cases, treated at the same time upon the expectant method . . .
>
> It is in proportion as we arrive at precision, in respect to the natural history of diseases, that this mode will be pursued with the greatest advantage. It is because we are approaching to that precision that I think it scarcely rash to predict, that in fifty years the art of healing will be grounded on many exact rules, which we and our predecessors have not known. These rules will not be brought forward as derived from grand principles of physiology, or pathology; they must be deduced from the aggregate of careful, faithful observations of individual facts, made by men of enlightened minds.[12]

This was a hand across the years to Edsall, like his predecessor a teacher of clinical medicine in the Harvard Medical School, and like him seeing patients in the same wards in the same building of the Massachusetts General Hospital where Jackson had worked almost a hundred years before.

In 1912, as now, the Harvard Medical School was one of the top schools in the country, though not as explosively fertile a place as the Hopkins, or one with as lustrous a tradition as the University of Pennsylvania. It had been forced through the improvements of the late nineteenth century – including reform of the curriculum and raising of entrance standards – under the leadership of Harvard's remarkable President Charles Eliot, and with the help of such doctors as Calvin Ellis and James C. White. In fact, under Eliot's leadership, the Harvard Medical School was setting the pace among medical schools in raising requirements.

In the years before Eliot, the school expected students to study for two years, but offered only one year of lectures, repeated. The medical faculty ruled quite independent of the university, paid the school's expenses from student fees, and divided any surplus among themselves after the usual custom—a proceeding which led Eliot to call them "a sort of trading corporation as well as a body of teachers."[13] Any gaps in the student's education were supposed to be taken care of by an apprentice system in which the young doctor worked for three years with some regular practitioner; this system was thoroughly outmoded by 1869 when Eliot became president.

Later, in writing to Edward Everett Hale about the accomplishments of his forty years as president of Harvard, Eliot put first of his nine points "The reorganization and ample endowment of the Medical School."[14] His first step in this campaign was to treat the medical faculty's invitation to attend meetings as an invitation to preside, which he did from then on. As Dr. Holmes put it, "Our new president, Eliot, has turned the whole University over like a flapjack. There never was such a bouleversement."[15]

The opposition was as stiff-necked and vituperative as medical rearguard action usually is. Once when the Corporation of the university had been called into the fight, Dr. Henry J. Bigelow, leader of the opposition, exclaimed: "Does the Corporation hold opinions on medical education? Who are the Corporation? Does Mr. Lowell know anything about medical education? or Reverend Putnam? Or Judge Bigelow? Why, Mr. Crowninshield carries a horse-chestnut in his pocket to keep off rheumatism! Is the new medical education to be best directed by a man who carries horse-chestnuts in his pockets to cure rheumatism?"[16]

In response to Eliot's suggestion of having written examinations for the M.D. degree, Dr. Bigelow snorted, "I had to tell him that he knew nothing about the quality of the Harvard Medical students; more than half of them can barely write. Of course, they can't pass written examinations."[17]

Eliot later described to Edsall the system of final examinations which Dr. Bigelow thus defended—upon which the credit of a Harvard-trained physician rested.

Nine professors met in the examining room and nine students were admitted, each taking his seat at one of the nine tables at which the professors were sitting. Dr. Bigelow struck a bell and each professor at once fell to work questioning the student before him as rapidly as possible. At the end of just five minutes the bell was struck again and rapidly the students were hurried to the next table – like a progressive euchre party as Mr. Eliot put it – and so on, each five minutes, until at the end of forty-five minutes the students were hurried from the room and, in Mr. Eliot's words "the professors walked – or I had rather say rushed" over to a centre table, Dr. Bigelow struck the bell and called the name of a student, whereupon the cards came forth, plain white on one side and with a large black ball on the other, being thrown down on the table by each professor. There was no discussion or other waste of time. If the student got five or more white cards he passed and went forth into the World as a physician. If he got five or more black balls he did not. It was undoubtedly an expeditious method but it did not last long after Mr. Eliot learned its character.[18]

The reforms went through one by one. In 1871 a three-year graded course was introduced, lengthened to four years in 1892. Also in 1871 the finances of the school were put in the hands of the university treasurer and the faculty were voted salaries.

Another portent that year was the creation of a new physiology laboratory under Dr. Henry Pickering Bowditch. Five years later Bowditch was the professor of physiology, and in 1883 became dean of the medical school for ten years. This was a landmark, as Bowditch was not an eminent clinician but had made his fame in the physiology laboratory.

Edsall used to tell another story that showed how Eliot's remarkable leadership continued:

Years later, at a time in fact when my contemporaries were already junior teachers, another incident occurred that Mr. Eliot told of with gusto. It showed that a strong hand was still necessary to surmount the resistance to change and the resentment at giving up established opinions. Mr. Eliot proposed to the faculty that a course be established in the subject, then quite new, of bacteriology, and that Dr. Ernst, a colleague of many of the men still in this faculty, be put in charge as he had had foreign training in the subject. This was violently opposed, especially by most of the clinicians, but Mr. Eliot essentially forced it through. The professor of pediatrics of that time was particularly intense in opposition. Soon the diagnosis of diphtheria by means of the bacillus came in and Dr. Ernst was the only one in the community who could do this. Shortly after, diphtheria antitoxin came in and again Dr. Ernst was the only one who could prepare antitoxin. One day going into Dr. Ernst's laboratory Mr. Eliot met the professor of pediatrics coming out. Mr. Eliot said "Doctor isn't this a peculiar place to find you?" The professor replied angrily "Yes, but you can't get your damned diagnosis anywhere else nowadays and you can't get your damned treatment anywhere else." [19]

The school began to look for professors who were the best men to be found — even outside of Boston. Thomas Dwight was brought to be professor of anatomy, after not only the U.S. but Europe had been canvassed. Dr. William T. Councilman (pathology) came from Hopkins and Dr. William T. Porter (physiology) from the University of Michigan. Such progress was made more easily in the laboratory branches at first, since the scientific teaching did not require hospital connections.

A letter which Eliot wrote in 1907 to Henry L. Higginson, fellow of the Corporation, considering the question of a new dean for the medical school, shows his leading spirit at work:

> We have an opportunity now to make a new declaration of what we expect from that School by appointing a new kind of Dean and letting him select the new kind of Secretary. We want to have a School which is going not only to train men learned and skilful in what is now known and applied, but expectant of progress, and desirous to contribute to new discovery. We want to have the whole atmosphere and spirit of the School a hopeful and expectant one as regards preventive medicine and medical and surgical discovery. To this end the administrative officers of the School ought to be men sympathetic with all laboratory research, and desirous of combining laboratory research with clinical research and hospital practice. To get this sort of man, I think we shall have to take as Dean and Secretary men whose chief interest lies in medical and surgical progress rather than in the cautious application of what is now supposed to be known. In all probability both men will have to be under forty years of age.[20]

The dean appointed to fill this bill in 1908 was Dr. Henry Christian, then only thirty-two, Assistant Professor of the Theory and Practice of Physic at the Harvard Medical School, who remained as dean until he became chief of medical service at the new Peter Bent Brigham Hospital.

At his last meeting with the medical faculty, held in May 1909, Eliot was greeted by a surprise testimonial and an address by Dr. Frederick C. Shattuck. In reply President Eliot spoke of his own efforts on behalf of the medical school, calling it "the most constructive part of my work . . ." He was ready to pay tribute to the Harvard spirit of the old school, hard though he had fought to reform it. "Sometimes we think critically of the old Medical School as a private venture," he said, "and indeed it was an establishment in which the principal teachers had a small pecuniary interest in the days when it was possible for a medical school to have a divisible surplus; but this interest was not their main interest. They were men of public spirit who meant to promote medical education, and to make the Medical School successful by training in it a large number of skilful practitioners . . ."

In a prescient look to the future Eliot said, "Sometimes I think that the coming twenty years will see a marvelous progress in medicine, like that in surgery during the past twenty years . . . New masteries of vital processes are almost in clear view. Money is going to be poured out for the promotion of medicine, and especially of preventive medicine . . ."[21]

Eliot did not go out of medical education when he was succeeded by Lowell as president of Harvard. He continued as a matter of course to act as elder statesman in the medical school, and through his position on various Rockefeller boards his influence was if anything wider than before. He became a member of the General Education Board in 1908, served on the International Health Board from its inception in 1913, and became a trustee of the Rockefeller Foundation in January of 1917. (He resigned all three of these posts later in 1917.)

As the quality of Harvard medical education improved, the proportion of students from outside New England rose, until in 1909 Flexner found that only 53 per cent were from Massachusetts, though 69 per cent came from the New England area.[22] Enrollments were still feeling the effects of Eliot's latest recommendation, that all candidates for admission must have not only a bachelor's degree but know something of physics and chemistry. Still, when I entered in 1910, there were fewer applicants than there were places, so if one had a college degree and the necessary science courses all one had to do was just go. George Minot, the discoverer of the cause and cure of pernicious anemia, is a case in point. After graduating from Harvard College, Minot didn't know what to do. He told me that one day he met Frank Rackemann and said, "Frank, what shall we do?" Frank replied, "I haven't the faintest idea; why don't we go down to the medical school and see what that's like?" They went over to the medical school and by the end of the morning had enrolled. It was as casual as that.

Another good example of the provincialism of Boston in those days around 1910 was manifested in Francis Peabody's doubts as to how to pursue his career. He had just finished his internship at the MGH and went to see Dr. Edwin Locke to discuss whether it was safe for him to leave Boston and go to Johns Hopkins. If he did this he was afraid Boston would forget him. Dr. Locke urged him to go; he said Hopkins was so advanced in its point of view that he could get only good from his experience. Peabody went there for two years and then went to the Rockefeller Institute, in the new hospital. He returned in 1913 with the opening of the Peter Bent Brigham Hospital, where he was the first chief resident and a wonderfully mature, inspiring resident. He had tremendous interest and insight, and certainly his sojourn in Johns Hopkins and the Rockefeller did him good. He returned to a bustling Harvard, at just the right time for rapid promotion, a promotion he richly deserved. Francis Peabody was gifted with a great talent for social relations — all of his qualities were of the highest degree. He was a man after whom many young men patterned themselves — one of the greatest of his generation.

Some time ago I called on Abraham Flexner in order to get some information from him about Walter Cannon and David Edsall in relation to the development of modern medicine in America. To my great surprise, I found he refused to say one word about Cannon, Edsall, or Harvard. I couldn't understand why because in one of his books I remembered that he thanked Edsall in the preface for a great deal of advice. In the course of the conversation it finally came out that from his point of view Johns Hopkins was the only school in America that taught medicine at the beginning of the twentieth century, and that Hopkins deserved all the credit for the modernization of medicine in the United States. By 1910 Hopkins had indeed seeded professors all over the United States. The point is, and I think it is well taken, that at the time I began medical school in 1911 Harvard was still a local school largely dominated by local people with limited points of view (though there were some notable exceptions).

Flexner was being perfectly consistent when he told President Lowell that "In any real sense of the word, there is no such thing as the Harvard Medical School,"[23] and backed this up by saying,

"The Harvard Medical School is not really a part of Harvard University in the same sense in which the Harvard faculty of arts and sciences is a part. It runs itself on the basis of seniority. I will give you an illustration. Professor Roach,* your professor of pediatrics, has recently died. I will name his successor."

"Why, that has not even been considered," replied President Lowell.

"Oh," responded Flexner, "it does not have to be considered; it is settled in advance. His successor will be Professor X." †

When President Lowell said he had never heard of him, Flexner said, "These matters are decided at the Tavern Club in Boston by a small group of prominent and busy Boston physicians."

Flexner considered Dr. Morse a good doctor, though with no reputation as a scientific pediatrician. He was nevertheless duly appointed.

Of course, another reason Flexner looked down on the Boston school was that he was a very vigorous proponent of the German style of doing things. German methods had had a great success at Johns Hopkins, but Hopkins was founded when German influence was at its height, while the Harvard Medical School grew on foundations which owed much more to French and English methods.

When Edsall arrived in 1912, the school boasted some twenty-eight professors, with a further group of thirty junior men and 125 instructors, lecturers, and assistants, all this to take care of 290 medical students, 218 summer students, and 156 graduate students. Edsall as Jackson Professor of Clinical Medicine and Christian as

* He meant Thomas Morgan Rotch.

† Dr. John Lovett Morse succeeded Dr. Rotch in 1915.

the Hersey Professor of the Theory and Practice of Physic were the top professors in a newly-united department of medicine, which also boasted such luminaries as Dr. George Sears, brilliant teacher and diagnostician, head of medicine at the Boston City Hospital; Dr. Richard Cabot; Dr. Elliott P. Joslin, expert on diabetes and later head of medicine at the Deaconess Hospital.

Many of them were great teachers. From my point of view, everything that I took in my first year of medical school was exciting, and I was quite prepared to continue each course that I took as my life's work. For histology I had Dr. Charles Minot, who improved the technique for making histological slides. He was very able, as was the professor of anatomy, Thomas Dwight. Dwight was a wonderfully dedicated anatomist who loved to lecture and dissect a cadaver and then pretend to cut himself. At this point he would turn to the class and say, "When you cut yourself you always want to suck it." When he brought his finger to his mouth, two or three in the class always keeled over, much to his apparent pleasure.

Pathology was exciting largely because of Professor William T. Councilman who was a marvelous teacher. Councilman, who was Hopkins-trained, had a wonderful way of communicating his enthusiasm until you felt that he wanted everybody to become a pathologist. One of the great errors of my youth came when I attempted to answer an oral question put to me by Councilman during a pathology examination. It was the shortest examination I ever had. It was Councilman's habit to throw a slide on the blackboard and then call on a student for identification. He stuttered very badly, and I remember that during this particular examination he flashed a slide on the board, called on me, and stuttered, "What's that?"

I stuttered in reply, "That's the liver," and the whole class just roared with pleasure because I stuttered just the way Councilman did. Unfortunately he thought that I was imitating him, and what was worse, I was wrong, and he got very angry. "Liver?" he stuttered. "Never!"

The funny thing is that I have since lectured in that lecture hall countless times, and whenever I come to the word liver I stutter. I have not stuttered over the word liver in any other lecture hall.

Otto Folin had charm. He was a wonderful teacher and like Stanley Benedict of Cornell opened up biological chemistry in a quite marvelous way. His previous training had been that of chemist in the McLean Hospital. Very early in life he had had a malignant tumor of his parotid gland, and during the operation to correct the condition his facial nerve was severed, leaving him with a facial palsy, so that he spoke out of one side of his mouth. He was shy and spoke with a marked Swedish accent throughout his life.

Another of the chemists who taught with Folin was L. J. Henderson. However, they didn't get on and in time Henderson left the Medical School and returned to Harvard College. In later years Edsall wrote of Henderson: "One of the most brilliant men Medicine has produced in this country . . . first demonstrated the acid base balance (acidosis etc.) and other fundamental facts."[24] At the MGH Henderson had an excellent influence on young physicians such as Walter Palmer, Arlie Bock, and James Gamble, and he stimulated them in the direction of their very exciting experiments in acid-base balance.

The person who most excited me during my first year in medical school was Walter Cannon. Cannon was a pure product of America. When he finished his medical school training, he almost at once began teaching physiology at Harvard, first as an assistant to Bowditch and later as a replacement of Professor William T. Porter. Henry Pickering Bowditch had been professor of physiology since 1876. In 1871 he had started the first physiology laboratory for students in the U.S.[25] He was dean of the medical school for ten years. He had demonstrated the "all or nothing" law of cardiac contraction and also the Treppe phenomenon in cardiac muscle. Dr. J. Collins Warren said of him, "Dr. Bowditch was the first professional medical scientist in the faculty, the first to devote his full working day to investigation and instruction in the newly formed department of physiology, the first teacher in the Medical School who did not choose to practice medicine."[26]

After about 1900 Bowditch did little teaching. In his late years he spent a good deal of time taking photographs of various types of people and superimposing one photograph on another in an effort to find an American racial type or a European racial type and so on. Professor William T. Porter was really the professor of physiology at the Harvard Medical School. But Porter was such a strict disciplinarian and gave such low grades to students that sometime in 1903 the students rose up in arms. Their unanimous choice was Cannon. The revolt was so serious that the authorities had to accept it, and soon after Porter was made professor of research physiology and Cannon was appointed as professor of physiology (1906). From my first year in medical school on I saw a good deal of Porter. He was a superb experimenter and a very careful operator. Everything that he did was immaculately and beautifully done.

At the time I came to study with Walter Cannon in 1911 he had finished the period of his early work on the gastrointestinal tract, and he was beginning to study the effect of adrenalin on it. This was really the beginning of his interest in the emotions in relation to the sympathetic nervous system.

In my second year, the class took a course in medicine with Richard Cabot, who taught by the case history method, the Clinicopathological Conference which he had introduced. He would present a case history, and then on the basis of the evidence, he would call on a student to make a diagnosis. By the time everyone in the class had performed once on these diagnoses, Cabot knew us all by name, a not inconsiderable feat since there were about 85 students in the class. He never forgot those names.

The statistical techniques and physical diagnosis in which Cabot specialized made him a very good teacher for young students. There were and always are leaders who are good for young students and people who are good for older students. Cabot was a very good teacher for younger students. David Edsall, on the other hand, was excellent for house officers. Richard Cabot was always sure he was right. Edsall was never sure he was right. Edsall approached medicine speculatively and was always a little hesitant. "This is probably so and so, *but* it might be thus and so." Cabot was never doubtful. Because he was a man who was blessed by knowing and believed that he always ought to be forthright and honest, Cabot told the patient his prognosis. "You'll be dead in a week or you'll be well in ten days." Occasionally his prognoses were wrong and as a result such patients regarded him with less than fondness.

Cabot, however, was superb in coordinating history-taking and physical diagnosis. There were few laboratory tools in those days, and laboratory work was little more than blood counts and urinalyses. Physical examination, on the other hand, was most important, and we concentrated our efforts on learning physical signs, percussing and listening. We knew our physical examinations better in those days than we seem to know them now. The minutiae of physical examination today has been taken over by various refined and exquisite laboratory tests, X-ray, fluoroscope, etc. and as a result the doctor has become less skilled in physical diagnosis.

To a degree student life in my day at the Harvard Medical School was different from what it is today. For instance, students had no dormitory to live in and getting food was quite a scramble. But though living conditions were hard, they didn't seem to matter because everybody in medical school was very ardent.

The trouble with the curriculum in those days was that it was too full. Students went to four lectures every afternoon from two to six and just sat while four people lectured at them one after the other. The lectures were unrelieved by demonstration, what demonstrations there were usually occurring in the morning. One of the great things that Edsall later did as dean was to eliminate that block of four lectures so that the student never had more than two an afternoon, and had at least two free afternoons a week to do what fancy or virtue dictated. It was a more difficult reform to accomplish than outward appearances would seem to suggest. Everybody who has a chance to lecture in medical school wants to continue to do that for the rest of his life.

I remember with a good deal of dislike my third-year course in clinical medicine, before Edsall and Christian had had a chance to reform it. The student soon learned that for many of the lectures all he had to do was to take Osler's textbook to class and underline the passages in the book which were paraphrased.

As Edsall complained to Christian at the start of that school year (1912–13), "My personal teaching I find is now *all* clinical lectures and I wish to subtract some of these and add ward rounds and section work."[27] But he was quite prepared to take things slowly. He wrote in the same letter:

I have not yet even attempted to plan any changes in my work, for next year. Conditions are quite different from those I am accustomed to and I should like to get opportunity to watch things working before making any major suggestions . . . Since you said we could plan for a general rearrangement of the whole medical teaching during next year I thought it would be unwise to make changes now and then change again next year.

The only things I see now that do appeal to me as desirable to alter are the large number of lectures and large sized classes, the absence of any noteworthy amount of anything of the distinctively ward-clerk type of work . . .

I am thinking also of doing some work myself in physical diagnosis in the last half of the coming year and trying to coordinate the work of the different men a little better and to introduce some *normal* physical diagnosis before they get to the pathological.

As to a chairman of the department — I don't want it, now, at any rate. Executive functions don't appeal to me greatly anyway . . . If you will suggest someone whose usefulness you know in such work (I do not know the men well enough) I will support him if the matter comes up before your return.

Edsall's teaching methods were always undergoing revision. In the fall of 1918 he was writing to Christian:

I have been adopting a new scheme with the students this fall, or rather I had better say I have tried it on them the one time that I met them and have told them that I was going to continue it further, namely, I have assigned before the hour a definite amount of reading for them to do in Osler and have said that I might refer them to other books and then have conducted and plan to conduct an hour that is partly lecture and partly recitation but in large part really a conference, my purpose being to make them read more than they do on the one hand and on the other hand to try to clear up misunderstandings and so forth that they have in their minds that are hard to get at otherwise. I have a feeling that in this way I am able to do them more good than by simply lecturing to them . . .

I began with them on typhoid fever and asked them to read that carefully and at the same time to read over carefully a few other things, chiefly tuberculosis, septicaemia and pneumonia, as being the most nearly related to and confounded with typhoid and the ordinary diseases, and necessary for them to comprehend in order to comprehend typhoid. I shall go on with typhoid for an hour or two to come.[28]

It is no surprise that Edsall was still using typhoid fever as his base line in the infectious diseases. A thorough understanding of typhoid gave a doctor a good grounding in those days when typhoid was still the chief summer disease. It was beginning to drop off sharply, however, with the purification of water supplies. The Mass. General Hospital had 106 cases in the summer of 1910, 61 the summer of 1912 when Edsall arrived. In any case Boston, with its better water supply, had a smaller incidence of cases than Philadelphia: in 1910 Boston had 79 typhoid deaths or one for every 8481 of the population; Philadelphia the same year had 272 deaths, or one in every 5658.[29] Edsall mentioned that in his work at the Episcopal Hospital in Philadelphia he had seen "enormous numbers of acute infections, especially in typhoid fever . . . In four and one-half years there were, for instance, 3,677 cases there of typhoid fever alone."[30]

Good clinical judgment was life-saving in typhoid, when prompt surgical intervention was called for on very obscure manifestations, like rupture of intestinal lymph nodes with minimum clinical manifestations in a somnolent patient. Edsall's vast experience made these lectures very exciting and valuable.

Having seen typhoid in Philadelphia, Edsall had seen the disease at its worst, for there it was a continual epidemic from the contaminated water supply. He had instituted several innovations in treatment. The main one was a change from the accepted starvation diet, to give the patients a bland high calorie diet. This originated with Edsall in Philadelphia and at the same time with Dr. Fred Shattuck in Boston. The low calorie diet was cruel for a patient who was using the energy needed for high fever. Roger Lee once told me that the hunger he experienced when he had typhoid fever accounted for his stealing food from the diet kitchen.

David Edsall himself had typhoid in Philadelphia in the early 1890's. He used to dream of eating delicious meals, and would awaken to find a nurse at his bedside, bringing him a glass of boiled milk.[31] During his convalescence, he was so starved of salt that he would sprinkle it on his bread and butter until it was white with it — and this tasted delicious to him.[32]

Both Edsall and Christian were effective in their teaching of such clinical problems, but Edsall's lectures were the more dramatic. One thing he did was to distribute miniature copies of clinical charts to illustrate the differences and emergencies encountered with typhoid fever. These were photographs of actual patients' charts, and the pattern of chill and fever lines made a lasting impact on the students.

CHAPTER 11

Tragedy, 1912

EVERYTHING augured well that fall of 1912, and it seemed as if Edsall was ready to settle down to his new career at Harvard. The future seemed bright. This, however, was not the case, and Edsall's first burden at Harvard was to deal with a personal tragedy.

Margaret had been overjoyed at his Harvard appointment. She wrote to her dear friend Mary Kirkbride, "It seems too wonderful to be true & if anything could give me joy now, that would."[1] (Her sister Eleanor had died the month before.) She went on, "Some day I should like to pour into your appreciative ears the long story of how things have been managed at St. Louis—but that is not for publication! It does seem strange that we should have jumped out of the frying pan into the fire, but at least it looks as if we were to have some certainty and peace at last, after three most racking winters of change and complicated professional relations."

After a relaxed and happy summer, the family moved to 80 Marlborough Street in Boston in October, a handsome house in a street of handsome houses where it seemed half the doctors of the medical school had their homes. All three children became ill that fall, Richard nearly dying of croup and John and Geoffrey having a strep infection. A letter which Margaret Edsall wrote at the end of October to Mary Kirkbride recounts some of her troubles. It also shows the spontaneous warmth which so drew people to her. After speaking of some sad news of Mary's with ready sympathy she wrote:

> I am about tired out myself or I should have written before.
>
> David sprained his ankle just before we moved & I had a lot of extra work. Libbis * left me just before to be married—I had several changes in my household & 2 weeks ago Monday, after a year without croup Richard had another of his croup spasms but I got a doctor in to resuscitate him. Since then John has had a 10 days' illness with

* The children's favorite governess.

streptococcus throat & Geoffrey a week of it — both better, but not out yet. And I have tried to get the house in order.

I may be too tired to think or to write straight but I am never too tired to love you all with all my heart & to enter keenly into your sorrows & joys. With a heart full of love to you all,

MARGARET [2]

Next Geoffrey fell ill again and this time the diagnosis was diphtheria — a word of terror in those days, and particularly since Edsall had lost a young brother from that cause, and his little namesake, his brother Frank's child, had also died of it. Geoffrey was treated promptly with antitoxin and was on the mend when his mother became ill of pneumonia.

During the anxious period that followed, Dr. Sam Hamill wrote to Locke from Philadelphia:

I telegraphed you to send me these daily bulletins because I want to be constantly informed and because I had not the heart to ask poor David to do this. I have not even had the courage to write him as the receipt of messages of sympathy can only add to the weight of his awful burden . . .

I wish you would tell him that I have transmitted all of your messages to Pearce, Nicholson, Taylor, Pemberton, his brother Henry Edsall, Jopson, and some other of his friends who are constantly inquiring of me . . .

Pearce and I discussed the advisability of going on to Boston but decided that it would be unwise as we both felt that in similar circumstances we would prefer to be left alone. It is entirely unnecessary for me to tell you how deeply I feel for David in his overwhelming grief. I had hoped that his trials had ended. He seemed so happy and so full of hope when I saw him in Washington a few weeks ago. He really seemed a new Edsall. To think, that just as he was emerging from these years of disappointment and depression, he should experience this sorrow, which eclipses all the others, is most distressing. Should Mrs. Edsall recover, the effect of the intense anxiety that he is experiencing is almost sure to break his health. Should she die, I fear the result. I hope you will keep a very close watch upon him. He has not much nervous resistance. He had such fits of intense depression when in Philadelphia as to sometimes alarm me. The statement in your second telegram of yesterday that he is now a wreck did not surprise me but has greatly disturbed me . . .

It seems strange that in this city where David has so many active & passive enemies that he should have so many real friends.[3]

Despite all that Locke and his colleagues could do, Margaret Edsall died on November 19.

Looking back on her illness years later, Locke called her illness influenza with pneumonia, "one of those toxic cases that just went." Even today, he said, even with antibiotics, he thought she could not have been saved.[4]

A week after Margaret's death, Edsall wrote to Mrs. Locke, "It seems a little thing to thank you for the chrysanthemums when I have the great cause for thankfulness that I have that Mrs. Edsall was in hands that I so trusted, and had such devotion, that I am spared any thought that even the slightest thing might have been done better or more quickly. I am grateful for the freedom from that thought."[5]

Edsall took his children back to Milton where Margaret's mother and sisters could help him. He managed to carry on with his work – in January he was lecturing to the third year students on pneumonia – but he was a grief-stricken man for many months.

His friends were very worried about him. Hamill wrote anxiously to Locke from Philadelphia:

> I am sorry that you could not have gotten David away for a few days during the holiday season. As I wrote you, Howland and I tried to do so, but he felt that he could not leave his family. As soon as I can put things into shape, I want to go to Boston and spend a few days with him. I do not see, at the present moment, any possibility of this being at an early date . . .
>
> I hope that you are able to see something of David. I think he needs your friendship, and I should like you to be as close to him as I was when he was here. He is a man who needs someone to whom he can freely open his heart at times, and from whom, when the blue devils are thickest, he can receive help and encouragement. I know that he feels the lack of this in Boston, because he has frequently written me long letters in which he said that it would be such a relief to him if he had someone to whom he could talk intimately. Edsall is a man whom you will never get close to unless you go more than half way. He is absolutely undemonstrative, and even when you know him as well as I, you will never see the slighest demonstration of affectionate

friendship, such as the average man would show to his fellowman. I do not mean the palavering kind of demonstration, but the manly type – if that conveys anything to your mind. I received a long letter from him this morning, for instance, in which he pictured his life problems as I question if he would do to any other man, and yet there was not the slightest expression of the appreciation in which I know he holds my friendship, or the slightest suggestion of affection.

I mention these things in order that you may never misinterpret his attitude, and that you may be somewhat more aggressive in your attitude toward him. I know that he has a very profound regard for you, and I hope and believe that you can give him something which he very much needs in his present life.[6]

CHAPTER 12

The Jackson Professor at Work, 1912–1923

EDSALL's greatest drive was to convert medicine to a study of the mechanisms of disease — the physiological approach to medicine. This was one of his important contributions and he was one of the pioneers of that movement in the United States. Before World War I the Germans, French, and English were, of course, preeminent in physiological medicine. Indeed, Europeans like Müller, Rubner, Kraus, Bayliss, and Haldane helped develop the new era of physiological medicine while America (at the beginning of the 20th century) was still devoted to the anatomic and pathologic approach. Edsall, however, was one of the lonely ones who tried to shift the focus of attention to a study of the mechanisms of disease.

"We have come much more generally to recognize diseases . . . as made up of a group of disturbances of function," said Edsall, discussing this question as early as 1912 before the Massachusetts Medical Society.

> [We] have recognized that these disorders of functions must be clearly appreciated individually before we can comprehend any disease as a whole. It is through this change of conceptions that general physiology, physiological chemistry, the study of the normal reactions of immunity and similar questions have become so lively a part of clinical investigation. Not many years ago the physiologist was, so to speak, the mere acquaintance of the medical and surgical clinician. Now in his daily work the physiologist is quite as important to him as the pathologist . . .
>
> Indeed in nearly all conditions pathology itself now approaches physiology so closely that, at the present day, pathological investigation begins rather at the point where physiological processes become deranged than where disease terminates in death . . .
>
> The influence of the newer point of view upon the methods of treatment may be illustrated by so familiar a disease as typhoid fever. When I studied medicine I was, essentially, taught certain routine

methods of treating typhoid fever, which were varied in character somewhat according to the severity of the disease but were really directed chiefly against the disease as an entity. Now we are taught to treat the patient's powers of resistance, his excretion, his disturbances of circulation, digestion and metabolism but, as to the disease itself, only certain characters and accidents that are somewhat peculiar to this disease.[1]

Such ideas were not only important in influencing the direction of work in the hospital, but extended to the courses and the general point of view of the department of medicine at the school. Before Edsall came to Harvard, most of the clinical teaching was taken directly from Osler's textbook. The discussion of disease on the ward was largely a textbook discussion. Edsall didn't think much of that approach; it was his view that no one ought to give a lecture that could be found in a textbook.

He had a very clear idea of what constituted good teaching and had stated this idea in telling fashion years before coming to Boston in his presidential address to the American Pediatric Society in May 1910. After discussing the state of pediatrics in general, he went on to the question of the education of pediatricians — remarks which had a wide application.

The chief essentials for making well trained men and for attracting first-class men into a subject as their own work are that the men shall do most of the work themselves and not merely be told about it; that they shall be made precise and critical in their methods, and that the craving for a sound understanding of the reasons for things be satisfied by having them clearly comprehend the relation between the fundamentals and the practical things.

In all these ways general medicine and general surgery have been quite transformed within two decades in a large group of our best medical schools, and now, instead of the instruction being as it was earlier, almost wholly didactic and at best given in the form of clinical lectures to large classes, students do most of their work in wards and dispensaries, either in small groups or, still better, as individuals, acting practically as temporary assistant internes. Precision and a critical point of view are given them through the fact that purely observational methods, which always have an indeterminable amount of error associated with them, are supplemented by the employment of laboratory methods and instruments of precision, which not only give added information, but have the invaluable element of demanding accuracy and precision in their use and thus training into more precise habits.

The craving for a fundamental understanding of facts that are rather baldly stated in textbooks is, in some places at least, being vigorously met by cooperative teaching between clinical and anatomical, physiological, pathological and other departments.[2] *

A good deal of this had yet to be developed at the Harvard Medical School, as can be seen from Edsall's comments to Christian at the end of Chapter 10.

I was a third-year student the first year Edsall taught, 1912–13, and I retain a memory of endless lectures in the hospital amphitheatre. Inside of two years, Edsall had students participating on a much more exciting basis. The department of medicine announced:

The general plan of the work will be to instruct the students during the second half of the second year in the methods of history-taking, physical examination, and clinical laboratory technique, to drill the students in the third year as assistants in the Out-Patient Departments of the hospitals, and in the fourth year to give them continued contact with patients by having them serve as clinical clerks in the hospital wards.

Much of this work will be conducted in small sections, while lectures and clinical lectures will be given for the class with the view of presenting to the students a more comprehensive knowledge of medicine.

Students were now given a thorough grounding in "the taking of histories, methods of physical examination, and in the examination of urine, blood, sputum, and gastric contents." These and other basic tools of research were emphasized:

Students will be instructed and exercised in the chemical, microscopical, and bacteriological methods used in the practice of medicine. It is expected that each student by frequent opportunity will attain the necessary proficiency to enable him to utilize these methods in the diagnosis and prognosis of disease.[3]

Edsall was further concerned to make the various elements of the student's education meaningful as a whole, particularly to over-

* From Annual address of the President, American Pediatric Society, May 3, 1910, J. Pediat. 47: 806–816, 1955.

come the tendency to regard the laboratory groundwork as something separate from its clinical application. By 1919–1920, the second-year course in physical diagnosis had been aligned with the anatomy course, and third-year students were having clinico-pathological demonstrations weekly.[4]

At the hospital Edsall was also a leader in making changes. From the beginning he had a seat on the General Executive Committee, a new administrative body which had been set up in the same adventurous spirit that led to his appointment. As Edsall recognized, "In this hospital, with its very old traditions, with everything long established on the more usual lines of administration, but so well administered as to have served as a model for many other institutions, a very radical reorganization was carried out by mutual desire of the trustees, the superintendent, and the staff, through which important powers and responsibilities have been given to the staff representatives . . . the change has been to the very great benefit of the hospital."[5] *

This General Executive Committee started functioning at the time of Edsall's arrival, and has met weekly ever since. The first six members were: Dr. Maurice H. Richardson, chairman, Dr. Francis B. Harrington, Dr. Samuel J. Mixter, Dr. Algernon Coolidge, Dr. David L. Edsall, Dr. Richard C. Cabot, and secretary ex officio Dr. Frederic A. Washburn, the hospital superintendent.[6] They dealt with many important matters, including, that first year, the reorganization of the medical wards and the system of visiting physicians. The east and west services were set up each with a chief, and under him two visiting physicians who served six months each, as did the assistant visiting physicians. The next year saw the appointment of a paid house physician on the East Medical Service, Dr. Walter W. Palmer. This was a position of a new type, for not only was the house physician supposed to help and direct the house pupils, or interns, in the absence of the visiting physician or the chief of service; he was largely engaged in research. This same year, precedent was broken further by assigning third-year students to the outpatient clinics.

For a long time the MGH took nobody except Bostonians as

interns. It was said, though I have not confirmed this, that until about 1903 or 1904 all interns at the MGH put on dinner jackets for dinner. At least the atmosphere was such that one could believe it. With the advent of David Edsall, however, the procedures for choosing interns changed. At first most of the interns who were chosen were from Harvard, but representatives from other schools were later chosen as well, because Edsall believed that the MGH should take the best from other schools.

The examination, however, was a terror, and I think that it was meant to be. For example, when I was ushered into the examination room, the first thing I saw was a tremendously long table with a huddled group of examiners at one end and an empty chair at the other, and the chair was a trick — if you were lucky, someone had warned you not to sit down. If you were unfortunate and sat down, you were immediately asked, "Who told you to sit down?" I'd been warned not to sit down, and I waited until I was asked to take a seat.

There were five doctor examiners on my board. The first was Dr. Herman F. Vickery, a small bearded man with a humorous look and a nose that twitched. If you didn't know Dr. Vickery well, you never quite knew if he was laughing at you or not. For all of that he was an excellent clinician. Another member was Dr. Frederick Lord, who knew pneumonia and lung diseases from the clinical as well as the bacteriological approach extraordinarily well. Dr. "Big Bill" Smith, Dr. George Sears of the City Hospital, and Dr. Edsall rounded out the board.

It is interesting what chance does in an examination. The evening before the exam I was browsing through Forschheimer's text on pharmacology in medicine and by chance came upon asthma. Because my mother suffered from asthma, I thought I had better read about it. And it turned out that the first question I was asked to discuss was asthma. I gushed until they told me to keep quiet and changed the subject. I think that my disquisition on asthma made a good impression.

The thing that really clinched my appointment, I suspect, was not the examination at all, but the fact that I had made an interesting diagnosis in the outpatient department about a month before. One day I had turned up there before anyone else and came face to face with the first patient. I took one look and said to myself, "Good gracious, he has a leonine expression." I took some scrapings from his nose and sure enough Hansen's bacillus turned up in the microscope. I had discovered a case of leprosy (the only case I have seen in Boston). By the time a clinic physician came in, I had made the diagnosis. My diagnosis made a great impression because this particular man had been a patient at the Brigham for over a year, and they had not made the diagnosis. In those days the competition between the Brigham and MGH was intense.

The internship was a long appointment which lasted for eighteen months. It began with being a "sub sub-pup" (or pupil), then advancing to "sub-pup," and finally "pup." These positions, devoted mostly to laboratory work, were followed by the junior and senior positions. The senior's word was law, and everybody under him obeyed him implicitly. Serving directly under the senior was a junior who was on duty every other day. Those two men had direct charge of all the patients. The pup was usually in charge of the work in the outpatient department, and the sub-pup and the sub sub-pup did what they were told by the pup in the outpatient department in the morning and spent their afternoons doing the laboratory work in the wards. This work consisted of urine analysis and blood counts, and in retrospect the number of routine bloods and urines which we did daily on patients was excessive. The great increase in laboratory determinations for the diagnosis and prognosis of disease had hardly begun in those days. Today a house officer is responsible for one ward. We five took care of a whole unit: two wards, and single and double rooms for the very sick.

There were two services, the east and the west, and the choice of service was important because it showed the kind of visiting physicians one wanted to work with. The west at that time was chiefly clinical, while the east was dominated by Edsall who was more interested in research and physiology. There was great rivalry between the house pupils of the two services, so much so that each service had its own distinctive design of tie, and they continually vied with each other as to which was best in patient care and diagnosis.

Medicine was entirely different from what it is now. When I began, most of the patients suffered from infections: pneumonia, typhoid fever, diphtheria, meningitis, rheumatic fever and the like. Occasionally we would see a patient with typhus fever, which we called Brill's disease. These were people who had had typhus in their youth and now had a recrudescence that Hans Zinsser later showed was the old typhus infection harbored in the spleen. We saw many cases of primary anemia. Of course in my day there was no adequate treatment for these, and I remember George Minot, who was ahead of me, writing in one of his notes: "This disease has something to do with food, and we must find out what it is." Later, of course, he made his great discovery of the cause of pernicious anemia that won him a Nobel Prize.

When I began my internship, bacteriology held the greatest attraction for the young physician. Its development held out great promise. For example, in 1910 Bela Schick had worked out the toxin-antitoxin preparation for immunization against diphtheria, and a short time later the Rockefeller Institute developed a serum for meningitis which they claimed lowered the death rate from one hundred to fifty per cent. In spite of all this ferment, there was precious little we could do for our patients in the way of therapy.

I had not yet taken my degree when I became an intern, for I had a chance to start my internship in February 1914 and Edsall arranged for me to postpone my last four months of school which were to be spent chiefly in pathology. In the beginning one did not live at the hospital as there was no room, and I lived with Lawrence Lunt, Carl Binger, and Phil Pierson in a little apartment which overlooked the river, near the site now occupied by Phillips House. When the weather got hot and close we set up a tent on the roof. That set a trend and a short time later some women who lived at the settlement house next door, the Elizabeth Peabody House, decided to do the same thing. In the end almost everybody in the neighborhood tented on the roof.

When I became a junior, I followed custom and moved into the house officers' flat in the hospital. Discipline at the hospital was rigorously enforced by Colonel Washburn, the hospital superintendent. Interns, for example, were not allowed to have side pockets in their trousers because the colonel thought it was unseemly for house officers to stick their hands in their pockets. He only allowed rear pockets.

The one place where the colonel emerged with an unblemished victory was in the matter of routing house officers out of bed. In those days each house officer's room had an instrument called a "whosit." A "whosit" was a hollow tube that ran from the flat to the hall below, and a nurse needing advice would whistle up the tube, and invariably the house officer would call back, "Whosit?" The nurse could then tell him it was time to get up or a patient needed attention. All house officers had their beds placed so that they didn't have to move to answer the "whosit." After answering yes they tended to go back to sleep. When the colonel became aware of this practice, he had all the beds moved across the room from the "whosit"; to answer it one now had to get out of bed and walk across the room which made it extraordinarily difficult to walk back to the arms of Morpheus. Internship was no sleeping matter. But if you didn't like the nurse, you could pour water down the "whosit"!

At the end of my internship I went to New York to work under Eugene DuBois and Graham Lusk, still without a degree. In the year that I and the others were away Edsall kept his eye on us. There can be no doubt that his plan was to bring us all back as researchers and clinicians in a new full-time unit and in that way revamp the medical service at the MGH. This is very much what everyone practices now but nobody did in the years before World War I. I don't know how Edsall learned that I had profited from my year with DuBois, or how he divined that I wanted to get back to the MGH, but DuBois and Lusk wrote a final exam for me which made them laugh and Dr. Edsall informed me that the Harvard Medical School had awarded me my M.D. degree. He then promptly offered me one of the new residencies in medicine at the hospital which I just as promptly accepted. I had learned by this time never to question opportunity.

The residency was something new that Edsall had started at the MGH. The first medical resident was Walter Palmer, known to his friends as "Bill." He was interested in chemistry, an enthusiasm he just couldn't seem to transmit to the interns. The next two were Paul Dudley White and myself. Edsall had sent us away to get special training and now that we had it he wanted us to run the medical services. But that was easier said than done, because the interns objected very strenuously, claiming that it took away authority from the senior who traditionally rose from sub sub-pup to become absolute monarch of the service. In practice, although the residents were nominally in charge of the service, intern opposition was strong enough to prevent them from doing any clinical work on the wards. I found that when I went on the wards to advise brash young house officers I was met with less than enthusiasm, and I soon restricted myself to making ward rounds with Edsall.

The interns were not the only ones to object to the residents. Another was Dr. Willey Denis, the resident chemist. Willey Denis was a large woman, an excellent chemist who later in life was appointed professor of chemistry at Tulane. When I knew her, her pet peeve was residents. One of the first things I did when I became resident at the MGH was to put in for a desk in the chemistry laboratory, not knowing that Willey Denis didn't like residents cluttering up her laboratory. I doubt that it would have made any difference, but the very first day I was in the laboratory I started a kjeldahl analysis to determine nitrogen metabolism. I stepped out of the laboratory for a moment, clean forgot about the experiment, and went off to a matinée. When I finally got back, I found that Dr. Denis had turned off the kjeldahl. She further allowed that I was a pain in the neck.

Three days later a good-sized pig appeared in the laboratory, so placed that the front of his cage faced my desk. For the next few weeks as I worked in the laboratory that pig grunted in my back. I weathered the incident because I felt I deserved it, and in time the pig was removed. Eventually Dr. Denis and I became quite good friends and wrote several papers together.

One of Edsall's great strengths was in selecting promising young men and sending them away for postgraduate research training. His primary idea was to help build research competence in various fields so that when these young men completed their graduate studies they could carry on as his staff at the MGH. I think it's fair to say that Edsall was not satisfied with the training which was then available there. Many of the young men Edsall chose for postgraduate training at this time later became distinguished investigators and clinicians. Both Howard Means and Cecil Drinker, who in later life made several distinguished contributions to physiology, were initially sent by Edsall to Denmark to study respiratory exchange with the great physiologist August Krogh. It was, for example, under Edsall's guidance and urging that George Minot went to Europe to study blood, a training that laid the ground work for his later famous studies in pernicious anemia. With Edsall's encouragement, Paul Dudley White went to England to work with Sir James McKenzie and Sir Thomas Lewis on problems of the heart. He returned to Boston with an Einthoven string galvanometer, a very complicated and cumbersome machine designed to take electrocardiograms. White, of course, has continued his studies to this day and has had a most distinguished career in cardiology.

Not everybody went abroad. Harry Newburgh studied physiology and pneumonia with Porter at the Medical School; Jim Gamble and Bill Palmer remained in Boston to work in physiology and chemistry with L. J. Henderson. Frank Rackemann was sent by Edsall to work with Longcope in New York in the then newly developed field of allergy. The vistas were truly wide, and Edsall men were seeded widely. Oswald Robertson, for example, went to work with Rufus Cole at the Rockefeller Institute on the bacteriological aspects of pneumonia, while Carl Binger was sent to study pharmacology with Abel at the Hopkins, and then to the Rockefeller Hospital to investigate therapy in pneumonia. In all this process Edsall concentrated on getting research done on important diseases. Stanley Cobb, for example, took his internship with Harvey Cushing although he never planned to be a surgeon. To Edsall's mind it was good preparation for Cobb's eventual pursuit of neurology and psychiatry, which he then took up at Johns Hopkins, working on psychiatry under Adolph Meyer and with Howell in the physiology of the nervous system.

Although there were many clinicians in America at that time, few were active in academic medicine or interested in developing research. There was an urgent need for men who were specially trained, and were also willing to devote their lives to teaching and research. Edsall could institute such a program at the MGH because he saw the opportunity for careers in academic medicine and knew where proper training was available.

The MGH already had a tradition of laboratory investigation, though few of the leading clinicians had been concerned with it in earlier years. To my mind this began when J. Homer Wright came to Boston, after completing his studies abroad, to become pathologist at the MGH. There were sixteen benches for staff members in Wright's laboratory in the pathology building when it was completed in 1896. Each staff member who wanted to do some research had to pay twenty-five dollars a year for the privilege of using a bench. I might add there was a good deal of scrambling for benches and for the chance to do some work under Wright, who, albeit very shy, was a very brilliant man. He devised a technique for counting platelets in the blood and several stains for facilitating blood examinations.

Medical research, which had been advancing by tiny steps, was ready for an almost explosive growth in 1912. Looking back on this period twenty years later, Edsall commented on

the swiftness with which medical knowledge has since then progressed and the extent to which it has in recent years successfully sought the aid of diversified disciplines, some of which disciplines seemed only a few years ago to be almost or quite alien to actual medical interests. It may certainly be said that in no form of intellectual activity has there ever been more rapid increase of effective knowledge in a half-century, or greater variety of new forms of attack. The result has been an entire transformation of the character of the practical aspects of medical work and an even greater change in the character and the variety of investigative activities and of the personnel engaged in this investigation. Until very recently, speaking in the usual historical sense, serious devotion to research in medicine was looked at askance by very many of the medical profession itself . . .

And it should be held in mind that the purely clinical viewpoint was almost wholly dominant until quite recent times, even in those subjects that are now commonly termed the "medical sciences." It is not very long since that there were in this country almost no persons who devoted their lives to the medical sciences then, even as teachers. The medical sciences were taught almost wholly through lectures by clinicians, which they gave as a collateral part of their activities while striving for consulting practice and often merely as an opening toward teaching positions in the clinical branches.

It is not to be wondered at that with that viewpoint of the medical profession itself, and with almost no provision of funds or facilities for serious research, the impression became strongly stamped upon the minds of many university men in other faculties that medicine was an almost pure vocation — a very valuable one, but entirely utilitarian in character and unsuited for men with distinctly scientific gifts and interests. This impression remained long after medical conditions had greatly altered, and it has not yet wholly vanished.[7]

In another speech Edsall commented on the scientists' side of this: "I was told by various men in medical science with some bitterness that the atmosphere of almost any medical school and medical faculty was impossible for them to develop in, that to work in them was, as one able man put it, a prostitution of science, and they urged repeatedly that the medical sciences be made part of the general college faculty and be divorced entirely from the medical schools except for the necessary teaching."[8]

That Edsall was one of a little understood group of new men, a new breed in medicine, is obvious even in Richard Cabot's comment to Christian after Edsall became dean. Cabot called Edsall "one of the best practitioners I know — despite his strong talents for investigation & for public health work."[9]

Edsall was indeed creating a new breed of men in medicine, and at the start the funds for this kind of work were extremely limited. There were the Dalton Scholarships, dating from 1891, and there was the Walcott Fellowship for a research worker in medicine which had been endowed by Dr. Frederick C. Shattuck in 1910. The hospital provided one salary for a resident physician, and on Paul Dudley White's return from Europe set up another for him.[10] But even by the time Edsall left the hospital in 1923, he had only $13,000 to support all the work that was being done.[11] By contrast, the MGH in the mid-60's spent some $10,000,000 in medical research, and there were 840 investigators and clinical and research fellows.[12]

Despite the minimal funds, Edsall gradually had an active research program going, based on hope and enthusiasm instead of grants and salaries. As he described this development to the faculty when he became dean of the medical school:

When I came here [to Boston] I had one full-time man on my staff, the Walcott Fellow; there were none others who definitely gave more than a small fraction of their time to their hospital and school work. Since then I have had no new salaries from the school but the hospital has added a full-time assistant on salary and three residents on full time to Dr. Cabot's service and mine. These men all have the privilege of teaching and by agreement with the authorities have at least half their time free of routine duties to devote to research. I have arranged this in writing, not as an underhand method of securing the services of these men in increasing the force of my department but by the frank and I believe entirely sound argument that that method secures much abler and more efficient men than do positions that do not permit of teaching and research training, an argument that I have found the authorities increasingly ready to appreciate and act upon. I have also consistently combined, whenever possible, small salaries from the school and small sources of income from the hospital on one promising man, rather than using these salaries for several men, thus securing for one man a total that would make available a larger amount of his time and more opportunity for his development. Next year we shall have there in medicine seven or eight junior men, I hope, giving from half to the whole of their time to their hospital and school work and several of them besides the residents will have the privilege of rooms, board and laundry in the hospital if they desire. They are mostly very inadequately paid and are held chiefly by the precarious bond of devotion to their work, but it has at least sufficed to hold them and conditions are improving.[13]

A glimmer of hope for adequate funds for full-time salaries, laboratories, and all the rest appeared early in Edsall's years at MGH although at first it seemed not to concern developments at that hospital. The Rockefeller General Education Board, having assisted the Johns Hopkins and the St. Louis medical schools in putting a major part of their medical teaching on a full-time basis, was interested in supporting similar developments at Harvard.

The phrase "full time" was for some twenty years a battle cry. Edsall's first efforts in that direction at Penn. had helped to bring about his downfall there. Yet full time was really the recognition that medical teaching and the advances in knowledge meant that a top professor in a clinical subject such as medicine, surgery, or pediatrics would have all he could do to teach, conduct research, and run a hospital service, without the heavy demands on his time which a busy practice entailed. If scientific medicine was to advance, the men who were to make that advance had to be paid an adequate salary, not one that could equal what a top consultant could make in fees, but at least one that would enable a man to maintain his family above the poverty level.

Edsall had approached the problem gradually, half time at Penn., half time when he came to Boston, though he found that in actual practice his work at MGH was nearer the full-time standard. In the same way, Cushing and Christian, starting with a half-time proposal in forming the plans for the Brigham, had gone on to something broader. As Cushing described it in a later Annual Report, "We were the first clinical teachers, so far as I know, who desired and were permitted to give their undivided attention to the work of a teaching hospital and to confine their professional activities within its walls."[14] * They were still working out the details of this system in the early years at the Brigham, when the Rockefeller proposal, including the strict Rockefeller provisions for no fees to individual doctors, came into the picture. In 1913 and 1914, discussions centered around the departments of pediatrics, obstetrics, medicine and surgery. Since Mr. Abraham Flexner, the chief architect of the Rockefeller system of full time and at this time assistant secretary of the General Education Board, had the fixed impression that the Brigham was the only hospital that could be considered as a university hospital, this naturally involved Drs. Cushing and Christian.

Both thought it over, and both turned it down, Christian not so adamant as Cushing but liking the proposal no better. Cushing said of this later: "Coming from a race of general practitioners, the intimate and confidential relation between doctor and patient — one of the most precious things in Medicine — was in my blood and I could not look upon the cold institutional program with any great enthusiasm, much less with any expectation that it would serve to make something out of me that I was not already."[15] *

In April 1914 Cushing wrote to President Lowell:

> It is my impression that the reasons for the adoption in Baltimore of the proposal from the General Education Board were based upon the feeling, chiefly on the part of the heads of the preclinical departments, that the school was getting into a rut, from which it should be extricated

* From Fulton, John F., *Harvey Cushing*, 1946. Courtesy of Charles C. Thomas, Publisher, Springfield, Illinois.

no matter how seriously the method adopted might temporarily wrench their machine. Unquestionably this feeling would not have arisen had the heads of the clinical departments been given the opportunity some years ago voluntarily to place themselves on a whole-service basis similar to that on which my colleague and myself now serve in the Brigham Hospital.

It seems to me that the clinical departments of the Harvard Medical School, on the other hand, by their own initiative give promise of getting out of the rut into which they had fallen, and the injection at this time of a new element into the situation might seriously complicate it. Were we at a standstill or in financial straits we might have to ask others to help pull our load, and to accept their terms no matter how experimental they might appear to be. On the other hand, were our engine moving more smoothly than it is, we might otherwise be justified in venturing to test it on an unbroken road.

Not without various difficulties we are slowly working out our own problems at the Brigham Hospital — problems which concern the continuous whole-time service of the clinical chiefs — problems therefore which are somewhat idealistic and which moreover are novel to those who are concerned with the hospital administration. The difficulties are gradually being smoothed out, but the experiment, from all points of view, cannot as yet be deemed a complete success. Before we have quite found ourselves, on this basis, it is proposed that we make a still more radical experiment insofar as our relation to the institution is concerned . . .[16] *

Lowell replied on April 14, "Your position seems to me perfectly reasonable; and while it will prevent our applying to the General Education Board now for a gift for full-time clinical professors, it may not prevent our doing so at some future time . . ."[17] * (This reply casts some doubt on the statement sometimes heard that Cushing's rejection of the Rockefeller full-time grant caused a permanent "coolness" between him and the administration, particularly in view of the fact that another alternative was at hand. Harvard never did adopt the Rockefeller version of full time despite further years of tentative negotiations.)

When Cushing and Christian had refused to abandon their newly created system in favor of a more rigid way of doing things,

* From Fulton, John F., *Harvey Cushing*, 1946. Courtesy of Charles C. Thomas, Publisher, Springfield, Illinois.

Edsall wrote to President Lowell suggesting that the MGH could be used instead, for medicine and surgery. This required a considerable persuasive effort before Mr. Flexner could be convinced that the MGH would do at all. It was, he thought, too far away from the school. It was controlled by an independent board of trustees and had no properly established administrative link with the school. It did not have the proper tradition. And the situation in Boston was impossible anyway, the relations between the school and its many hospitals being too tenuous, complicated, and impermanent.

At this point Edsall committed to paper a few of the things he as an outsider had learned about the situation in Boston, and the MGH in particular. These he sent to President Lowell as a summary of points which it might be helpful to bring to Mr. Flexner's attention.

(1) Abraham Flexner had the feeling some years ago, that he expressed to me (and this was I know a somewhat widespread feeling outside Boston) that there were so many conflicting interests here and so much conflict of opinion, that it would be almost impossible to organize coherently all the hospitals etc. connected with the School especially because the School has no absolute control of any of the hospitals and not even an open affiliation with some of them. I feared this myself when I came here, but the thing that has surprised me most has been the remarkable progress toward cooperation and cohesion in such varied institutions and the loyal interest in and support of the School when it has been doing such radical things as it has in many ways. The result of the difference between this situation and that in Philadelphia or New York, for instance, (with which situations Flexner, I am sure, classes this to a large extent) is that here Harvard University is the Alma Mater of nearly all the men on the various hospital and other boards and [they] have Harvard's interests at heart, while in Philadelphia the University of Pennsylvania has very little place in, or interest from, the community and the same is largely true in New York. Of the 36 managers or trustees of the Presbyterian Hospital in New York I think none are Columbia men and none are in any way connected with the government of Columbia.* The contrast with the M.G.H. where practically all the staff and most of the Trustees are Harvard men, and

* Not 100 per cent true, but the contrast with Harvard was nevertheless striking.

where many of the influential authorities are connected with the governing bodies of Harvard is evident. The lack of an open contract with the M.G.H. becomes therefore of much the same significance or lack of significance as the fact that there are no binding agreements between many of the German hospitals and medical schools, – i.e. their interests are much the same and their governors are much the same or have the same interests. Hence there is little possibility of a rupture.

(2) As to the M.G.H. as a teaching and research hospital. I had been "warned" by various interested friends that the traditions and customs of the place and the rigid discipline of the superintendent would make it necessary to be very cautious and restricted in the things I attempted to do there that were contrary to precedent. So that I did go slowly for some time. As a matter of fact, I have this year made a good many radical changes, nothing I have asked for has been met with any but a warm spirit of cooperation. I have met with no restriction and now students are used more freely in that hospital and are in some ways a more definite part of the organization of the hospital than in any other hospital I know, including Johns Hopkins, and there is no hospital in which more teaching is done or more encouraged, and the attitude toward research is more generous and encouraging on the part of administration and staff and trustees than in any other place unless it be Johns Hopkins and the Brigham – and not greater in those places. I had a good many fears at first whether I should be able to approach an organization and general scheme of teaching and research that would be what seems to me right. Now I think only the money and a little patience are necessary. It may be worth while to note that the M.G.H. is by many people in this country looked upon as the model hospital and that Dr. Smith, the superintendent of Johns Hopkins, has modelled his general administration there after that at the M.G.H. Dr. Washburn, the superintendent, as well as his assistant, I have found to be deeply interested in teaching and research and extremely broad and far-sighted men.

(3) There is a good deal to be said in favor of having a full-time plan run in a hospital where there is also a service run on the usual plan alongside it. The chief criticism of the full-time scheme is that it will lead to an unpractical spirit – that clinicians will become academic prigs I believe Dr. Osler * put it. This is the greatest possibility of harm

* Osler had written from England to Hopkins' President Remsen before that school accepted the full-time system, "I fear lest the broad open spirit which has characterized the school should narrow, as teacher and student chased each other down the fascinating road of research." (Osler to President Remsen, Sept. 1, 1911. Privately printed. [HMS]) His opinion on the matter was even more definite a few years later: "The burning question to be settled by this generation

in the plan, I think, and it would seem to be pretty certain of being avoided if the men worked freely in their own way but alongside of others who are fine men but not on that plan . . .[18]

Flexner finally was brought to explore the possibility of using the MGH as the teaching hospital for full-time medicine and surgery. He said that of course the school would have to have the right to nominate the chiefs of service, spelled out in a fifty-year contract.[19] Dr. Walcott, president of the MGH, said such a contract would be "highly desirable."[20] Flexner pointed out that of course no full professors could receive any fees for private practice. This was agreed to.[21] He also wanted to be able to move the full-time professorships to the Brigham at a later date, if this became feasible.[22] The delicate negotiations continued through 1914, and finally were in such promising shape that President Lowell made a formal approach to the MGH in a letter to Walcott, Sept. 15, 1914.[23] In this letter he alluded to the negotiations still unsettled on the subject of a full-time surgical service, but outlined a proposal to add a third professor of surgery there.

Dr. Samuel J. Mixter was the head of the West Surgical Service, Dr. Charles L. Scudder head of the East Surgical. Dr. Maurice Richardson, the eminent surgeon-in-chief, had just died. It was apparently clear to all concerned that neither Dr. Mixter nor Dr. Scudder had any wish to become an academic surgeon on a Rockefeller no-fee basis. The proposal therefore was to appoint a third

---

relates to the whole-time clinical teacher. It has been forced on the profession by men who know nothing of clinical medicine, and there has been a 'mess of pottage' side to the business in the shape of big Rockefeller cheques at which my gorge rises. To have a group of cloistered clinicians away completely from the broad current of professional life would be bad for teacher and worse for student. The primary work of a professor of medicine in a medical school is in the wards, teaching his pupils how to deal with patients and their diseases. His business is to turn out men who know how to handle the sick. His business, also, is to bring into play all the resources of the laboratories in the investigation of disease, for which purpose he must have about him active young men . . . His business, further, is to get into close touch with the profession and the public, and with both to play the missionary; and this he can only do if engaged part of his time in consulting practice. There always have been of choice whole-time clinicians . . . By all means let us have them in the special hospitals attached to institutes of research, as in the Rockefeller; but spare the medical schools an experiment, which may be successful now and then, but which – from my point of view – can not but lower in type and tone the work of the clinical professoriate." (Osler, "The Coming of Age of Internal Medicine in America." *International Clinics*, 4, 25th series (1915):1.)

surgeon, give him one-third of the beds, and eventually work back to two services as one of the other two places became vacant. But the surgical staff met in November 1914 and turned the whole thing down.

A letter from Jerome Greene, a powerful trustee of the General Education Board, to President Lowell just at this time made it plain that even if the MGH surgeons had agreed to the plan, it might not have gone through. Lowell quoted to Edsall, Greene's statement that he did not see any possibility of the MGH plan working out. It might have been acceptable if the Brigham had not been in the picture, but the Brigham was right there next to the school and was the obvious choice for the Harvard University hospital.[24]

A pause followed this stalemate. In June 1916 the question was taken up in a new way, the negotiations being put in the hands of President-emeritus Charles Eliot, with a committee consisting of W. S. Thayer, Reid Hunt, Walter Cannon, and David Edsall.

A new proposal was made in December of that year on a provisional basis, this time including seven clinical units: two in clinical medicine, two in clinical surgery, and one each in pediatrics, obstetrics, and diseases of the nervous system. This drew a response from Dr. Wallace Buttrick, the secretary of the General Education Board, which made clear how far apart still were the Harvard and Rockefeller points of view.

The General Education Board has interested itself in establishing full-time departments of medicine, surgery and pediatrics [wrote Dr. Buttrick] on the general theory that such full-time departments would practically determine the character, organization and conduct of a medical school. These full-time departments have involved the ultimate *suppression* of the part-time departments . . .

Dr. Eliot's memorandum suggests a different type of organization, viz., the continuance of a number of part-time clinical departments, as at present, and the addition thereto of full-time departments in the same fields . . .

It is questionable whether in this form the full-time scheme could achieve its purposes. The full-time scheme aims at several objects — increased activity in research, better care of patients and students, and the banishment of professional business from research and teaching clinics.

The full-time scheme was, in a word, meant to end one order of things and to instal a new order of things. Dr. Eliot's proposition continues the old order and adds the new order, in so far as medicine, surgery and pediatrics are concerned.

Dr. Eliot's scheme does not even give the full-time unit a commanding position as against the part-time units. For example, the department of medicine might consist of four units, of which one would be a full-time unit. The full-time professor might not be chairman of the department; he might have no greater weight in the faculty than a part-time professor; he might not represent his subject on the administrative board which is the real governing body. Thus Dr. Eliot's plan would not transform the general character of the medical school, as the institutions previously reorganized on the full-time plan have been transformed.[25]

It was not until April that Dr. Eliot and his committee had prepared a fuller explanation of how far their adherence to full time extended. Wrote Dr. Eliot to Dr. Buttrick in defence of Harvard's multiple system of hospitals and clinics:

It is of course more difficult to develop the full-time policy in a large school possessing valuable connections with many hospitals than it is in a new and small school which is closely connected with only one hospital. It will take some time, as well as much money, to apply in the Harvard Medical School the full-time principle as widely as the authorities of the School desire and mean to apply it. Even in the Medical School of Johns Hopkins University the prohibition of private practice is applied to comparatively few of the teaching staff today.

The School has no desire to maintain clinics in the major subjects not led by full-time professors. On the contrary, it would apply the full-time method in all the important clinical branches and units . . .

The Harvard Medical School, however, firmly believes in using several hospitals instead of one, and many hundreds of beds instead of from one hundred to one hundred and fifty in one University hospital. It believes that in a large medical school, which admits classes of more than one hundred men, more than one clinic for Medicine and more than one for Surgery are indispensable.

As for the proportion of full-time to part-time men, he stated clearly:

While the Harvard Medical School desires to put full-time professors at the head of all its clinical teaching, it expects to continue to employ a large number of part-time teachers, in order to utilize its extensive clinical and laboratory facilities, and give amply individual instruction in the great variety of subjects with which it deals . . . There are more than three hundred lecturers, instructors, and assistants, most of whom are private practitioners as well as hospital practitioners . . . The existence of this useful body of instruction given by part-time teachers, who are familiar with practice in homes as well as in hospitals, does not interfere in any way with the control of all the principal clinics by full-time professors . . .

In closing, Dr. Eliot reaffirmed his university's independent tradition with the comment:

No "transformation" of the Harvard Medical School and no suppression of any of its parts or departments are needed or desired. Its history is honorable, and its present condition effective. It now asks from the General Education Board (1) aid in establishing that great improvement in medical education, the full-time professorship without private practice, and (2) in building up and maintaining a well-equipped Department of Preventive Medicine and Public Health.[26]

When it became clear that the answer of the General Education Board would be No, the application was withdrawn in May 1917. A number of years would pass and views would change before major help to the medical school came from the Rockefeller boards.

Edsall, having come to the welcome conclusion that "money and a little patience" were the chief things needed in solving his problems in building up research at the MGH, found that extra amounts of patience would be needed. Until the completion of the Moseley Building in 1916 the space problem was nearly impossible to solve. When Paul White arrived with his Einthoven string galvanometer the only space which could be found was, in his words, "a little closet next to the syphilitic bathroom."[27] Then Dr. Means came back with his apparatus to measure respiration. Means, like myself, had received training in metabolism, except that in addition to working with DuBois he had studied with August Krogh in Denmark and at the Nutrition Laboratory in Boston with Benedict. I suppose that Edsall from the beginning had visualized us as a team, or at least had hoped that we would make a good team. He was right, and from those early days forward Howard Means and I were always close friends and associates. After some scrounging Edsall presented us with a tiny cubbyhole underneath the outpatient amphitheater. The cubbyhole was walled off by sheets and just big enough for a simple Benedict machine to do basal metabolisms on a single patient. Doing metabolic research in the beginning was not simple. When we began, for example, there was no way of collecting 24-hour urines or weighing patients accurately, but within a very short time Means and I confirmed the standards we had learned from DuBois and set to work.

When the administrative offices moved to the Moseley Building, Edsall managed to keep the space at the Blossom Street end of the Bulfinch Building for research. Howard Means wrote about this at the time:

> The organization of a laboratory for clinical investigation was made possible in the spring of 1918, when the old offices of the Administration on the ground floor of the Bulfinch Building were vacated. The collection of equipment having been started as early as the spring of 1916, reconstruction was immediately begun. In this way six rooms became available, which now constitute a fairly satisfactory laboratory for about twelve workers. Before the organization of this Laboratory clinical research had to be conducted in any corner that the investigator could find.
>
> The old cashier's office has been converted into a very good chemical laboratory, and to it has been added a portion of the hallway opposite. Dr. Howland's old office has become a bacteriological laboratory, and Dr. Washburn's a blood laboratory. The old stenographers' office and the visitors' room are now used for metabolism work. Near by under the west end of the portico is the cardiographic laboratory . . .
>
> The chief idea underlying the establishment of the Medical Laboratory is that it shall be primarily a place for original work in the field of clinical investigation. Routine work is not done there except in the case of certain highly technical tests which cannot be done elsewhere . . .
>
> From the time of its organization to the present [1921] a large number of workers have contributed to the progress made by the Medical Laboratory, and the collected papers of the Laboratory now make a series of seven fair-sized volumes . . .[28]

Some of the more important of these studies included Arlie Bock's work on blood gases with Professor L. J. Henderson; pathology of the blood by Dr. Minot & colleagues; metabolism of bile pigments by Dr. Chester Jones; study of thyroid disease by the thyroid committee involving medical and surgical services, x-ray, and lab measurements of respiratory metabolism under Drs. Means and Aub; clinical calorimetry of infants by Dr. Fritz Talbot; studies of bacteriology of pneumonia by Dr. Lord; blood and urinary chemistry in diabetes and nephritis by Dr. Reginald Fitz; and laboratory work for Dr. Rackemann's Anaphylaxis Clinic.

The year after Edsall left the hospital he managed to find money to add an ell to this section of the Bulfinch Building, to build the famous Ward 4 and other much needed facilities. The story of Ward 4, its significance to medical research, and Edsall's role in creating it, has been well told by Means.[29]

Although I was a supernumerary as a resident, and no great discoveries came out of the research, my work gave me great joy. Most of us had this feeling, and I can testify that the morale among us was very high. Our training had allowed us to do research work on patients which had not been done before in the hospital. One must keep in mind that the amount of research then going on among clinicians in America was very small indeed. The average physician reporting his cases to medical societies was content with mere clinical description.

Very little attention was paid to the physiology of disease. As a matter of fact, there were few interested societies to which one could report such findings. One was the Society for Experimental Biology and Medicine, the "Meltzerverein" in New York, so called because it was organized by Dr. S. J. Meltzer of the Rockefeller Institute. However, it was a dangerous forum for young investigators to attend because if one had the temerity to present a paper, Dr. Meltzer would more than likely jump up and cry, "Interesting, interesting, but when I worked on that problem thirty years ago . . ." I only add that under such treatment young egos collapsed quicker than punctured balloons. Meltzer, however, was one of the great pioneers and had done some physiological work in many diseases.

Actually, however, there were enough new problems for all. The age of physiological medicine was just beginning. By current standards much of our early work was crude. We certainly didn't have many of the exquisite and refined research techniques that exist today. We had one great advantage, however. We were the pioneers. There was little work by others which needed reference.

How much of this development at the MGH was due to Edsall? In his book, Dr. Rackemann says, "There is no doubt that it was Dr. Edsall who did more than any other man to promote clinical research at the Massachusetts General Hospital. It was he who wanted to extend and improve the training and experience of his young men. It was he who helped to start their investigations and who encouraged them, and finally it was he who 'sold' the idea to the Director and to the Trustees of the Hospital, so that his young scientists could find little places here and there to set up their laboratories and have a little money for apparatus and supplies."[30]

Dr. Means agrees with this too. "Of course, bits of research had been done as extra-curricular activity by members of the volunteer staff, long before Edsall. But he can be credited with introducing clinical investigation at the Hospital in a professional sense. That was one of the things for which he was brought to Boston."[31]

And the director of the hospital gave official recognition to this when he wrote in his annual report on Edsall's retirement from the hospital: "Dr. Edsall's broad knowledge of organization of schools of medicine and hospitals has been of great service to us. His wide learning and commanding personality attracted promising young men to him. He built up from nothing our medical research laboratories and started them upon their career of achievement. The historian of the future will credit him with giving the Hospital a new and broader conception of its functions."[32]

CHAPTER 13

The Industrial Disease Clinic, and Other Research, 1912–1923

ONE of the most far-reaching of Edsall's innovations was the establishment of an Industrial Disease Clinic at the Massachusetts General Hospital. Edsall had long been persuaded of the importance of industry in relation to disease. It seemed impossible to him that the surroundings and activities which occupied a man all his working hours should be without importance to his health. Further, he was well aware of European experience with industrial poisoning, though the relevance of this experience to the U.S. was then overlooked by most industrialists and many members of the medical profession. Even in 1917, speaking before the Johns Hopkins Medical Society, Edsall knew that industrial disease was regarded as his hobby-horse:

"If in a brief time I can give a superficial suggestion of the general relations of medicine to industry, and can combat what I think I may fairly call the common academic aloofness of medicine toward these things . . . I shall have accomplished my main purpose . . .

"In bringing in one of his pet views in a discussion before the British Medical Association, Clifford Allbutt once announced, 'I will now trot out my old jade.' Dr. Miller intimated, when he asked me to come here, that I might be permitted to do the same."[1]

The medical profession, in general, had recognized few instances of industrial poisoning, and in any case most doctors were unaccustomed to thinking of such hazards as problems in public health. With his strong interest in preventive medicine, Edsall set out to change this situation, beginning in his first days at the MGH. Facts were primary, and he picked an ideal method of gaining them.

The Industrial Disease Clinic was set up, first under Dr. Harry Linenthal, then under Dr. Wade Wright. In the beginning, patients were supposed to be referred to this clinic from the rest of the hospital, but the numbers referred were disappointingly small. Nevertheless, the hospital's method of record-keeping was changed to include information on a patient's work, with a description of exposure to harmful materials, dust, unusual strain and other details. Edsall often cited Dr. Alice Hamilton's statement to him early in 1914 that among the many hospitals she had visited in this country, she had found only two in which the records could be counted on to contain statements that were of any value to her in her work on industrial diseases.[2]

Within six months, despite the imperfect working of the referral system, they turned up 276 cases out of 1507 male medical admissions to the outpatient department, "in which industry was either the predominating cause of ill-health or was a very important cause or was very important in treating the patient. In a large proportion of these instances the relation of industry would have been entirely overlooked without this routine detailed inquiry."[3] Cases of poisoning from benzine, brass, naphtha, turpentine, and benzol were uncovered, which had been wrongly diagnosed as anything from "apprehension" to "appendicitis."[4] And of course there were the cases of lead poisoning, which yielded very satisfactory statistics, since the presence of lead could be so clearly identified. So many cases of lead poisoning were found that the hospital issued the famous "lead slips," "Advice to Persons Working with Lead" and "Precautions for Printers," with suggestions ranging from "walk to work" to "insist that floors should not be swept during working hours." These slips found their way back into many industrial establishments, and caused an improvement in working conditions well beyond the health of the individual patient. As Edsall reported, "In a great many instances the manufacturers have been more or less ignorant of the dangers to which they are subjecting their working people, and they are often very willing when these are pointed out to them to do a very great deal to prevent them."[5]

In fact, the employers were sometimes ahead of the doctors. Edsall cited the case of a patient with hemorrhages which he suspected were caused by benzol poisoning from the dope used in making artificial leather. Edsall visited the factory and there he found, in his words:

> The exposure to benzol was very severe and the ventilation very bad. The manager called in the chemist and we went over the matter together, and as a consequence he told the chemist to get rid of the benzol completely if possible and said that he would look into a proper ventilating system. But he also told me that he had had a man die some months before with hemorrhages, and though he did not know that benzol was seriously poisonous he had wondered whether the dope might not have had something to do with the man's death and had asked the doctors in the town who had had charge of the man if it had and they had said "No." He was a humane man and openly resented the fact that this advice had led him unknowingly to expose others unnecessarily. I could multiply similar illustrations.[6]

The early accomplishments in the clinic whetted the group's desire for more. They knew many cases were escaping them because even among MGH physicians, "few of them understood what the hazards were and little was accomplished."[7] Then through the generosity of "a friend of the hospital," Dr. Frederick Shattuck found them enough money for a medical fellow in industrial disease and other expenses of the clinic,[8] and they put a social worker at the front door of the outpatient department where she checked every single patient admitted and referred the likely-looking cases straightaway. A brief sample from her instructions is enlightening: she sent *all* ammunition workers, brass foundrymen, bronze workers, buffers, grinders, polishers, filers, chemical workers, chippers, cigar makers, and so on down the alphabet; and others if their work put them in hazard, such as farmers "if accustomed to using insect sprays (lead arsenate) etc."; stokers and firemen, also coal passers "if story of high heat"; upholsterers and mattress workers "if work is said to be dusty." (Another interesting sidelight is the notation: "IN GENERAL, no children, unless illegally employed.")[9]

Now the clinic began receiving thirty or more patients a day, and was turning up as many cases of lead poisoning in one year as the hospital had formerly treated in five. Miss Susan Holton guarded the door of the outpatient clinic, where she soon developed a high degree of skill in detecting which patients might have an industry-related ailment. She was followed by Miss Alice Sinclair. By the time the system was well shaken down, they were seeing nearly 20 per cent of the hospital's outpatients (between March 1916 and March 1917, 5121 patients came to the Industrial Disease Clinic [10]).

And their usefulness was not confined to the hospital. Miss Helen Bradfield, as their chief social worker, did a great deal of visiting in factories and homes to check on conditions.

Now they were really getting the information needed to take step one, to prove that industry "is a very important factor in general medicine . . . There has been a widespread, passive feeling among medical men that this is not true . . . It is difficult to prove . . . if one takes as the basis of his search the usual medical records, for generally nothing is said in them that really touches the point." [11]

"One of the most influential things leading me to institute this method of inquiring into the effects of occupation on health," Edsall told the Massachusetts Medical Society in 1914, "was the desire to provide material that would be useful in making laws and special rules relating to occupation and its effect upon health. Many such laws and rules, like many others of those laws of social bearing that have become so frequent in recent years, here and abroad, have been made, not only in this country but abroad, upon very slender information, because information was not available in any accurate form but the laws had become necessary." [12]

Edsall welcomed the opportunity he saw for medicine to improve the general circumstances of life, and to bring an element of helpful reasonableness into an area of strife between employer and worker. When a few years later, as consultant for the Industrial Accident Board of Massachusetts, he was confronted by the complications of compensation cases under state law, he had to admit that from one point of view all disturbed health in industry might be subject to compensation, though he recognized that this would probably lay an impossible burden on industry. "Probably the best available way that we have to cover these things," he said in 1916, "is by something that spreads out over all industry and is aided by the State — that is, by health insurance or something like it." [13]

Wade Wright was the devoted doctor in the clinic who spent full time on finding industrial disease in the outpatient department.

With the advent of the war and the forming of the Base Hospital, the clinic was temporarily closed, for Wade Wright joined the Mass. General unit and went to France. After the war it was revived again, organized jointly with the medical school, but after Dr. Wright left in 1924 the work gradually took a different form. The Division of Industrial Hygiene had been started, the School of Public Health was functioning, the *Journal of Industrial Hygiene* was spreading the word, and the need for the clinic's pioneering efforts had passed.

Edsall's industrial disease work had many ramifications, state and national. As Edsall described it in 1915 to Dean Bradford, who was considering Harvard's facilities in relation to an Institute of Hygiene proposed by Abraham Flexner:

"We have already had a study done for the Bureau of Labor Statistics in Washington . . . [One] is now about to begin in cooperation with the Bureau of Child Labor and others are in contemplation. We are also constantly in touch with the State Board of Labor and Industries, the State Health Department and the Industrial Accident Board. The latter Board is constantly sending us cases of industrial accident or disease for "impartial opinions" in settlement of claims for compensations. In fact this is, I think the only hospital that has, anywhere, by agreement, put its facilities at the service of the State Board in order to aid in just settlement of these claims. This has been going on for more than a year, and is constantly increasing in frequency of use by the State Board. I have been asked by the Board of Labor and Industries to act as a strike mediator in an important strike where the workmen claimed damages to health from the process and after visiting the factory and examining the men gave a report which was accepted by both sides and the strike was called off." [14]

Edsall's interest in these questions led him not only to be frequently called in consultation; he was for a number of years a member of the state Public Health Council and the Committee on Preventive Medicine and Hygiene. He was also appointed to the federal Committee on Industrial Fatigue (a national defense subcommittee) which was established after the U.S. entered the war. Of course, wartime conditions made these industrial problems acute, especially in the manufacture of munitions, where frequently we imported a process without knowledge of the necessary safeguards to the workers.

One wartime contact, which was to bear great fruit for Harvard in later years, was his working with Alice Hamilton. He had first met Dr. Hamilton while she was at Hull House, when he asked her to take him on a tour of the Pullman Works and the Old Dutch Lead factories in Chicago. But apart from that and a conference on various industrial poisons in 1914 when Dr. Hamilton was working on rubber factories, the two had seldom been able to get together. When during World War I Dr. Hamilton was working on poisoning in the munition industry for the Federal Government, she asked for a committee of medical men to act as consultants, and she was given Dr. Edsall and Richard Pearce (then professor of pathology and research medicine at the University of Pennsylvania). She would report to them the dubious cases she found and they would find some local specialist who would examine the case and report back to her.[15]

One of the most curious problems in which Edsall was consulted occurred in the Franklin, New Jersey, plant of the New Jersey Zinc Company, near his old home in Hamburg.

Edsall described events in his speech at Johns Hopkins on "Medical-Industrial Relations of the War":

> In a very large plant engaged in producing raw material that is used to an enormous extent, as benzol is, in war as well as in peace, the company officers and the plant physicians have been much worried by the fact that there have been a considerable number of cases of an extraordinary and in some instances very grave disease in their men, and they did not know how to stop it. This disorder is most interesting clinically. It nearly always begins with a peculiar gait which one writer has aptly termed the "rooster gait." [16]

The men affected would tell a story of going up hill on their toes and being forced to run downhill faster and faster to keep from falling forward, and they had to climb stairs on all fours. One patient reported that

> any influence, such as walking down a steep incline, which interferes with the protection which his gait offers him against propulsion, brings propulsion out. Recently, for instance, he lifted a box and started to

walk forward. This weight brought him up on his toes and he at once commenced to run and finally fell. The patient rides a bicycle, but must have help to get on, and always ends his ride by falling off.[17]

Other manifestations were mask-like faces and monotonous voices, muscular twitching, cramps in the calves, and stiffness of the leg muscles. Occasionally patients had uncontrollable laughter, or less frequently, crying.

Edsall pointed out, "Direct physical examination shows almost nothing and there are not any focalizing signs."[18]

He told how the company had hired "a clinician well trained in the study of nervous disease" to discover the nature of the disease, and had finally decided that it was most likely to be manganese poisoning, since manganese was present in the ore handled by the plant. As Edsall recounted, "He made a valuable scientific report, but the company officials were annoyed because they said his report was simply an interesting medical paper and the thing that interested them was the economic and human matter of the prevention of the condition."[19]

The Franklin mill's concentrating process required the ore to be ground and then thoroughly dried and screened and passed through magnetic separators so that iron and manganese were collected on a separate belt. In order to make this separation, the whole process had to be bone dry and was therefore very dusty. Edsall was sure that manganese dust was in the air, since the crude ore contained 9 per cent manganese, but most of the local investigators thought that the magnets which separated the ores produced the symptoms. An elaborate study was made of the magnets, some of which were sent to Harvard to Cecil Drinker's laboratory, where they provided the first project for a new research department in applied physiology. Even Tyzzer's waltzing mice were involved in the research, but the magnets produced no symptoms and had to be exonerated.

Edsall was called in because the chief physician of the plant had been an intern under Edsall and knew he was visiting nearby. Edsall said:

I took Dr. Cannon and Dr. Wade Wright along with me and we went first to see the plant. I only epitomized what they felt at once when I wrote that further study of the nature of the disorder would be interesting and profitable from a scientific standpoint and should be done, but that the imperative practical fact obvious to anyone who entered the plant was that the dust hazard was simply amazing and the dust contained at least one substance known to be poisonous and in all probability the actual cause of the condition, and the real thing to do was to have engineers devise means for reducing the dust – a difficult problem, but not more so than others that I had known to be solved. An executive officer of the company replied that they had had such plans under consideration for some time, but that he did not believe that they would prevent the disease. In view of my letter, however, he had ordered that they be taken up actively at once. The chief physician wrote me that he had been of my opinion for a good while, but had not been able to convince his superiors and thought that an outside opinion was necessary . . . They have nearly completed a system to control the dust at a cost of over $300,000, and are now interested in having a careful study of the etiology and treatment of the disorder and they have given us at Harvard a generous budget for a two years' or longer scientific study of the whole question.[20]

The results of this study appeared over the signatures of David Edsall, F. P. Wilbur, and Cecil K. Drinker in *The Journal of Industrial Hygiene* in August 1919.[21] Another paper on this subject by Edsall and Drinker, entitled "The Clinical Aspects of Chronic Manganese Poisoning," was published in a volume honoring Sir William Osler on his 70th birthday. This volume of papers by Osler's pupils and co-workers was put together by Dr. Welch and an Anglo-American committee and published under the title *Contributions to Medical and Biological Research.*[22]

The studies in the laboratory gave a full picture of the operation of manganese as a poison, its curious symptoms arising from the poison's action on specific basal ganglia.[23] As a result, not only was manganese poisoning prevented in all the New Jersey Zinc plants, but its incidence was reduced throughout the country when these findings became known.

One of the first things Edsall did in his own research after settling down in Boston was to get the work he had done in the Carnegie Nutrition Laboratory ready for publication. He gave the Shattuck Lecture in June of 1912 on this subject, and a paper entitled "The Efficiency and Significance of Different Forms of

Respiration"[24] before the Association of American Physicians at their 1912 meeting. J. H. Means remembers the impact of this paper, and said it "heralded a growing interest on the part of clinicians in respiratory physiology that was to last for a decade or more."[25]

At the Carnegie Nutrition Laboratory, Dr. Benedict had developed an apparatus which measured the oxygen intake, the CO_2 output, and the total air inspired. Various other instruments enabled Edsall to measure the pulse rate, blood pressure, and CO_2 of the alveolar air of his subjects. As a result of experiments with different types of breathing, Edsall arrived at some tentative conclusions as to the meaning of various types of breathing seen in disease, and a doubt remained in his mind as to the usefulness of certain standard forms of treatment. He said to the Massachusetts Medical Society:

> We certainly need to determine as far as we can, in any pathological disturbance of respiration, when the disturbance is associated with exhaustion of the respiratory centre, and when it is, on the contrary, due to over stimulation of the centre. The treatment would, in the two cases, be the opposite. I think we may reasonably suspect that we have in many instances been making an error entirely similar to that which was made for years in treating the circulatory failure in acute infectious diseases with the mistaken idea that it was primarily due to excess of blood pressure whereas it is usually due chiefly to quite the contrary condition. We may, that is, in respiratory distress, equally well have been dealing at times with a condition directly the contrary of what we thought we were treating, and may, especially, very often have been treating an overexcited respiratory centre by still further exciting it. Likewise it appears to me that the effects of the respiratory stimulants need, in purely experimental studies, as well as clinically, to be investigated to determine much more accurately . . . their effect . . .
>
> I think all clinicians will admit, that . . . we can do very little to relieve respiratory distress . . . No drugs are more sadly disappointing when we most desperately need them than those that are classed as respiratory stimulants . . .
>
> It is quite possible that we may find that some drugs that are called respiratory stimulants actually lower the efficiency of the function of respiration in some circumstances of disease and even in normal con-

ditions, and the contrary may be true of some so-called respiratory depressants.

His experiments with 15-minute periods of normal, superficial, and slow deep breathing showed him how deceptive subjective impressions, and even clinical observations, could be. He continued in his speech:

In most instances we had the impression of having breathed less in total amount during the period of superficial breathing than in the other periods. In every instance quite the contrary was true. The record always showed in the neighborhood of 25 per cent. more total (tidal) ventilation in this period than in the normal period and the lowest amount of air used was always in the slow deep breathing; as much as 25 per cent. less than in the normal period and as much as 40 per cent. less than in the period of superficial breathing. Evidently, then, from the standpoint of gas exchange the slow, deep form of breathing is very economical of effort while the superficial form is extremely wasteful.

Also to his surprise, he found that no matter whether his subjects breathed normally, or superficially, or as slowly and deeply as possible "without distinct discomfort," "the lowest amount possible without producing distinct respiratory distress"—"the real amount of air that reached the alveoli was almost exactly the same in all instances" and the oxygen absorption and CO_2 output were also the same, whatever the rate of breathing.

Nature requires of us only a definitely essential amount of alveolar ventilation, and if other conditions remain the same this amount is the same, but the number of breathing efforts required to accomplish this ventilation varies extremely widely with different forms of breathing.[26]

He soon followed up the question of drugs with J. H. Means, and their paper "The Effect of Strychnin, Caffein, Atropin and Camphor on the Respiration and Respiratory Metabolism in Normal Human Subjects" appeared in *The Archives of Internal Medicine* in December 1914. Although worked out on only two subjects, the results seemed worth stating:

1. Strychnin: No action on the respiratory center or circulation, possibly an increase in metabolism.

2. Caffein: Stimulation of respiratory center, moderate increase in metabolism, no effect on circulation.

3. Atropin: Marked increase in metabolism, in doses large enough to cause the classical action on the circulation. Marked drop in the calculated values for alveolar carbon dioxid and rise in alveolar ventilation. Whether these latter effects are due to stimulation of the respiratory center or to changes in the dead space is not clear at the present writing.

4. Camphor: The results not agreeing, no conclusions are justifiable.[27]

He worked again with Dr. Means in reporting a case of family periodic paralysis, which harked back to his earlier experience with Flexner and Mitchell in Philadelphia.[28] He gave a few miscellaneous addresses where invited, and made a number of outlines of ideal administrative and teaching situations in medicine, for instance "Relation of the Staff to the Administration of Hospitals,"[29] in which he described the workings of the MGH General Executive Committee, given to the Philadelphia Pediatric Society, or "Movements in Medicine"[30] in which he discussed such things as medical insurance and relations between clinical and laboratory branches in teaching, before the Massachusetts Medical Society. But most of his publications for this period related to industrial medicine and the work of the hospital in this field.

During Edsall's first decade in Boston, the medical atmosphere was gradually changing. Though retired, President Eliot kept track of everything to do with the medical school which he had, in a sense, created. When Eliot heard that his friend Bishop William Lawrence was ailing, he wrote, "I hope that you are employing the most sensible physician in Boston, and that he is not an old man. The bright young fellows have had a much better training than the old ones received."[31]

Edsall had an opportunity to see prominent patients while he was building this modern scientific unit at the MGH, for he had an office at the Blossom Street end of the Bulfinch Building during most of this time. Other men like George Minot who wanted particularly to see patients also had simple offices for seeing private patients. It didn't take long for Edsall to get a practice of important people. His large figure, his deep voice, his obvious kindliness, his authoritative way of speaking – and also his mastery of clinical medicine – produced a prompt feeling of security and confidence. He was always interested in a person's whole environment and personality, whether of a simple ward patient or an important private patient, and this endeared him to people. Their confidence was well deserved for such qualities added to wide understanding made for excellent analyses in diagnosis. He was an outstanding clinician.

CHAPTER 14

Edsall Makes a New Home, 1915–1928

When Margaret Edsall died in 1912, Edsall was left with three small boys to take care of. The first thing he did was to abandon the Marlborough Street house and move back to the Tileston family home in Milton. There his wife's sister Amelia, five years younger than Margaret, ran his household and took care of the children. Amelia got on well with the children but she was not a good housekeeper and after a year this arrangement was given up. Margaret had always felt concern about Amelia. As Harriet Mixter wrote later, Margaret believed Amelia "had unusual talents, but so far life had been unsatisfactory to her because nothing really big enough had offered her an adequate outlet."[1]

Amelia later found scope for her restless energy in nursing and work with refugees during the war. She went to Serbia where she worked devotedly for five years. In 1920 she died in Belgrade of pneumonia, the third of the sisters to meet a tragically early death. She was, however, honored by the nation. Her funeral procession in Belgrade was over a mile long,[2] and she was posthumously awarded the Order of St. Sava by Alexander I of Serbia.

Edsall next turned to a friend of his wife's, Elizabeth Pendleton Kennedy, known to her friends as Pen. Pen was a Virginian, and related to all the blue-blooded old Virginia families, but she had made a rather unusual life for herself, leaving home as a young woman to work for Doubleday Page in New York as a reader. She was an interesting and cultivated person, a combination of southern aristocrat and New York radical, interested in the labor movement, woman's suffrage and Negro rights. She had known Margaret Tileston in Radcliffe, and often visited the Edsalls in Philadelphia and on Cape Cod. Early in 1914, Edsall persuaded her to give up her job and move into the big house in Milton to take care of them all.

Her stepsons remember that at first this was a distinct improve-

DAVID LINN EDSALL

Young David Edsall

Dr. Edsall in Pittsburgh

David and Margaret Edsall
in Philadelphia

Alfred Stengel in 1910

William Pepper, Jr.

Courtesy of University of Pennsylvania

Ivan Petrovich Pavlov with Walter Cannon at the International Physiological Congress held at the Harvard Medical School in 1929.

At the opening of the Peter Bent Brigham Hospital, 1913. Left to right, John L. Bremer, Theobald Smith, Walter B. Cannon, Harold C. Ernst, Milton J. Rosenau, David L. Edsall, Charles S. Minot, Sir William Osler, Harvey Cushing, W. T. Councilman, Henry A. Christian, S. Burt Wolbach. *Courtesy of Harvard Medical Archives*

The Edsall and Houghton families in China, 1926. (Pen Kennedy Edsall seated front left.)

a) Carl Binger, b) Francis W. Peabody and George R. Minot, c) Stanley Cobb, d) Paul Dudley White and J.C.A., e) Edward D. Churchill and James Howard Means.

a)

b)

c)

d)

e)

(Left) Louisa Richardson shortly before h
marriage. (*Photo by Frizell. Courtesy*
Radcliffe College Alumnae Association
(Top) Grandfather and grandchild. (B
tom) David Edsall about 1926.

ment. The house ran smoothly, the cook cooked better meals, and tact and good humor governed her relations with the boys.

In June of 1915 she and Edsall were married. The marriage was in many ways unfortunate. Pen had been leading an independent life for many years. She was forty-five when she married, the same age as her husband. She found the conventional society of Milton very trying and after two years the family moved to Cambridge. But by this time Pen was obviously an unhappy woman with a vengeful tongue. Edsall sent his son Richard away to boarding school because she was so harsh with him.

The family always wondered if the automobile accident in which she and Edsall were involved had contributed to the change. At the end of their first summer, which was spent in Ogunquit, Maine, Edsall and Pen were driving back to Boston on the Newburyport Turnpike when they struck a car emerging from a side road. This car was driven by Richard T. Crane, Jr., who was somewhat hurt, but Crane's houseguest, Benjamin S. Cable (Assistant Secretary of Commerce and Labor under Taft), was killed. Edsall himself was not badly hurt, but Mrs. Edsall was thrown through the windshield, receiving cuts on her face which left permanent scars. She also injured her leg. She was lame from that injury and had pain from it for a long time afterward. Dr. Geoffrey Edsall remembers, "As a child I always felt that the accident was responsible for the change in Pen's personality. At any rate she complained a great deal about the pain in her leg, the difficulty it caused her, and she limped rather obviously for many months afterwards. She seemed to be quite miserable during this whole period and never quite became her bright self again." [3]

She always maintained that her husband had not been at fault in the accident, but it was a bad enough shock without that. The case came to court but no blame was placed on either party and both drivers automatically had their licenses suspended for six months. This was a considerable hardship for Edsall, whose style of life did not include a chauffeur. Further, Mr. Crane claimed that the accident had indeed been Edsall's fault and that he had behaved "very badly." His side of the story ran all over Boston. Edsall was always at a disadvantage in this kind of situation, as his instinctive response to such stories was a dignified silence.

In the fall of 1917 they moved to Cambridge, living at 68 Sparks Street for five years and then at 50 Concord Avenue for several years after that. At this time they had a more congenial social life among university people, but his friends remember how they suffered for Edsall when Pen berated him in public. Like most social southerners, she demanded intense attention from her male associates, which represented the southern tradition of courtesy towards ladies. This was far more intense than what one saw in Boston. When her husband became head of the Harvard Medical School, she cherished her position as Mrs. Dean, and was angry with her husband if medical school professors were not properly deferential to her. On one occasion, just after she had made a prolonged stay in Virginia, when I was still living in their house, we went together to a faculty tea. I thought it would not be proper to be too attentive to her at the tea but did the best I could in an inconspicuous way. But she expected more attention and respect than it occurred to this faculty to give her, and she scolded us both unmercifully at home that evening.

Edsall weathered this type of behavior as best he could, and it was not until his sons were nearly grown that he could bring himself to speak to them of the problems at home which he feared would separate him from his children.

Unhappiness at home made Edsall more dependent upon a few close friends. He used to enjoy long walks with many of these and his summers were enlivened by these rough tramping expeditions. From his early years in Boston, his ties with some of his close friends grew out of such prolonged walking trips.

One of these friends was Carl Binger, whose relationship with Edsall went back to his close early friendship with the youngest sister of Edsall's first wife. Edsall took Binger and me as medical interns in his early years at the Mass. General Hospital, a wonderful fifteen months of medical training under a brilliant clinician. After that Binger went to the Rockefeller Hospital where he investigated various types of the pneumococcus in the pneumonias, and the effects of physical therapies, but that too did not hold his attention for long and he went to Dr. Adolph Meyer in Baltimore to begin his lifelong study of psychiatry.

On one of the mountain rambles they used to take in their holidays, Edsall lost his cane down a sheer cliff and Binger tried and failed to recover it. But the next year they took the same way and Binger tried again and this time found the cane wedged in the cliff. You can imagine the triumph and elation. Their love of rough climbing brought them close together in the summers, but their diets were not as varied as usual and a mutual dislike of peanut butter came from their rather monotonous meals.

Many of Edsall's friendships were cemented during such rugged walks in the White Mountains. Walter Cannon was a favored companion and they had many hilarious stories to tell after their trips. They came one evening after a hard day of walking to a swanky summer hotel where they had promised themselves a comfortable night. They had only their walking clothes which the headwaiter thought unsuitable in the dressy dining room, and he therefore fixed an isolated table on the porch surrounded by a screen. Of course they were promptly recognized by some of the hotel guests, who were horrified by their treatment and insisted on their eating in the dining room, even though they both preferred the porch. Their walking clothes were most picturesque in that environment.

Another friend who shared Edsall's country pleasures, if not his mountaineering jaunts, was L. J. Henderson. He and Edsall frequently spent weekends in each other's summer homes, discussing university and life problems. He was a favored advisor and stimulating friend.

Edsall loved wild unspoiled hilly country. In such surroundings he had grown up, and there he felt at home. He cared less for the high mountains, and never felt really at ease beside the ocean. In Hamburg, and later in Millwood, Virginia, and Greensboro, Vermont, he had houses in the high hills which pleased him best. One of the most beautiful of these summer homes came through his wife, Pen, who had lived in the valley of the Shenandoah as a girl. In 1920 the Edsalls bought a house in Millwood, Virginia, to use as a vacation home. Millwood was a rural community about ten miles from Winchester and perhaps sixty from Washington. It had the well-groomed, long-farmed look of that part of Virginia. The Shenandoah River ran two miles from the house, the Blue Ridge Mountains rose invitingly just across the river, and the long high ridge of Massanutten Mountain could be seen further west from the fields around the house. Mrs. Edsall felt at home, back among her Virginia cousins, while the young people picnicked and played tennis and went canoeing on the river.

After they had first settled in at Millwood, Edsall wrote to his friend Dr. Shattuck:

> The boys were never so happy in their lives with so many delightful things to do that they do not know which way to turn at any moment. It is an almost unique persistence of the charming old county life with almost all the lovely old plantations still occupied by the descendents of those who built the places a century or century and a half ago. And almost every step you take you tread on history . . . "Saratoga" (where Mrs. Edsall's mother was brought up) was built by General Morgan with his Hessian prisoners as laborers and named for the battle. We look from our porch directly into Ashby's Gap which Moseby used constantly in his raids . . . One of Mrs. Edsall's aunts said she counted 63 fires one evening while Sheridan was raiding about here.[4]

In Cambridge, too, Edsall was very fond of walking in all weathers, and would tramp for miles around his neighborhood, where the handsome old houses stood among generous gardens and lawns, shaded by elms and other large trees, or he would stride out into the suburbs which then had many truck gardens and small farms. He had a great love for flowers and for the beauties of nature so that he found the gardens in the suburbs of Boston particularly beautiful and worthy of long walks. And these walks he was prone to take alone while he pondered over the problems he had to deal with after he became dean of the Medical School. He frequently pointed these walks towards the homes of his friends where he would stop for discussion and advice about some problem involving Boston and Harvard personalities and customs. His first wife's cousin, the Unitarian minister Henry Wilder Foote, had a house in Belmont on the side of the steep "bel mont" ridge (not far from the house where we now live, then owned by the famous printer and typographer Bruce Rogers) and Edsall frequently walked out in that direction. He sometimes carried home a beautiful bunch of lilacs he had picked along the road (he was meticulous about asking permission) and was acquainted with our luxuriant hedge of old lilacs before we were.

Other particular friends of those days were Bliss Perry, professor of English at Harvard, Felix Frankfurter of the Law School, and Alfred North Whitehead, the famous English philosopher who had come to the Harvard faculty in his advancing years and who was so affable and wise.

These were scholars with broad knowledge who liked good conversation just as did Edsall, and who gave good advice in relation to university policies as well as world affairs. Devotion to them and respect for them and their wives was mutual. When he had had a particularly unpleasant encounter, his restless reaction would direct his steps after dinner to some friend's house where good advice and interesting mental distraction were always available to cheer him.

A man of deeply warm feelings, which he could express only to a few intimates, his relationship with his children was of deep concern to him. His sons were in their late 'teens or twenties before he found himself able to speak frankly to them about the conditions of their home. In 1926 he and Pen set off for China, Edsall being sent by the Rockefeller China Medical Board to advise on the work of the department of medicine in their prize medical school, the Peking Union Medical College. On the long boat trip across the Pacific, Edsall read the recently published novel by Warwick Deeping, *Sorrell and Son*, and wrote back to his youngest son Geoffrey asking him to read it too. The fictional Sorrell reared his son alone through desperately difficult circumstances, but the relationship between father and son was as ideal as Deeping's richly Victorian talent could make it. Breaking his usual reticence, Edsall wrote:

> It is fair to say that with conditions as they have been at home it was in some ways far more difficult to keep myself out of the conflict than it was with Sorrell where the cause of the conflict [the wife] had gone. At any rate Sorrell did not care more or make more effort (except in a purely physical way) than I have for you boys and shall while I last. It may give you a clearer idea of the intense desire of parents to have their children avoid the futile and unhappy & mistaken things that have passed in review before them in their longer experience. Waste of time and effort and of mind is the great source of regret in later years to those capable of doing worthwhile things.[5]

He asked the Old Corner bookstore to send Geoff a copy of the book, "chiefly," he wrote, "because it pictures the kind of relations I want to have with you and the kind of attitude that most intelligent young men get after they have tried out the temptations of life — i.e. that 'the job' and the people you love are the only things that matter much."

The six months in China made a great impression on Edsall, and were one of the high points in his life. He was immensely attracted by the Chinese people and enjoyed them thoroughly in all possible ways. He was delighted with their wit and never tired of telling the Chinese version of George Washington and the cherry tree, which tickled him because it showed, among other things, the Chinese capacity for laughing at themselves. The story takes its point from the fact that the Chinese traditionally hate to say anything that will upset anybody or hurt his feelings. Therefore when little Washington George (as the Chinese called him) was confronted by his father with the challenging question, little Washington George burst into tears and sobbed, "I am sorry, Father, but I don't know what is the matter with me today. I cannot tell a lie." [6]

Edsall's letters home were filled with a traveller's impressions of Japan and China. After a brief time in Shanghai he wrote to his eldest son, John:

> The rickshas are thousands — you pay 5¢ (gold) for a 10 minute ride & so on up to about 25¢ for an hour, & it is quite fascinating to watch the lope of the various ricksha men. Most people like the ricksha. I don't much as they are cramped for me, and I am so long & heavy that I can't sit back comfortably as I tend to upset it backward — or it feels as if it would do so. They bear more pathetically heavy burdens in this country than in any I ever saw. At any moment on a busy street, you see men staggering along with a twisting half-trot with a split-bamboo pole over the shoulder and slung from each end a heavy load — as, for instance two or three bunches of bananas. I should think the human frame would not stand it many years.[7]

Another set of medical traveller's impressions can be seen in his photographs from the trip, in which temples and Chinese babies share the honors with impressively high mounds of human excrement.

Arrived in Peking, the Edsalls were guests of the Medical College's director, Dr. Henry S. Houghton, staying in his beautiful Chinese house, and attending many a formal banquet and entertainment. Among these festivities was a luncheon given by President

Tsao on Thanksgiving Day, and another given by W. W. Yen at which Edsall met four ex-premiers, the secretary of education, and the secretary of the interior. Edsall made some notes about a banquet he attended at the home of Dr. Wu Hsien * on October 10, 1926,[8] where the guests were regaled by more than thirty-one different dishes including old eggs, sparrows, fish tripe, shark's fins, bamboo mushrooms, "eight-precious" pudding, and two kinds of soup for dessert. Fortunately Edsall had a cheerful curiosity about new foods and a cooperative digestion to cope with the many dishes cooked in heavy sesame oil. He enjoyed the hot wine which opened the meal, though warning that if you drank it hot your cup would be promptly refilled, while if you let it cool it developed a bitter or pungent taste and was "not very nice." However, "caution leads," he added, "to allowing it to cool."

He found the Shantung cabbage "delicate and delicious," the aged eggs to resemble a mellow cheese in taste, but shark's fins (a notable delicacy) "almost tasteless with a rather repulsive insipidity . . . Their reputation as a delicacy seems to depend merely upon costliness."

He remarked that dining on mice was a "foreign fiction," but he did not look forward to sea slugs, which "are said to be fearful and wonderful things to contemplate swallowing."

His technique with chopsticks appears to have been adequate to the occasion, though he said that trying to hold a large item firmly enough to take bites from it led you to crowd it hastily into your mouth for fear of disaster, and as for pursuing slippery items such as bamboo shoots or abaloni, "it is a great indoor sport for a tyro to chase them around a plate with chopsticks. The great trick with such is to hold your chopsticks vertically and pinch them from each side. If you tackle them obliquely they shoot away at once."

Dr. Francis Dieuaide, acting head of the department of medicine when Edsall was there, recalls, "He was a great 'hit' with the Chinese who were deeply impressed by his imperturbability and wisdom. They saw in him both mental faculties and physical attributes ascribed to one of their ancient philosophers. (Parenthetically, the latter reference was to the broad, high forehead in addition to the height.) It was said that Chinese staff members told the authori-

* Dr. Wu was a notable biochemist who took his Ph.D. with Otto Folin.

ties they hoped to have a permanent professor who would be just like Dr. Edsall." [9]

The department of medicine was then undergoing a reconsideration of the extent to which it should try to cover the range of disease with which it was faced. "I believe Dr. Edsall felt fairly broad coverage was desirable," Dr. Dieuaide stated. He continued, "Dr. Edsall's visit was most stimulating to the Department members and . . . his advice about the work of the Department was exceedingly helpful. More than this was true, for in his short visit he won the highest respect and endeared himself among the staff . . . Dr. Edsall actually attended ward rounds and gave a number of lectures throughout the visit. He especially enjoyed the direct contact with teaching and the hospital service (possibly his last such contacts?)." [10]

Meanwhile the boys were growing up. John, the eldest, had planned to become a doctor from the time he was six or eight years old, and followed a nearly straight line to arrive at his goal, entering the Harvard Medical School in 1923. He had every encouragement to study abroad, and spent two years in England working in biochemistry in Hopkins' laboratory at Cambridge. The first summer he and his classmate Jeffries Wyman went to Graz, which was still a mecca for medical students. John Edsall and Wyman each lived with a different German family, read and talked German constantly and really gained a grasp of the language. They had letters of introduction to distinguished men there: Pregl, who won a Nobel Prize for his work in microanalysis, Fritz Reuter, the professor of legal medicine, and Otto Loewi (also a Nobel Prize winner) who took them climbing in the mountains of the Tyrol around Obergurgl.

When John Edsall came back from England, he started working with Edwin Cohn on muscle fiber (this was in his third year in medical school) in some of the precious free time which Edsall as dean had managed to clear for students. He followed this up even more intensively in his fourth year. He had gone as far as taking the examination for an internship at the MGH and winning an appointment when he realized that he was much too interested in research to spend twenty-one months becoming a skilled M.D., and thus an eminent scientist in biochemistry was born.

Richard, the second son, took longer to find his feet. As a young man, he aimed at a literary career, and went off to Greece to work on poetry and drama. Later he went into advertising, and from there into market research in Canada, where he managed a very successful enterprise. He died in May 1967.

The youngest son, Geoffrey, seemed all set for a career in geology, but a stiff summer course in chemistry at Berkeley in California with Latimer and Hildebrand sparked his interest in the biological sciences, and after a brief time at Stanford he came back to Harvard to go into medicine, graduating from the Harvard Medical School in 1934. Public health claimed him early in his career, and he is now professor of applied microbiology at the Harvard School of Public Health, and superintendent of the Institute of Laboratories of the Commonwealth of Massachusetts.

CHAPTER 15

The New Dean, 1918

EDSALL spent the war years at the Massachusetts General Hospital, working and teaching doubly hard. The hospital had sixty-seven of its staff in military service and the medical school had kept less than 50 per cent of its teaching force. The MGH provided the nucleus for U. S. Base Hospital No. 6, which in France grew to 100 wards and nearly 5,000 beds; preparations for this were begun more than a year before the U.S. entered the war. The Harvard Medical School formed Base Hospital No. 5, organized as a Red Cross Unit under Dr. Harvey Cushing, which served with the British Army, caring for 45,837 patients in nearly two years' service. Many Harvard and MGH personnel also served with the Harvard Surgical Unit, attached to the British Expeditionary Force from June 1915 to January 1919.[1]

Those men who stayed at home were teaching accelerated courses and doing considerable war work besides. Early in 1918 an outbreak of scarlet fever in the hospital closed the MGH wards to all but emergencies for nearly a month. The fall of 1918 saw the outbreak of the influenza epidemic, so they went on an emergency basis again. Even the 200 surgical beds were turned over to the influenza victims, over 800 cases being admitted to the hospital. The senior medical students were excused from school to help Boston's hard-pressed doctors. The school itself was closed for a week at the height of the epidemic.

At the MGH, the new laboratories in Bulfinch which Edsall had won on the completion of the Moseley Building were operating, mostly on war work. Edsall himself had an increasing load of consultation and committee work for state and federal government.

Early in 1918, Dean Edward H. Bradford of the medical school was reaching seventy and was ready to retire, and the university was looking for a new dean of the medical school. Dr. Christian wrote to the retired President Eliot at some length in April, discussing possible men, and arguing against having anyone too closely associated with the Brigham Hospital, though he added, "I do not feel that with the right type of man that the Brigham Hospital or any other hospital needs any particular friend in court."[2]

Christian's first choice was Walter Cannon, but Cannon was in France, and was known to be against the idea in any case. Further, said Christian to President Eliot, "He is opposed I believe in certain circles for a reason which I think may interest if not amuse you; namely, that he is too representative of your own ideas and policies."

As to Francis Peabody, Christian wrote, he "would make an admirable Dean but . . . for his future development it is better for him not to be in such an administrative position."

Then he named Edsall and said, "I think he presents many good qualities. He is not systematic nor a very good administrator but he has fine ideas about the development of medicine and I believe that he would be a good man."

In any case, he hoped the new dean would "be progressive and acquainted with the new medicine . . . Above all things I hope that we will not get a clinician with narrow training however effective he may have been in Committee work and administrative details."

President Eliot wrote back, "I, too, hoped to see Cannon made the new Dean; but his absence for the war makes that selection impossible now . . . Dr. Edsall would in my opinion be an excellent Dean from every point of view; but he certainly would have to give up some of the public services he is rendering. I hope the choice will fall on him. He would need, of course, all the assistance which a good Business Manager can give."[3]

Edsall's experience with administrations was not such as to make him look with optimism at the idea of his being dean of the Harvard Medical School and his first reply was a flat "No."[4] Once he had been persuaded to change his mind, he wrote to President Lowell[5] to make sure that various arrangements would be satisfactory to the administration before he was willing to accept. His first point was that he should have an assistant to handle the details of the dean's office — the interviews with students, correspondence, and similar day-to-day matters. (The assistant was found in Dr.

Worth Hale, who had become Harvard's assistant professor of pharmacology in 1913 after working with Reid Hunt in the Public Health Service. He was appointed as assistant dean and remained Edsall's right hand throughout his administration. One of his most important influences on the school was his control of admissions.[6])

Next Edsall arranged that his school secretary as well as his consulting office were to be at the MGH.

According to the usual custom, the position of dean of the medical school was considered to require half a man's time, and for this Edsall asked a salary of $2,500, not to replace what he could make by consulting work but to make it possible for him to give up outside practice.

He pointed out that under the MGH rules, he would be retired at sixty, but at sixty he would still be educating his children. He asked to be allowed to retire at sixty-five instead of sixty (if he was in good health) as he wished to be able to devote his major efforts to academic medicine and the dean's office, and not worry about making his practice financially productive.

His next two points concerned the governing structure of the medical school, which he felt needed reform, and adherence to the principle of securing the opinion of the Committee of Full Professors on important appointments.

Lastly, he was willing to give up all his outside war work, with the exception of his position as consultant to the U. S. Public Health Service in industrial hygiene. He felt in that case that things might come up which he was better fitted to do for the government than anyone else, and they were of such importance that he ought to do them.

Edsall's appointment as dean was announced in June 1918. He was chosen because the university and the school wanted to move forward – if they had not, the revolutionary changes that were made relatively peacefully in the following fifteen years would never have occurred. The ideas that Edsall presented to the faculty of medicine in his long confidential statement of April 7, 1919, had been ripening since he had watched age and seniority dim the lustre of the University of Pennsylvania's medical school and since he had survived and learned from the storms of the St. Louis venture. Every speech and every paper he had written on the subjects of administration, education, and the advance of medicine, had developed the principles he now stated, which went as far back as his presidential address to the American Pediatric Association in 1910 and his lecture "The Clinician, the Hospital and the Medical School" at the Aesculapian Club in 1912. Since 1912 he had been putting his principles into practice, fostering the growth of scientific medicine at the Massachusetts General Hospital despite the difficulties of the war.

Now for the third time Edsall had the opportunity to lead a developing situation in a medical school, and this time he determined to move with care and to carry his colleagues with him. His first faculty meeting was taken up with emergency measures such as the closing of the school for a week because of the influenza epidemic, and it was not until April 7, 1919 that he was able to address a faculty meeting presided over by President Lowell and attended by some thirty doctors and Dean Eugene H. Smith of the dental school. Distinguished founders of the medical school looked down from the walls as Edsall rose to make his major policy address.[7]

Edsall saw before them a period of rapid change, in which changes in medicine itself, in the importance of preventive medicine, and in possible socialization would bring about decided transformations of familiar practices. He also foresaw that Harvard would be challenged by its strongest rivals, a challenge he welcomed since it also gave the opportunity for the Harvard Medical School to achieve "actual supremacy." He continued:

> My chief hesitation in addressing you in the way that I am about to do, is that it may seem to indicate that I consider that the Dean does or should control and direct policy. On the contrary, the Dean has only one vote when it comes to an actual determination of policy, but he is charged with the duty of a more serious and general consideration of policy than are other members of the Faculty and he necessarily has the ear of the administrative authorities of the University more than do others. He is popularly thought to have, and to a limited extent undoubtedly does have, unusual opportunities to exert influence as compared with other Faculty members and he may readily seem to use his influence wrongly and for his own ends unless his purposes are known and approved of; but on the other hand, results are laid at his door whether he advocated the things that brought them about or not.

With a glance at the lessons learned in Philadelphia and particularly in St. Louis, he stated:

> I have seen and felt elsewhere very serious results from secret academic diplomacy in others, especially at times when important changes were being undertaken, and frankness seems to me much safer for the School, whatever its influence upon one's own plans.

The first point of policy he took up was the choice of new men, particularly young men below the rank of professor. He spoke of Hopkins as outstanding in this respect, citing the fact that it

> had a small freshly chosen and very remarkable faculty, and the faculty there has always given peculiarly jealous care to the choice of important junior men as well as professors . . . Positions of any importance are rarely filled there without repeated, unhurried and very frank discussion by the senior faculty of those under consideration. This is not always true here . . . We have had a kindly but dangerous shyness of what seems like criticism of candidates . . .

The next key point he took up was the allocation of funds:

> There has been so large a growth of special departments and, owing to important hospital connections, so large a number of major clinical departments that a large budget has been so stretched to cover many objects that it covers almost all thinly. Distinctive results in almost any line have been extremely difficult . . . At Hopkins, their smaller budget expended upon fewer objects did distinctly better . . . Some extra financial difficulty will always be imposed upon us by the very extent of our opportunity but some things not now necessary have been done and time and a consistent policy can concentrate our efforts somewhat.

This was to be a cornerstone of his later policy.

He next touched on the difficulty of organizing such a school in the most effective way. Although as he said, "The relations with the hospitals have been progressing very well in recent years . . . The very large number of laboratory and clinical departments, the numerous scattered hospitals, and especially the fact that the latter are managed by Boards having at most no organic connection with the School makes cohesive and well-systematized effort extremely difficult here." He went on, "I believe there has been less real contemplation and study of general policy here by the Faculty or Full Professors than by similar controlling bodies in any important school," and with a prescient look forward he added, "In large part it has seemed to me due to the form of organization of the government of the School, which, in order to get business done, almost forces the Dean to be an autocrat or to depend upon the Administrative Board as an oligarchy." Various re-distributions of power were later made among Administrative Board, Full Professors, Executive Committee, and Dean, but none ever solved this problem.

He then took up future developments in medical education, where preventive medicine seemed to him the most important:

> The teaching of medicine, by and large, is still almost entirely teaching the care of the individual sick. A department has been established here and in some other schools dealing with preventive medicine but preventive medicine is of course as diverse and varied in its applications as is the treatment of diseases of various kinds . . . I believe that a new point of view and a new atmosphere in medical schools in this matter is well-nigh necessary and is one of the most important things to be considered. It is probable that the first medical school that . . . consciously and systematically emphasizes this side of medicine . . . and makes them actual living parts of all important medical activities will within a few years largely serve public health and largely increase its own prestige . . . Most medical graduates now know much less and think less about their practical responsibility in this work than they do of the application of pathology, anatomy, chemistry, etc., to their clinical work, and yet none of these has even now more insistent practical importance.

He foresaw the possibility of "interesting and probably radical results" in a new relationship between the doctors and the public, and felt that "particularly if public control in any very extensive form comes about, the preventive side of medicine will become vastly more insistent and perhaps the dominant side." Among other reasons for this was the straightforward fact that "if the care of health becomes, so to speak, a direct tax instead of an indirect one as it is at present, economic pressure and desire for a reduction of costs to the general public and to industrial and other establishments will make an insistent demand for prevention."

As he looked forward to the development of public health work, he was concerned to stick to the essentials as he saw them.

If one means by a School of Public Health new buildings, new and separate departments and other large new ventures, I should be opposed to any such undertaking at present. A school to my mind, however, is fundamentally not buildings and not a separate organization but leaders in certain subjects, facilities for and coordination of their work and a collection of disciples. Such a school we could by the proper development of the suitable fundamental departments have here as a part of the Medical School.

The public health developments already in use or available were indeed remarkable: departments of preventive medicine, pathology and bacteriology, physiology, biological chemistry and pharmacology in the Harvard Medical School; special facilities in comparative pathology, tropical medicine and industrial medicine; cooperation with the university's department of sanitary engineering among others in Cambridge; state and municipal health departments to provide material; affiliations that covered contagious diseases, mental hygiene, and other needed elements—all this a few years earlier had forced the Rockefeller committee ("which did not recommend that we receive money for a School of Hygiene") to admit that Harvard possibilities were finer than anywhere else. This dream became a reality within a few years, however, with the generous assistance of Rockefeller funds.

Next he took up various problems looming on the horizon, among them the fact that more young men of promise, and more money, were going into the clinical branches of medicine than into the fundamental sciences. The scientific branches had been the first to achieve academic salaried status; now the clinical branches were getting ahead, perhaps too far. "Ten years ago in many laboratory branches there was a very considerable choice of desirable men . . . Now the choice in almost all such branches is much narrower . . . There is a threat in this, in the future, of really serious damage to medical teaching and to medicine in general unless this tide is turned." Keeping a balance in this and similar situations was to occupy him throughout his time as dean.

Next he brought up another potentially dangerous situation: the competition from other medical schools for good men.

In recent years we have had here an extremely heavy loss because we could not offset the inducements offered. At least three other losses have very recently been or are now possible here for the same reason. Any one of these would be a grave blow. Also several men have been suggested by members of the Faculty as perhaps very desirable men to secure within a few years. I know that in none of these instances is it at all probable that the inducements that we have just now to offer would be sufficient to secure them. Inability to secure or to hold the personnel desired is the worst misfortune that could befall us.

One means of making sure that this disaster should not overtake them lay in the management of their money. "Of any funds available for the School a considerable sum should be kept constantly mobile, being applied for temporary purposes that are desirable, but not mortgaged to any department or departments. When need arises for large funds to secure or to hold any important man or to carry on some extremely important piece of work there would then be a reservoir to draw from." The DeLamar funds were the major source which allowed Edsall to carry out this policy from the beginning, giving him room to maneuver, but, as he noted later in his speech, DeLamar funds "are intended to be applied, not to the general needs of the School and not directly to routine teaching in any departments, but to increase of knowledge regarding the origin, cause and prevention of disease and regarding nutrition."

About another financial policy he also wanted no misunderstanding.

Actual inefficiency and suffering should be relieved whenever possible but when this is once done each step should be to make something that will be distinctive and that will add to the prestige of the School, and this should be done only in as many ways as the money available will do it thoroughly well. Dispersion of even a very large sum over many objects shows very little result and often brings discredit. Devoted to one or two or three objects and well employed it brings impressive results and great credit. Done in the latter way, I should feel much optimism in going after further funds to cover the objects still needing development. Done in the former way, I should not care to have any responsibility in the matter . . . I think the principle is absolutely essential to follow one thing at a time and make it really first-class and to add no new ventures not absolutely necessary until the essential things have all gradually been made such that we can be really proud of them.

From there he went on to a description of full time, Harvard style. He was against a rigid system, and noted that even at the Hopkins a more flexible system might soon be put into operation. He proposed "to progress by partial steps, putting individual departments or parts of departments into proper form step by step," and cited his own experience and that of Dr. Christian in developing considerable research on very small sums.

Through all this ran the inspiring view that the school could not only be outstanding but preeminent. "We have here all told finer potential opportunities for teaching and research in almost all the clinical branches than exist at present elsewhere or are likely to be brought together . . . We have also a peculiarly fine opportunity in public health teaching and research . . ." As to the school's clinical relations, "Any one familiar with the conditions elsewhere will I think agree that properly coordinated and with proper and clearly defined purpose and general atmosphere it is possible to make these relations more effectual than those anywhere else, but," and the but was important, "I think we need to determine clearly and wisely what general guiding purposes we have in view and not develop casually and separately in the different hospitals and different departments."

Discussion followed, which even the dry style of the official minutes shows was lively:

It was agreed that salaries in the laboratory departments are so low that it is extremely difficult for a man to start a career in this field, and consequently good men are not attracted. If there are not good laboratory men, clinical medicine also will suffer. *President Lowell* feels these salaries should be raised. *Dr. Cushing* thinks that this difficulty could be met by having laboratory work done in hospitals. *Dr. Cannon* considers that it would be a grave error to have all laboratory work done in connection with practical problems; it has been brought out by the war that men in pure sciences have a wider and more penetrating knowledge in many cases than those in applied sciences.

Dr. Rosenau and *Dr. R. C. Cabot* expressed their approval of the Dean's desire to have the School notable in certain directions rather than fair in all. *Dr. Christian* agreed, provided the fundamental training of medical students throughout all departments is up to a given minimum standard.

Dr. Christian proposed that the School should make a large plan for raising and expenditure of money, and present it to the community.

In answer to a question, *President Lowell* stated that the Corporation would welcome such a plan.

On motion of Dr. Ernst it was VOTED that Dr. Edsall's report should be printed and confidentially distributed among the Faculty.[8]

CHAPTER 16

A Stormy Beginning, 1918–1923

THE TIME IMMEDIATELY FOLLOWING the war saw tremendous expansion of medical teaching, a ferment of change, experiments of all kinds — the expression of dissatisfactions which the war had brought to a head. It was not an easy time to be the new dean of an ancient but forward-looking medical school.

"Just now most of the other important places are in a perfect welter of contention," Edsall wrote to President Lowell in May 1921. "We have escaped it here and it seems to me that at the moment it would be most important to escape any sense of contention and especially just as the American Medical Association is coming here and there would be such splendid opportunities for gossip."

"I cannot quite see what is the matter with Medicine in general at this moment throughout the country . . ." Edsall continued. "Part of it is due to the influence of Mr. Flexner I am sure, but there is also more than that and I think it is in part the neurotic effects of the war . . . I have never seen anything that even approached the present state of excitement and unrest and unsteadiness in most places. I heard a very distinguished man say the other day that Harvard seemed to be the only medical school of importance in the country that was not having a brain storm."[1]

But things were calm at Harvard only by comparison. Ten years later Edsall could look back on this period and see what a razor's edge he had traversed. "Many of the senior staff, and most of the junior staff, were in a state of dangerous dissatisfaction and depression," he wrote in his 1927–28 Dean's Report, and the school was "in so precarious a condition that it was quite uncertain whether it would not go downhill rather than forward. The funds provided for the very able personnel to do their work were, in a number of instances, not more than one-half of what other schools were providing, sometimes less than that. There was . . . strong effort made on the part of a large number of other schools to take away the major number of the distinguished personnel in the School . . . The School had been in such financial condition that it had even been obliged to allow its plant to run down to an extent that has necessitated expending within the past five years . . . about $200,000 to rehabilitate it . . . The condition at that time was really rather alarming."[2]

Perhaps the most dangerous of these attempts to entice away the leading figures of the Medical School came in 1920 when the Mayo Clinic tried to get Walter Cannon and Otto Folin. They made the most tempting offer they could, much larger salaries, ample provision for research under most favorable conditions, no necessity to teach or serve on committees. If these two had gone the situation at Harvard might have been serious, as others would have followed their lead. But after months of consideration Cannon decided to stay at Harvard and Folin stayed with him.

Cannon wrote to Dr. Rowntree in refusing the Mayo offer:

> At Rochester, I should be concerned with research on clinical problems by means of physiological methods. The emphasis would be on applied physiology, and the persons most likely to be influenced would be your selected graduate students who might have their thinking directed towards conceptions of disturbed function as they considered disease processes. In my position at Harvard, I should continue a somewhat more free range in physiological research, whether practical or not, dependent on my personal interest. Connection with a teaching institution would permit me and the group associated with me here to use this place more effectively than a great clinic could be used, in developing and training teachers of Physiology to fill positions elsewhere . . .
>
> On the other hand, established habits of work at Harvard, the conveniences of libraries and laboratories already in working order, the unusual group of men not only in my own department but in the Medical faculty, long existent friendships and associations, — tended to keep me here.
>
> On thinking matters over with great care, the question settled down to this, where could I as a physiologist be of greatest service to medicine? You, perhaps, do not know the present status of physiology in this country. There are three main Chairs now vacant, — Hopkins, Pennsylvania, and Columbia, and there are only three places where new phy-

siologists are being developed, — Washington University at St. Louis, Chicago, and Harvard. If I should leave Harvard, there would be four vacant Chairs. The group here would be seriously broken, and for some time the probabilities are that only two places would be left where training in physiological teaching and research would be carried on. It is obvious that adequately trained men cannot be passed on to the clinical years in our Medical Schools, unless there are skillful teachers devoted to high standards occupying Chairs in the laboratory subjects. In the present critical situation of the pre-clinical subjects, it seemed to me not so important to medicine in general for me to support graduate teaching as to do my best to make provision for the training of undergraduates.[3]

Dr. Cannon's wife remembers that the Mayo offer was three times what Harvard was paying him, but it failed to tempt him. There is a famous story about Cannon that once when some doctors were discussing what they would do if they had all the money they wanted, Cannon said he did not wish for more. "I have all the money I want," he said. "Mrs. Cannon gives me $10 month and with that I pay my car fare, buy my lunch and have my hair cut."

"It does indicate," said Mrs. Cannon, "why the Mayo offer was refused, because he was perfectly happy where he was. Harvard did nothing except possibly hope he would stay, but for a few months or possibly years afterwards some mysterious person added one or two thousand dollars a month or a year to his salary through the University and we never knew who was the generous donor."[4] (I myself always thought the mysterious benefactor was Dr. Shattuck.) It was not until 1930 that Cannon's work received much more financial support; then Edsall was able to get a generous endowment from the Rockefeller Foundation.

People wanted to stay at Harvard because the surroundings were better than they were in other places. There was less fighting. If you worked here you got enough so that you could carry on. It wasn't luxurious, but professors are queer birds — they like to be around people they like. That was true at Harvard more than anywhere else. The Mayo Clinic was not an academic institution. The doctors had too many patients and spent most of their time looking at patients. No real research was going on at that time. That was the reason they wanted Cannon and Folin, to make it a great center for research and teaching, rather than just a big clinic. And they did do that. They attracted some good scientists, they built up the Mayo Foundation, and established effective relations with the University of Minnesota.

Cannon was one of the pillars of the Harvard Medical School. He was a modest person with no interest in personal power, but he was often considered the real dean of the school, particularly before Edsall came.

Dr. Cannon was not a cloistered scientist. His influence on his associates in the medical school was enormous. His considered opinions, his good judgment, and his courage made him valuable to committees, and he served on many. He represented the liberal group which was willing to experiment with education. In his judgment of people, he was not interested in safe personalities. Social charms were not enough for him. He sought outstanding thinkers, and was unprejudiced except against dishonesty and vulgarity. He helped produce the Golden Age when Harvard Medical School was expanding, and he was one of the small group which gave it greatness. It does not take many Cannons or Edsalls to make a school great — but it does take a few! [5]

In paying tribute to Cannon after his first twenty-five years as George Higginson Professor of Physiology, Edsall said, "I know of thirty-one persons, and there are perhaps more, who now have professorial rank in physiology or immediately related subjects and who sought their training with him. That tells its own story. They carry his influence throughout the several parts of this country, and to thousands of young students." [6]

The years had ripened Edsall's views on leadership and on building a staff. He fought for increased budgets because without them he would lose his men, but he fought for much more than that. He had a number of principles which he stated at one time or another on how to handle men. One was his avoidance of half measures, which he outlined to Dr. Charles Scudder. "To go half-way as you suggest and give a man a half-way position on trial I am entirely opposed to unless one is dealing with a very young and untried man or unless there is nothing else to be done. It means that one subconsciously admits such probable failure of the scheme that one is not willing to risk much on it. I have seen such a plan tried repeatedly and so far as I can remember without success. Success depends very largely upon giving the man and the venture that he undertakes free scope and authority, and being willing to take whatever risk there is after having made a choice."[7] A dean who talked about risk, and university trustees who backed him up — no wonder it is looked back on as a Golden Age.

Another principle was his firm adherence to merit rather than seniority. He wrote to Dr. Algernon Coolidge at one point:

> I had occasion ten years ago to go through, to my own great disadvantage, disturbances in two medical schools and hospitals which resulted in the very great loss of prestige of both those places. I resigned from one place chiefly because progress had long been impossible owing to advancement by seniority and I left the other place chiefly because of the general feeling in the group of men working there of friction with the trustees . . . [I] have therefore observed with a good deal of care what has been happening in other hospitals and medical schools . . .
>
> I feel sure that the continuance of the established principle of filling positions by merit rather than by duration of service and occasionally, when necessary, actually securing men from elsewhere, is an absolute foundation stone to success in the management of a hospital or a medical school, as it is in the management of any business. To adopt the other plan results inevitably in blocking the progress of and driving away the most able and intelligent group of on-coming men . . . Of course I need scarcely say that in matters of this kind any sensible and fair-minded man would, other things being equal or approximately equal, choose both a local man and a senior man.[8]

Another thing he had learned, patience. "I have come to have a profound respect for the ease with which irritated human nature can upset plans and institutions," he wrote.[9]

It was largely Edsall who established the principle that when you had a professorship to offer you didn't pick the next man in line; you looked all over the world and chose the man you thought was outstanding in that given field. Frequently people would say, "It's no use offering the professorship to X, he'll never accept the small salary that Harvard has to offer." This point of view never impressed Edsall.

This point was raised when Hans Zinsser was offered the professorship in bacteriology. Everybody said, "You can't get Zinsser. Why should he come up here? He is an absolute kingpin at Columbia." Edsall replied, "Well, all we can do is try. If we're going to make Harvard an outstanding institution, we have to make every effort to get the best man that is available for the job."

Knowing that Zinsser was burdened with administrative work at Columbia, Edsall devised a professorship for Zinsser at Harvard without any administrative burdens. It appealed to Zinsser, and he came in spite of the salary difference. Today Edsall's principle is firmly established at Harvard, and the university looks all over the world before an appointment is made, although sometimes without the thoughtfulness that Edsall exhibited.

The best man for the work—the right work for the man. Another example of bringing men and work together is the way Edsall persuaded the Rockefeller trustees to support a new line in psychiatry. Edsall was deeply interested in bringing psychiatric teaching into the hospital. He proposed a full-dress plan for improvements in psychiatry at the Harvard Medical School, only to find that Abraham Flexner would like to see an entirely different approach. So he sent Stanley Cobb, then a young assistant professor of neuropathology, to New York to see Flexner, with the result that Cobb spent some very fruitful time abroad, and returned from Europe trained in the modern treatment of psychoses and well oriented in the anatomy of the central nervous system. Dr. Cobb had already had an internship with Harvey Cushing and was well prepared to carry on the problem of psychoses and their possible relation to anatomy.

Flexner described how the Rockefeller General Education Board undertook to support this work when Dr. Cobb was nearing the end of his studies in Europe: "When he decided to return home, he notified Dr. Edsall, and Edsall once more came to see me. I asked him what Cobb wanted. Cobb had said nothing on the subject. 'Write him,' I suggested, 'and find out what he needs to begin with.' His needs were extremely simple: a dozen beds in the Boston City Hospital, a laboratory assistant or two, and unrestricted use of clinical material. The entire sum needed was appropriated by the General Education Board. Cobb returned and developed a splendid center of neurological study at Harvard."[10] (Cobb stayed at the

City Hospital for years, eventually in 1934 going to the MGH as head of an enlarged psychiatric department there.)

This was a part of many generous Rockefeller gifts in the fields of psychiatry, neurology, and neuropathology under Drs. Macfie Campbell and Stanley Cobb.

Throughout his administration, Edsall's relations with the Boston City Hospital were an example of his tact and patient determination. He did much throughout his deanship to make the hospital a true university institution. There were a few people there who were all in favor of what he wanted to do; these were Dr. Locke, who was physician-in-chief and a very close friend; Dr. G. C. Sears, who headed the first Harvard Service, 1915–1918; and to a lesser extent, Dr. Henry Jackson, one of the famous Jackson family. One of the things that Edsall tried to do was to get proper trustees appointed to the City Hospital so that he could have support for his ideas on an administrative level. When a good trustee came along, Edsall got things accomplished; however, when a political trustee was appointed, he got along much less well. On the whole, throughout his tenure as dean, the City Hospital developed well in relation to the Harvard Medical School. One of the reasons for this was the caliber of people whom Edsall placed at the City. Francis Peabody, the professor of medicine, was a brilliant clinician and teacher and was eminently successful in every way. He had a charming gracious manner and got along well with City Hospital doctors. Soma Weiss was tremendously helpful. He was obtained from New York where he had worked under DuBois. Everybody liked him, and while he was at the City Hospital, things went well and Weiss could get anything accomplished that he wanted. He left the City Hospital to become the head of the Medical Department at the Brigham Hospital. Stanley Cobb established an outstanding neurological unit there. Nevertheless, the City remained subject to trouble over appointments, particularly over staff seniority, which for long had been the technique for promotion. Edsall's success in getting staff appointments based on ability rather than age made a great improvement in teaching so that the institution became much more popular with Harvard students.

In a day when the bidding for men was rising at auction rates, money was a prime necessity. And between the Rockefellers and loyal Harvard benefactors, money was forthcoming. To illustrate the kind of starvation which typified the old way of doing things, in 1916 a committee on the medical school reported to the board of overseers, citing the department of pediatrics as outstanding among the clinical departments. The committee, whose chairman was Dr. Shattuck, wrote in their report:

> The amount and quality of work here, both in teaching and research, is quite out of proportion to its cost. Nineteen teachers are on the roll of the Department. Nearly half of these serve entirely without pay. Such self-denial on the part of the teachers cannot always be counted on, and we hope that in the near future $10,000 a year may be the annual expenditure on the Department instead of the present three thousand.
>
> In general it may be stated that while the School can, with close economy, pay, or nearly pay its way, the fructifying influence of a larger income is very desirable.[11]

In 1922 the fructifying influence of a larger income had hardly begun, and Edsall noted with alarm that the best medical schools were spending five to ten or more times what they had twenty years earlier, and the cost per student in some leading schools was $2,000 or over. The Harvard cost per student was low compared with its rivals. The dean had obviously often had to meet the argument that the school should stick to teaching and not spend so much money on research. He was at pains to point out in the *Harvard Alumni Bulletin* that the actual costs of research were small compared to maintaining a fine teaching staff, and should be considered an element in retaining that staff. But he felt, "The financial situation at present is by no means easy or adequate."[12]

Nevertheless, endowments were beginning their rocketing growth. And one of the chief of them was the gift of Captain Joseph Raphael DeLamar, whose princely gift came early in Edsall's administration. "The largest individual fund that either School has ever had," said Edsall in his 1927–28 Dean's Report, "left on such terms that it can be used very freely, and therefore with peculiar effectiveness, in advancing knowledge."

Captain DeLamar, born in Holland, had spent his early life at sea. He settled in the United States to build ships, then went West as a miner, and made his fortune as a mine owner and director of many other business enterprises. He might never have thought of endowing medical schools if it had not been for the suggestion of his lawyer, but as a result half of his immense fortune was divided among the medical schools of Columbia, Johns Hopkins, and Harvard. He was particularly interested in the relations between nutrition and health.

Some $2,000,000 came to hand at once in 1919, and as the estate was settled this amount grew to over $4 million. In twenty-five years Harvard's prudent investment and reinvestment had raised the capital to over $6 million.[13]

The DeLamar money gave Edsall the chance to distribute some unassigned funds where they were the most needed. He usually managed to hold a certain amount of DeLamar income in reserve, though the pressures on him to commit it all were very great. As an example of the freedom it gave him to put first things first, he wrote to Christian in 1922:

> When we get more DeLamar money in, if we do, it is my clear purpose . . . to urge very strongly that some of that money be applied to the development of research in medicine and surgery. I do not think it can be morally applied to any purpose but that (i.e. research) . . . and that opinion is based upon an interview with the lawyer who drew this will and who knew DeLamar's purposes perfectly clearly. As I said in the very earliest discussion of the DeLamar money and its uses, I felt that its application to the betterment of the conditions for research in the medical sciences was the imperative thing.

These sums had arrived in the nick of time. Edsall told Christian:

> Had the DeLamar income not been pretty generously used in that way and had I not been able to secure money from other sources, I think there is every reason to believe that we should not now have with us Cannon, probably Folin, certainly Drinker, Lawrence Henderson and some of the important younger men associated with these men, and very possibly also Reid Hunt. Some of it has been used – about $7000 – in connection with developing Peabody's salary and budget; otherwise we should certainly have lost him last year to Hopkins, as the City Hospital plan could not have been carried through. And some has been used in developing the Pediatric Department, which otherwise

could not have been done in such a way as to attract Schloss or any other satisfactory man. The situation was really dreadfully threatening at the time that the DeLamar funds began to come in and at the time when I secured some other funds . . .

The amount of pressure and effort made to get men of that type away from us has been perfectly extraordinary since I have been Dean.[14]

However, the larger income was soon forthcoming, and its influence was felt throughout the school. Between 1918 and 1928 the endowment of the Medical School increased from $4,300,000 to nearly $14,000,000.[15]

While Edsall was fighting to hold on to his professors, he was confronted, early in 1920, with strong pressure for a considerable enlargement of the school. An urgent discussion was carried on, much of it by letter, among President Lowell, former President Eliot, Drs. Walcott, Shattuck, Christian, and Edsall. The arguments advanced show the problems the school faced as Edsall took up the reins. The war ended with prospective students beating on the gates. In contrast to its prewar policy of accepting almost any qualified student, the medical school could pick and choose. President Eliot in his letter to Dr. Christian stated with cogency his view of the case for enlarging the school:

I believe it to be the *primary duty* of the Harvard Medical School to send out into the service of the community a larger number of well-trained physicians and surgeons than it is now sending out and a much larger number of health officers and sanitarians. There is desperate need all over the country for such physicians, surgeons, and sanitarians as the Harvard Medical School is now training with its confessedly inadequate staff and resources. The policy of the School should be not to keep the number of graduates down, but to increase that number as rapidly as possible.[16]

This was no visionary scheme, but based on plans for increased endowment, more beds for teaching, and full use of the buildings of the medical school. Eliot pointed out that these buildings had been planned for four classes of two hundred students each, though they were now felt to be cramped for classes of one hundred. But it was clear what he put first when he stated:

As a temporary measure the School might return to more lecture, book, and recitation work than we like, rather than hold down the number of students admitted to the School.

The school had recently increased the size of the freshman class from 100 to 120, and Christian felt even this to have been a mistake unless the number was to be reduced to 100 by the second year. He had his reasons as well:

> The Medical School is not adequately manned with competent teachers to give satisfactory instruction to the present numbers . . .
>
> I do not think that those who are favoring an increase in the number of students are very familiar with the actual conditions at the Harvard Medical School as they compare with facilities for teaching at other of the better schools . . .
>
> I have now been connected with the Harvard Medical School some seventeen years. During that period we have often been on the point of attaining first rank but we have always missed being more than a good school by reason of expanding with new departments or increased numbers of students when we were not financially able to do so . . .
>
> Again we have often been on the verge of an enthusiastic cooperative pull together; the real pull together has not actually come . . . I believe to expand the School now when so many of the teachers believe it unwise will again break up the spirit of cooperation that Edsall has gotten started . . .
>
> Finally, there is much reason to think that the increased numbers seeking admission last year may be a temporary condition . . . It seems to me wiser to wait until we know that the demand for entrance of more students is a permanent condition and then to go to work to try to provide endowment enough to give them not as good but better instruction than we can provide now . . .[17]

Edsall was heard from on January 23, and his was the view which prevailed. "The first duty of the School," he said, "is to make the facilities for teaching the number of students that we now have better."

Equally important, the teaching staff had to be strengthened by making positions more attractive, and by increasing the number of full-time positions. "More service will be done toward providing good doctors and much more will be done for the prestige of the School by training those men who will be the future teachers than by increasing the numbers of undergraduates."

Entire departments should be put on a full-time basis, Edsall went on, and the teaching of preventive medicine should be strengthened both directly in public health work, and throughout all departments.

To give increased numbers of students per se precedence over any of these things would I believe have a dreadfully damaging effect upon the School's reputation as well as upon its powers of accomplishing the best public service.

Furthermore, he pointed out the self-defeating likelihood that the best students would probably choose schools which kept their numbers down and standards high.

It was painfully obvious to the dean just how much in the way of additional endowment was required even to maintain Harvard's current position:

Judged by comparison with existing budgets in some competing schools, not one of our departments could now be made attractive enough to secure a distinguished new man or to prevent even threatening inroads upon our personnel without adding largely to its budget, in some cases very largely. Departmental budgets elsewhere go up to even $65,000 in one case, $70,000 in another, and a considerable number that I know of are *well* beyond $25,000. None of our departments has the latter amount spent upon it and most of them have about half that. Expenditures may be excessive elsewhere but we can not hold or secure men by telling them that other people are extravagant . . .

It must be recognized too that research in most lines of medical work is now extremely expensive . . . and research must be provided for with reasonable generosity because valuable men can not be secured or held unless this is done, because it is for the public good, and, quite as much, because one of our most useful activities is in training prospective teachers, consultants, research workers and other advanced men . . . Research as I look upon it is not merely interesting and capable of yielding useful knowledge; it is also an essential method of training men's minds and one that they demand.

If the money could be obtained, the dean knew of a more fruitful way to spend it than in enlarging the numbers of undergraduates.

Looking over the lists of individuals, I find that we have now over seventy young men on temporary appointment in the School . . . the most desirable sort of graduate students. It is now very difficult indeed to hold them here long enough to get the credit of having actually trained them here. They constantly refer to the better conditions elsewhere. Were money made available we could make them contented and the number would readily become 150 or 200 and we could in a few years become a chief source of supply of highly trained men for places in this country and elsewhere and thus meet a demand that is more insistent and more desperate than that for practicing doctors and quite as important for the public welfare.[18]

The school expanded gradually under Edsall's administration, but the growth was kept in balance with other advances. By 1922, there were nearly 500 students in the school. In 1923–24 it was decided to admit highly qualified men with advanced standing from other schools, and the limit on third and fourth year classes was raised from 125 to 135 men.[19] By 1925 the Harvard Medical School had 520 medical students, took 500 to 700 graduate medical students a year, also taught the dental students in their first two years, and counted in addition 300 interns and others doing postgraduate work – in all 1500 young men each year.[20]

Once in a responsible administrative position, Edsall found the cumbersomeness of the Harvard system more than an annoyance. The school was big, and with its many hospitals and virtually independent special departments, it had to contend even more than most with the apathy of professors and others towards the business of running the school. When he first took office, Edsall had three bodies charged with administration: a faculty council of six members, an administrative board of eight members (including President Lowell and the dean), and the committee of professors which consisted of Lowell, Edsall, and thirty-two full professors.

Many people thought that the committee of professors *ought* to run the school.* When upbraided by Christian for not making more

* As Eugene DuBois said in a speech on Edsall's retirement: "The problems of medical education are as old as the hills. The first reference that I have found to a medical faculty is in the twelfth chapter of Psalms. The prayer book gives this translation of the fourth verse: 'We are they that ought to speak, who is lord over us?' This surely must be a chorus from the faculty when King David was considering a change in the budget or the appointment of a new dean." (Eugene J. DuBois, "The Development of Clinical Subjects as Contributing to University Work." *Science*, 82 (Nov. 22, 1935): 473).

use of this body, Edsall replied by pointing out some of the practical drawbacks as he had found them.

"It is impossible to get a large body to go continuously week after week over many small details, discuss them properly and settle them," he wrote Christian in 1920. Even getting a reasonable attendance at meetings was difficult. Over one problem, Edsall noted, "it required a special second notice requesting a large attendance to get twenty men to be present, and at each of the three meetings I mention a considerable proportion of those present have left before the discussion was over even though a vote appeared to be imminent, because it was approaching 6 o'clock and they wanted to get home. I have no desire in the matter excepting to see some system adopted that will be really capable of carrying on an effective system of continued purpose and policy and a system whereby purposes and policies will not change from time to time in accordance with changing personnel present." [21]

Edsall felt it unfortunate that so much power lay in the hands of the dean himself, for this caused a radical shift in the direction of the school each time there was a change of deans. Also Edsall was not comfortable when his colleagues considered him a czar. Even in dealing with Dr. Christian whom he knew well and had worked with for years, Edsall ran into suspicions and fears that "politics" were controlling policy. Christian, who had been away, even accused him of avoiding consultation with the committee of professors because he was afraid of it; Edsall's reply showed that he had consulted them at every possible opportunity. He went on to mention

> the unfortunate tendency on the part of a few men in the School to believe always that there is something underhanded going on . . . My own attitude has been that I was very reluctant to take the Deanship and had no pleasure in it other than that it is an opportunity to try to further the activities of the School, that details of expenditure must be settled by the Corporation and that general methods of expenditure are wholly in their control but that if so important a thing as [the DeLamar] fund were to be administered for general purposes that were contrary to the desires of the great majority of professors the duties of the administrative officer of the School would be so unattractive that I should certainly not desire to carry those duties myself.[22]

In the year 1921–22 the faculty council was abolished, and a new administrative board was set up with a larger faculty representation. Members of the first board were Algernon Coolidge, Milton J. Rosenau, Harvey Cushing, Reid Hunt, J. L. Bremer, Walter Cannon, Macfie Campbell, Worth Hale, Burt Wolbach, Oscar Schloss, and Francis Peabody, as well as Edsall and Lowell.

In 1929 Edsall told me that he wanted me to go on the administrative board. More important, he told me very specifically when he asked me that I was not to be just his boy, that I was to express myself at any time in opposition and to really represent the best interests of the medical school. My understanding when I went on the board was that I never was to be influenced by his opinion.

After I had been on the board about four months, Dr. Bremer, who was a professor of anatomy and who represented the old Boston school, took me aside and said, "You know, Aub, you really ought to resign from the administrative board. You're making Edsall's life even more difficult. Everybody thinks that you are merely his mouthpiece and that your vote is the vote by which he's stacking the administrative board. If you wanted to be kindly towards him, I think you ought to resign." I thought that was a funny thing to say to me, but if a man like Bremer, who was on the board, felt that way I thought I had better resign.

There is no question in my mind that all through Edsall's life at Harvard, there were certain people at the medical school who did their best to make difficulties for him. Two or three years after World War I, I lived in Edsall's home for about six months. His wife had gone south for half a year, and they both asked me if I would come to live in their home. I considered it a great privilege and went. It was just at this time that Edsall was having a great deal of trouble with people at the Brigham. It was a very complicated situation. I think that basically it arose from the fact that both Henry Christian and Harvey Cushing had been trained at Johns Hopkins, where there was but one hospital in the university. If one was professor at that hospital he was also the head of that department in the university. They came to Boston with the idea that the Brigham was to be the university hospital.

As a matter of fact, Cushing had some justification for this idea when he came. Before he left Baltimore he asked President Lowell whether his professorship made him "the head of the surgical department." Dr. Francis B. Harrington replied for the president: "Your functions in the School would not be subordinate to any other member of the Surgical Department and you would be expected to take the leading part in the future organization and development of the Surgical Instruction. In other words you would assume a position similar to the one now held by Dr. Maurice Richardson." [23] In assuring Cushing that he would not be in a subordinate position, they managed to give him the impression that his was the leading role, a position that he assumed at once and never willingly relinquished, despite the explanations of deans and presidents.

Edsall, on the other hand, came from Philadelphia, where the medical school had many hospital resources. Because of his experience there he felt that a medical school should have more than one university hospital and should make every effort to have as much clinical material available as it could. It didn't matter to him which hospital or institution was involved: the point was it should be *available* to the medical school. In sum, his point of view was that the strength of Harvard lay in having several very good hospitals.

Their early work in Boston served to reinforce their divergent views. Edsall worked at the Massachusetts General Hospital, at some distance from the school, though relations with the school didn't seem to suffer on that account. All Cushing had to do was look out of his window in the Brigham to see the imposing columns of the administration building of the Harvard Medical School right across the way; it was natural to feel "closer." He expressed this in one of his annual reports at the Brigham: "Our proximity to the Medical School has from the outset obliged us to accept the unqualified role of a university hospital. This fact places burdens upon us which are very unequally shared by the other major hospitals with similar school affiliations but more remotely situated." [24]

At the same time Edsall was writing in the *Harvard Alumni Bulletin*: "We have such hearty cooperation in medical education and research from nine hospitals that there would be little academic advantage to the University even if it did actually own and administer them — but it is idle to think of any university owning and maintaining such a group as ours, for $50,000,000 would not represent the present-day cost of reproducing the hospitals and conducting them as they are now conducted."[25]

Both Harvey Cushing and Henry Christian insisted upon getting titular professorships in surgery and medicine for the Brigham. Although Edsall disliked that, he went along with it. However, when it came to taking his old chair, the Jackson Professorship of Clinical Medicine, and giving it to Dr. Peabody at the City Hospital, he balked, feeling and probably with some justice that because of Dr. Jackson's intimate connection with the Massachusetts General Hospital, it was a mistake to place the title elsewhere, as he pointed out to President Lowell.[26]

Edsall's relationship with Lowell was wonderful. Lowell had great confidence in him, and Edsall practically had only to say to Lowell what he wanted and it was accomplished. I think this arose from the fact that Edsall never wanted to get anything for himself and only thought of his ideal of what a medical school ought to be. His ideals and lack of bias toward people were so manifest that a man like Lowell, who was busy and didn't know anything about medicine, couldn't help but have confidence in this very big man — big in every way. Edsall never had any trouble with Lowell's successor James Conant although Conant paid more attention to the medical school because he was a chemist and thought he knew a good deal about medical schools and medicine. Perhaps when Conant took over, Edsall had reached the stage where all he had to do was present his case and let his prestige and status do the rest.

Edsall had only one objective, improving the medical school, and above everything else that meant getting the best person possible for a given job. Although Cushing and Christian were very opinionated and dogmatic people, they both had the same point of view toward men that Edsall had. I do not want to give the impression that Cushing's was the only opposition to Edsall in Boston and that Edsall was a shy violet when it came to a fight. Actually there was a great deal of opposition to Edsall in Boston. Moreover Edsall didn't avoid fights and he didn't compromise with regard to them. He knew what he wanted and went after it, although frequently it represented a battle. Alan Gregg once wrote me that

Edsall didn't always show the best judgment in that regard and that frequently he would just as soon fight as compromise. He did get into a good many fights, or at least some of his acts encouraged resistance. But as Edsall put it once, "in the words of a certain distinguished Irish gentleman, the best way to avoid a danger is to run plumb at it." [27]

CHAPTER 17

Reforms in Teaching

"THE dreadful burden of the student in *things taught*" — Edsall as dean continued to war against this form of creeping paralysis as he had when organizing his courses in medicine. He was equally on guard against things which "would kill the time and the individuality of the teachers, especially of those highly desirable men, the independent-minded scientists." [1]

The postwar climate of opinion favored reform. Those who had studied American doctors under the test of war conditions came back to their schools prepared for some radical changes. In his first Dean's Report, Edsall described how the graduates of even the best U. S. medical schools had been found to "lack training in logical reasoning and in application of detailed facts and methods though they often have a large knowledge of the details themselves. They very commonly lack any mental habit of employing their training in the medical sciences in solving practical problems and the details of knowledge seem often to have swamped the fundamental principles and sometimes . . . obscured the clarity of judgment." [2]

While this was no news to those who had been struggling for curriculum reform for a decade, the new impetus made it possible to carry through some sweeping changes. As early as 1913, a committee on curriculum, with Christian as chairman and including Cannon, Bradford, Coolidge, Graves, and Greenough, had proposed several liberalizing steps, and had recommended that "some free hours be left as far as practical for library reading, review, and voluntary courses," with the all-important proviso "that in order to bring about the changes the needed reduction of the time now allotted to the various Departments be made." [3]

This was finally achieved by the reforms of 1921–22, despite the sacrifice of what Edsall knew to be "so dear to the earnest teacher, the amount of time available to him for the exposition of his subject." [4]

> Of course, in getting into this sort of operation [Edsall said later] one always expects to meet with opposition from all sorts of persons, but, strange as it may seem, everybody agreed to the principle, but everybody objected to having his own time cut down. Dr. Cannon and I spent a great deal of time in preparing the way for a discussion of this matter with our faculty. Our faculty is a democratic body, and seventy-five or eighty men have a vote on all questions of policy. We thought we had a large job on our hands. We did what the king of France did, marched his columns of soldiers up the hill and down again because we found that nobody opposed the situation. We formulated surgical operations on the number of hours, and we eliminated 25 per cent of the fixed hours in all years.[5]

After two years' work and considerable horse trading behind the scenes, the departments agreed to accept an across-the-board reduction of 25 to 30 per cent in their precious time. The final committee report was passed without opposition.

"The freedom provided the student by this change may not appear great to those accustomed to the liberty now accorded . . . in the Arts and Sciences . . . This change in the curriculum is, however, a very decided and most radical one from the previous curriculum," the dean stated in 1922.[6]

In 1925 he reported, "even the poorer students have had a good deal of time to get over the exhaustion . . . produced . . . under the previous system. They do sometimes take exercise in their spare time now and they even read general literature occasionally, and a few things of that sort, but not enough to interfere with their successfully doing what we consider necessary for the medical degree."[7]

Looking back at the end of his career, he called this "the most important change of all that was undertaken."

> This was intended to overcome the great rigidity of the curriculum that made it impossible for the students to give special time and thought to matters that interested them particularly or that they needed to give attention to, and prevented any reasonable opportunity for advanced work in any line . . .
>
> A general change in the curriculum, which freed entirely three afternoons a week and which limited the amount of time during the day that might be required of them, has made great change in the students' attitude toward their work, in their whole method of working

as students of Medicine rather than merely as applicants for a license through examinations . . .

The effect of it is particularly shown in the use of the library, which immediately increased by nearly 50% and within three years increased by about 100%, and by the greatly increased numbers who have done extra advanced work . . .[8]

Following this reduction in the required number of lectures, there were two afternoons free of any compulsory periods, three if working on Saturday afternoons was contemplated. New voluntary courses appeared. And of course there were always the libraries. The students were too conscientious to go to the ball game often, or the matinee. Many were invited into a department to work in the laboratories and so got started in problems which fascinated them for years. Dr. Edsall's son John worked in Edwin Cohn's laboratory of physical chemistry, which beguiled him so much he gave up the idea of taking an internship in medicine and stayed in biochemistry as his life work as a professor at Harvard. In this way many students found special interests which influenced their future lives. The excessive drive of required lectures was much reduced, and the teachers worked much harder preparing their lectures so that students paid closer attention.

In medical teaching, perhaps more than any other field, a general raising of standards and increasingly stringent requirements for license by the states had combined over the preceding decade to create a stifling pressure. A movement was soon on foot in the Association of American Medical Colleges and the American Medical Association to modify the recommendations on which the states had based their rules.

As Edsall wrote to Dr. Cannon, asking him to speak for the dean on this matter at a forthcoming AAMC meeting:

As you are well aware in connection with the work on the Correlation Committee, any attempt to make the medical course more flexible and make it capable of meeting the needs of the varying types of students comes very close to difficulties oftentimes with these State Board regulations . . .

The most striking recent example [is] that of the Board of Regents of New York State, who have imposed regulations concerning both the number of hours to be taught in the subjects comprised under the medical sciences and also the years in which they should be taught. For example, if we were giving the adequate amount of pathology to meet their regulations we must still teach all that number of hours in the first two years or we do not meet their requirements. Wolbach, as you know, and some other pathologists believe that pathology should be spread out somewhat and should be brought more closely in contact with the clinical subjects. According to this regulation strictly interpreted, or indeed interpreted in any way so far as I can see, no pathology could be taught in the third year in connection with clinical medicine or in the fourth year without being added on to what is considered the proper amount of time to be given to pathology . . .[9]

Edsall was in a good position to help the movement along, as he was on the council of the AAMC and a member of the commission on medical education which they organized to study the problem. At the same time that Edsall was announcing to the Harvard community a new curriculum, with its increased free time, he was able to report a hope for the relaxation of the rigid requirements of the state boards, since the Association of American Medical Colleges had reversed their position and were willing to state their requirements in terms of mastery rather than hours and credits.[10]

One of the qualities which made Edsall a good dean was his habit of guarding against extremes. As he said in speaking before his confreres at the Massachusetts Medical Society, the lessening of regulations

should not be so hurriedly and hastily and extensively carried out as to lead either to confusion or a lowering of standards, or to throwing on their own resources, students unaccustomed to this to such an extent as to lead them to ineffectual effort. I have seen the latter occur to such an extent in some foreign countries that I am clear in my mind that over-supervision and over-regulation produce a better result than too little of these . . .

Nevertheless he continued to fight for the students' opportunity to choose what they would do with their time.

This, I am sure, is one of the most serious defects in our system, and one not sufficiently often appreciated. One of the most justified criticisms of the present day graduate is that very often he lacks initiative and independent and practiced judgment. I am not at all ready to say that he is more defective in this way than earlier graduates. We easily forget how very defective we were ourselves at that stage. But I do believe he is more defective than he would be if we had a less schematic system . . . that threw more responsibility on him instead of taking responsibility from him . . .[11]

("So fixed and rigid a four years' period of tutelage . . . does not present an altogether inviting prospect to the independent and inquiring mind," he said at another time.[12]) And at the medical meeting he continued:

It is undeniable that in all forms of teaching, the common danger is to force the student into such an accumulation of facts and experiences that the rationale of it all is . . . lost. It is true that sometimes it seems to have been felt that the laboratory is an open sesame, and it is forgotten that over feeding can happen with laboratory teaching as with any other . . . The purpose of experience in the laboratory — to help them comprehend the subject — has in such cases been often to some degree directly defeated by the methods used in the laboratory.[13]

Pioneer of "scientific medicine" as he was, he had often to combat fear and misunderstanding of the new methods. "Nothing would more rapidly degrade the medical course than to fear making it scientific in the true sense," he wrote in his first Dean's Report. "It demands a nice adjustment to make a background of facts and methods sufficient to give some skill in their use and to bring out clearly fundamental principles of thought and action, while at the same time avoiding such a multiplication of minutiae as actually to obscure the view of principles and confuse the students' judgment as to the essentials."[14]

Harvard went right ahead with the reforms it wished to make, and if government regulations interfered, they too were reformed. Ten years later Edsall observed, "This was the first school that did anything to overcome the cramping effects of legal and quasi-legal regulations, and undoubtedly its example was in part responsible for the rapid undertaking of similar freeing of time in a considerable number of schools."[15]

The reduction in hours did not stand as a solitary reform but was one of a group, all aimed at changing Harvard medical education in fundamental ways. Like all improvements in a living process, these were never finished, but went on year by year, and at the end of his tenure Edsall wrote, "Still more needs to be done gradually in this."[16]

Change in institutions is a subtle thing. Leaders like Edsall and Cannon were able to command changes which might have seemed revolutionary if less skillfully carried out. Edsall wrote:

> In the major changes it meant usually seeing large numbers of teachers who were involved and having long discussions with them and persuading them before the matters came up for a vote — Cannon particularly having shared to a large degree in all this labor of interviews of individuals in regard to the first main change in the curriculum . . . There was the most cooperative action on the part of practically all our colleagues in these matters, but they were busy with their work and the initiative in practically all these instances had to come from the Dean, and the general scheme had to be worked out by him.[17]

Other related changes included the development of comprehensive examinations; the use of tutors, English style, as Edsall had worked with his tutor at Penn.; changes in grading; the requirement of a thesis; a new system of elective courses and opportunities for individual work in various departments; and freedom for men in the top 15 per cent of the senior class to do independent work, perhaps at another school. There was also the opportunity to do summer work for credit, in order to clear some time in the following school year for individual projects.

Edsall had advanced ideas about the function examinations were supposed to perform:

> In my mind too many "tests" and partial examinations are frequently used in many places in this country . . . Unless they are very carefully used, they tend to hold the indifferent student to his work but at the expense of taking away from all students one of the important elements in their training; that is, a training in initiative and independent effort. It were better, I think, to let the poor student run more risk in order to give all the students more responsibility . . . a point that is, I think, better recognized in most European countries than it is in this country.[18]

There was a "general examination" in use when Edsall became dean, but it was chiefly a review examination on the work of the last two years, and was not fruitful in forcing the student to think constructively about his work as he went along. However, the finals given in 1920 were "altered in a very forceful way," as Edsall put it.[19] This new examination was conducted by a junior committee on examinations, under the chairmanship of Channing Frothingham, then instructor in medicine, and including Stanley Cobb, instructor in physiology, Elliott Cutler, instructor in surgery, George R. Minot, assistant professor of medicine, and Francis M. Rackemann, assistant in medicine. (These men, young at this time, all became leaders in the school in future years.) The senior committee (President Lowell, Dean Edsall, and several full professors) was somewhat concerned at letting the younger men handle it, but the events justified their decision. As the seniors reported in September:

> It may be regarded as extremely doubtful whether the older members of the faculty would conduct an examination designed to test the student's power of correlating his clinical and preclinical subjects as did the group of younger men with fresher experience. Furthermore . . . that this class worked hard, read incessantly, and particularly read back into their early work in the school, was apparently due not so much to the warning they received of the character of the examination to be given but to their conviction that in the hands of the Junior Examining Committee the program would be carried through with a searching literalness which could not be escaped.[20]

"This examination is a difficult one to carry out," Edsall stated in 1925. "It is, in fact, the most difficult form of examination that I have ever dealt with . . . We have teams of three to read each book, those three representing clinical branches and some one or other of the medical sciences . . . Trained examiners are needed in the work."[21]

A few years after the first trials, Edsall reported, "After this experience the statement seems justified that it appears to be accomplishing the most important object . . . At first the student body considered the general examination a severe and rather fearful trial . . . The performance of the individual student in his correlating examination is now usually good, and often extremely creditable."[22]

Five years after the examination was adopted, the students

requested that honors should be awarded solely on the basis of performance in that examination, because as Edsall said, "it went further than anything else in the course to determine who was an effective man at the end of the course."[23]

After a fifteen-year trial, he was even more confident of the usefulness of this particular device. He wrote:

> The results are sincerely gratifying as contrasted with the previous conditions during the 30 years that I have known something about them.
>
> It appears to have had a deep effect upon the attitude of the students in employing their medical sciences in their clinical thinking. This is conspicuously shown in the manner in which they have learned to approach this examination itself and in the standing they have taken before the National Board of Medical Examiners as contrasted with the students from other schools. It has had a strong influence also upon a large proportion of the Faculty in leading their teaching toward this point of view as against the tendency to have departmental teaching in compartments that have very little relation with each other.[24]

It was a great deal easier to lead the student to a correlating point of view than to do the same with the faculty. One reason for this of course was the increase in specialization, as Edsall described it in a speech in 1925:

> This accumulation of knowledge has been so great that it has forced into much greater specialization those who teach, and has lessened in some ways the capacity on the part of the teachers to deal with the relations to subjects other than their own . . . The obvious and often suggested way of meeting this situation is to have men of more general training in charge of the departments. Such a plan is, however, doomed to failure . . . However effective or ineffective the conditions were a generation or more ago, it is quite clear that with the present and probable future conditions of knowledge, versatility of the old type will but rarely escape being superficiality.[25]

Edsall sought approaches to a new kind of versatility, and one means was the arrangement of the courses themselves. Very early in his administration a number of cross-fertilizing experiments were tried:

> For example, second-year Anatomy (cross section and other topographical practical anatomy) was arranged some years ago so that it begins just before and then runs concurrently with normal Physical

Diagnosis. The student also studies Xrays of the chest and so forth in the anatomical course for the illumination that they throw upon living anatomy, and the instructors in the two courses keep somewhat familiar each with what is being done in the other . . . The effect has been to increase the student's tendency to think of many physical signs as expressions of usual but to some degree variable anatomical conditions, not as dogmatic sizes, positions and so forth of organs, rather arbitrarily set by books and teachers.

Among further things I might mention that Pharmacology and the earliest course on Medicine now have similar time relations and an effort is made in the clinic to develop simple typical disease phenomena as disturbances of physiological processes rather than as mere descriptions in a text book, and to have the student see how far one can or can not oppose or offset these disturbances by application of knowledge of the physiological action of drugs and so forth.[26]

A few more years' experience in these efforts bore fruit in a paper which Edsall gave before the Interurban Clinical Club in 1924, "The Question of Correlation and Opportunity of the Superior Student with Comments on the Comprehensive Examination and the Tutorial Method." David Riesman, in writing the history of the club, reported this talk in greater detail than any other in the book.[27]

Here Edsall discussed all these newer forms of instruction in which he was deeply interested. He spoke particularly of the course in medicine which was given to bridge the gap between physiology and medicine (the course given by Means, Minot, Aub, and Chester M. Jones), and the course intended to correlate surgery and anatomy (which he said the students regarded as "very interesting entertainment, although somewhat of a stimulant"). He pointed out that the important thing was for the student to correlate — he must be the one to do the work.

In another speech in 1925 he said:

I have repeatedly seen a look of surprise on the faces of practitioners as well as of students when they heard for the first time the wholly truthful aphorism of Dr. Cushing that medicine and surgery are the same subjects differing only in their therapeutics. I am often reminded of a story current in Harvard College of a student who learned in a course on government of a conspicuous statesman named Hamilton, in a course on economics of a man named Hamilton who was an able financier, and in a course on history of a man named Hamilton who was killed in a duel with Aaron Burr; and the student was quite astonished but wholly pleased to learn that these were all one and the same person.[28]

The word "correlation" proved to be a catchy one, and Edsall became exercised when people believed it to be a system. "It is not a 'system,'" he replied to Abraham Flexner, who had asked him how it worked. "It is the way I think and the way any one trained at all in sound diagnosis and therapeutics thinks." He continued:

> It is, as I conceive it, almost the contrary of a prescribed system. As affecting the course of clinical medicine as we are carrying it out, if properly done, it will require that the student get cohesion and systematic knowledge not in lectures and prescribed exercises but through guided reading of text books and articles . . .
>
> The only new element is the presence in the clinic in a very few exercises of a physiologist, a pathologist, or a pharmacologist, one, two or three of these men at a time, and a discussion and give-and-take between them, the clinician and the student . . . Clinical and laboratory teachers increasingly speak a common language but this pervades their teaching less than it does their work and the student apparently sees the relation with the laboratory branches become distant when he leaves them . . .
>
> Dr. Mall expressed the same purpose to me somewhat quizzically ten years ago when he said that "the ills of anatomy would never be cured until the Professor of Anatomy was also Professor of Physiology." Some men spontaneously do this. Somebody — I think you — said that Dr. Welch never did this sort of thing, that he simply taught naturally. I always thought that Dr. Welch did this almost more than anybody else I ever saw in this country but he did it naturally.[29]

Edsall was speaking from his own experience when he said that most students, even the good ones, tended to take their ideas from the text or the teacher unless they were directly stimulated into thinking for themselves, "but many of them, even the mediocre men, soon get quite keen and alert and intelligent in doing it if started on that road. I know from the change that I see in them after a very few weekly exercises with this in mind." He warned, however, that they should be caught early and not left until the fourth year or internship.[30]

This was the kind of thing which led Walter Bauer to call Edsall a "true experimentalist," not only in science but equally important, in medical teaching.[31]

Curriculum revision was a continuing process throughout Edsall's administration. In the second year work, which covered the most advanced preclinical subjects, examinations were reduced from nine to five, a sign of improved coordination of the nine original subjects.

In the third year, Edsall wrote, "a very radical change was made in the . . . curriculum which had been the most trying of all years, with thirteen different subjects, thirteen different examinations, and no defined coordination as to the way that these subjects were related to each other in time, etc. The major subjects and minor subjects were coordinated together . . . and the examinations were reduced to four."[32]

As to the fourth year, arrangements were made so that the men, "when they had shown prospects of profiting by this, should be allowed to concentrate for at least half, and if desired, more of the fourth year in one broad line, being required only to take what was considered a necessary minimum of medical and surgical ward clerking work. A large group (all told about 15% a year, the number allowed by the Faculty) have profitted by this, and their comments and their subsequent records seem to show that it has been a great advantage to them."[33]

With all this modernizing of courses, the old-fashioned system of grading could not long survive. The faculty did away with the percentage grade system, and stopped giving out marks to the students.

"The custom has been established," wrote Edsall later, "to have personal comments upon individual students . . . This has gradually transformed the attitude of the teachers from the mere giving of a mark to one of attention to the individual's abilities and qualities in diverse ways. The combined comments that accumulate on the personnel records of the student have become very influential in the evaluation of the men."[34]

The tutorial system grew up in relation with the efforts to make best use of the fourth year free time for the ablest students. In 1922 the top 15 per cent were given special freedom. The following year Edsall presented to the administrative board his plan for a tutorial system resembling that of Oxford or Cambridge, in providing for the best development of the exceptional student. At the same time he submitted the proposal to the Rockefeller General Education Board. The General Education Board agreed to pay salaries and expenses for the tutors for two years. Dr. Alfred C. Redfield was accordingly appointed in physiology and Dr. Arlie V. Bock in medicine.

In the course of that experimental first year, Dr. Redfield worked with twenty-five students, some from the fourth year and some from the first and second, most of them with the highest records but some with the lowest. The fourth-year students were guided in their attack on a specific problem and in their general reading. Eight men from the first-year class were given what amounted to a seminar in physiology instead of the usual course. Their general accomplishment was more than satisfactory.

Dr. Bock started out with six men, with whom he conducted two bedside clinics and one evening conference each week, plus assigning them to laboratories at the Brigham, City Hospital, and the MGH. He considered assigning a course of reading but decided to leave requirements at a minimum, "with the idea of permitting them to find their own way of working things out, with the aid only of suggestions from me. I still feel the second policy is the wiser one for men of this age for they seem to appreciate being freed from the irksomeness of the usual programme, and in the long run, will cover all the ground and more than might have been required." He further reported to Edsall, "I . . . fancied that I could see development where I used to look for it almost in vain." [35]

After a year and a half of experiment, Edsall applied to the General Education Board for help in putting the tutors on a permanent basis, outlining his hope to have one in physiology, one in medicine, one in surgery, and one in pathology. If they could have Rockefeller money for half of the expense, he was ready to use DeLamar funds for the other half. True to his principle of not frittering away money, he had been hanging on to some unexpended income of the DeLamar fund, but "the tutorial method I think is the most important thing that has arisen." [36]

From these modest beginnings the tutorial system has expanded until every entering student has a tutor for his first year, and there are tutorial seminars through the later years.

When discussing the tutorial system after a few years' trial, before the Massachusetts Medical Society (1924), Edsall considered that about 20 per cent of the students would do well in this type of individualized work. Finding the right tutors was most important, for it is a rare man who makes a good tutor. The system when well done made the men formulate opinions rather than recite facts, and this required the type of teaching which Dr. Arlie Bock did so well. It became clear that to make this successful required men who enjoyed original correlation of symptoms and understanding of symptom complexes. The teaching matured students and enlarged their point of view toward disease.

Edsall fought hard for the fullest development of all levels of student. In particular, he was aware of the challenge and freedom needed for the flowering of the superior student, "the most precious material of all." "We must do justice to the mediocre and the inferior," he said, "we must not by doing justice to them do injustice to the most important material of all." [37]

At the same time the tutorial system was being applied for the benefit of the outstanding student, a system of advisors was being worked out for the whole school. This went through a number of vicissitudes, since it was found that not every teacher makes a good advisor. The provision of summer work for credit also began to develop in importance as the individualized work of the fourth year attracted more and more students. Beginning with a very few, the number working off course requirements in the summer, so that they could pursue some particular interest in the fourth year, grew until there were 100 to 120 men each summer, "an evidence of their admirable spirit since it means doing extra work," Edsall wrote.[38]

Work elsewhere was one of the lures. "Very early," wrote Edsall, "there were developed opportunities for men to do parts of their work elsewhere, for credit, under proper conditions, furthering thereby the movement of men from one place to another in order that they might acquire subjects, or contacts with men, that seemed to them to be particularly desirable for them. This has been much increased until now every year a noteworthy number of men do this." [39]

Edsall was a great believer in the broadening influence of travel. In his own career, travelling often abroad and settling in a university town to work in the laboratory of some respected scientist, he had learned to know well the brilliant men who populated the laboratories of Europe. His international acquaintance with ideas and people gave him the breadth of knowledge, rare in that period, which made his advice to young men so valuable. As professor and as dean he showed his approval of travel in sending his young proteges to work under the bright minds of Europe — and indeed of America. His recommendation became an open sesame to the better laboratories and started a most exciting year. Not only did he help find the place, but, when he could, he found the funds which permitted the travel.

In summing up all these changes made over nearly two decades, Edsall wrote:

> I think that these have been, all told, of very deep importance in relation to the character of the atmosphere of the School and the kind of training that can be given the students . . . they have taken more time and more effort than anything else that has been done. The separate things are a part of a whole scheme, as you will see, to make the course more flexible in regard to the individual powers of the individual students, and also to permit all students to have more time for the development of their own powers . . .
>
> The effect of these, I think, is typified by what the able presidents of the third and fourth year classes told me at the time that the main changes were being carried out . . . The result during their own period as students had been to change the attitude of the great mass of students from one of merely reading textbooks and doing what they were told, to one of reading a great deal of medical journals, of medical monographs, etc. and actually studying medicine instead of studying for examination in a penny-in-the-slot style.[40]

Even in the twenties Edsall had no doubt that the modern methods of education were producing better doctors. In 1924 he said to the Massachusetts Medical Society:

> Granting, as is beyond question, that the present day graduate is better trained in his scientific background, and more capable of using this in practical ways, is he actually better trained in the definite, practical things that he must know as a dependable practitioner, and also does he possess as much skill and experience in dealing with human

beings? These together largely constitute what is often spoken of as the art of medicine, something that is at times a bit sneered at by those of formulistic scientific temperament. The use of the word art for this kind of skill should not carry the thought of an impressionistic method that that word is often used to connote. Skill of this kind means rather thoughtful and informed and effective methods of providing bodily and mental comfort. It means often especially the ability to take off people's minds and bodies those loads that are not very definable or measurable, but that often determine the difference between health and unhealth, or even at times between life and death.

In part, the above question can be answered promptly and positively. In physical examination, in diagnosis, in general methods of careful study of serious disease, and in the methods of accurate study of mild disease, the present graduate is very much better prepared than he ever was before. He has had (what I think is rarely appreciated, excepting by medical teachers) easily four times as much practical training and experience in the medical course as was usually the case at the time that I graduated in medicine, and for a long time after. The amount in most places has still been increasing in recent years. In addition to this, many more men now take a hospital internship than formerly.[41]

With his usual sagacious view of things, Edsall was aware of certain drawbacks as well in the modern training, which he expressed in a speech before the Southern Medical Association in 1928:

Most of the clinical teachers would tell you that the student is taught very much more practically now than twenty years ago; he sees more of patients, has more contact with them, and comes through the course with much more practical training than he had two decades or three decades ago. On the other hand, we hear many practitioners saying that the graduate as he comes out now is less practical in his attitude and has much less poise and initiative and judgment. Some of that is the natural feeling that all of us tend to get that things are not done nearly so well now as when we were students; but part of it is true . . .

The great fault is in the hospital training. Some of our states require hospital training; some of our universities do, but whether they do or not public opinion requires it. Now, the hospital is a very much better place for patients than it was twenty-five years ago; it is a very much better place for the more mature staff, and it is a much worse place for the training of an intern than it was twenty-five years ago. This is because he is more part of a machine now; he is so much in the habit of calling for expert help in every respect, calling for the radiologist, the neurologist, and other specialists, that he gets out of the habit of making decisions for himself. He also gets so much accustomed to the elaborate apparatus that hospitals have nowadays that he feels lost without it when he gets out. The whole experience tends to make him just what he is said to be; he often comes out with less poise and less ability to adapt himself to the demands of general practice.

In the days when I was in the hospital there was far less apparatus in use; it had not come in. In those days it was mostly a matter of training in individual judgment and experience. The consultant then was a man of superior knowledge and judgment; now the consultant is the man with these qualities, but also with much knowledge of apparatus and the use of it. The hospital nowadays is not so good a training for the youngster in medicine. I suspect we may go back in some form to what they are trying in California and Michigan, something like the apprentice system; but I personally have felt that the methods so far devised are not satisfactory. We might devise a method of attaching the student temporarily to excellent practitioners experienced in teaching.[42]

CHAPTER 18

The School of Public Health

EARLY IN Edsall's administration, the possibility arose again that the Rockefeller boards might give strong support to a full-scale development in the teaching of public health. They had already assisted the founding of a school of public health at Johns Hopkins which opened in 1918 with Dr. William Welch at its head. As early as 1915 Abraham Flexner had been negotiating with Harvard in a tentative way for the support of an "institute of hygiene," and Harvard's resources had been thoroughly explored then. Harvard had been giving a degree of Doctor of Public Health since 1911, and its School for Health Officers, a joint project with MIT, was established in 1913. The nucleus of this school included the medical school's professor of preventive medicine, Dr. Milton J. Rosenau, Harvard University's professor of sanitary engineering, George C. Whipple, and MIT's professor of biology and public health, W. T. Sedgwick. They combined to give the first real course for health officers in the United States.[1]

At the same time, there were full-scale departments in the medical school in closely related subjects. The department of preventive medicine and hygiene had been started in 1909, the department of tropical medicine (at first a school) in 1913, and the division of industrial hygiene in 1918, while the university's department of sanitary engineering dated from 1911.[2]

Most recent addition to this group was the department of industrial medicine, under Dr. Cecil K. Drinker, which offered a graduate course to train physicians for industry, also the first of its kind. Said Cecil Drinker, in describing this development for the alumni, "The field of industrial health, one of peculiar importance in New England, remained without special attention until the spring of 1918. At this time, thanks to the foresight and energy of Dr. Frederick C. Shattuck, a fund of $125,000 was collected from New

England manufacturers to be spent upon the teaching of students and the investigation of problems in industrial health."[3]

In addressing the section on industrial hygiene of the American Public Health Association the same year, Drinker said: "What we shall endeavor to bring out is the service which a fundamental science can render in practical conditions. It is just such realization which is making the advances of present-day, scientific medicine, and from the outset it is desirable to turn the student of industrial hygiene toward similar wholesome methods of thought."[4]

In August 1919 a new journal appeared to give a voice to the many new developments in this field. This was the *Journal of Industrial Hygiene*, of which David Edsall was the American editor. This new venture was close to his heart, for his primary interest in research was on the diseases associated with industry. The journal had an international board of editors, one here and one in England. During the twenties Edsall handed the managing over to Cecil and Katherine Drinker, and from 1927 to 1960 this post was filled by Philip Drinker.

During Edsall's editorship, along with the research and editorial activity of the Drinkers, the journal largely specialized in industrial diseases, but in 1950 the original vigorous editors were retiring and the journal united with the AMA journal, *Occupational Medicine*, which had been founded in 1946. Its scope in terms of subject matter was thus greatly broadened.[5]

This was all part of a steadily growing interest at Harvard in public health, signalized by the fact that fifty-four men registered in the Harvard-MIT School of Public Health in 1919–20. As for research, most departments in the medical school had been doing work of public import during the war, and many continued this in the years following. In his 1919–20 Dean's Report, Edsall cited the fact that the professor of tropical medicine had spent the year in Geneva organizing the medical activities of the League of Red Cross Societies, various members of the staff had worked in Santo Domingo at the request of the U. S. Navy on the subject of yaws, the department of preventive medicine had been working with the state of Massachusetts on infected rats, pharmacology had worked on the treatment of syphilis and the next year made startling improvements in the manufacture of "606," the department of industrial hygiene carried out a number of successful investigations, various members of the pathology and other departments studied typhus in Poland—and this was merely a summary of the larger investigations.

In particular, Edsall's interest in the relations between industry and health never waned. "I am perhaps prejudiced," he wrote once, "for I have long been deeply interested in it and in the inadequacy of training of physicians in the relation between industry and health. I had been greatly impressed for a long time over the almost entire neglect of industry in the medical history even when such minor things as the amount of tea and coffee the patient drank were carefully recorded . . . I have a little of the feeling one has for a pet child. . ."[6]

One of Edsall's first acts as dean had been to invite Dr. Alice Hamilton to come and give the Cutter Lectures in Preventive Medicine and Hygiene to the medical students. Alice Hamilton was a distinguished woman who had opened up industrial medicine in America. She was without question the leading person in the field, man or woman, and Edsall wanted very much to persuade her to join the faculty of the Harvard Medical School. He thought he might have a chance to persuade her, since there had been a recent shake-up in the government bureau under which she had been making investigations for the U.S. Department of Labor. At the same time, the Retail Trade Board of Boston was talking seriously of a five-year program which they would support to study the health situation in their stores. Of course, there had never been a woman on the medical school faculty, and women were not admitted as undergraduates or medical students, but Edsall did not find that tradition binding.

Edsall was able to advance to President Lowell both the necessity of using the Retail Trade Board's grant creditably, and the acknowledged fact that Dr. Hamilton was the best person in the country to head such an investigation.[7] Lowell was willing to consider the best person in the country, even if she was a woman.[8] Edsall then broached the matter to Dr. Hamilton and was able to assure her that

> the Committee in charge of the Industrial Medicine courses is, as a whole, most desirous that you should do this, if the work is actually

started, and I have also taken the matter up with the President and Fellows of Harvard University and have had word from them that they would be very glad if this plan could be carried out by you, and that they would be ready to receive at once from the Faculty a nomination for your appointment as Assistant Professor of Industrial Medicine . . . In case this goes through, our desire would of course be to have your advice and assistance not merely in this work but in regard to the general activities of the courses in Industrial Medicine . . .

May I say also that I have a very real pleasure in the thought that whether you should come to us in this or not, there is I think a very pleasant compliment in the fact that the President and Fellows have immediately acceded to this because of the service that you have rendered to the country already, although I think it is the first time that the proposition has ever come up to have a woman appointed to any position professorial or other in the University. Aside from my very great desire that we may be able to secure you for this work, I desire it also for the reason that I think it would be a large step forward in the proper attitude toward women in this University and in some other Universities.[9] *

Dr. Hamilton asked for a somewhat freer arrangement under which she could work for Harvard half the year and continue her own investigations for the government the rest of the time, and Edsall was glad to agree. Dr. Hamilton sent his letter on to her sister Edith, with the happy note at the foot:

Isn't this wonderful. Just as I had made up my mind that I had lost my chance by being too demanding this comes, doing away with

* After Alice Hamilton arrived and established the fact that she was not going to undermine the medical school by her sex, the question arose again about admitting women to the school as medical students. Edsall would have approved of such a step but most of the faculty would not. The admission of women students to the Harvard Medical School had to wait until World War II, when the number of applicants to the medical school was much reduced by the war. In 1945 the question arose at a faculty meeting, where there was considerable resistance. I was strongly in favor of it and brought up the suggestion. Finally, on the basis that this was a democratic institution and the number of women applicants was increasing while total applications were falling off, taken with the feeling that women had as much right to a medical education as men, it was decided to accept a few women. This caused a serious commotion; one of the professors at the meeting developed a paraplegia that night. From this time on, women were accepted in gradually increasing numbers. They did well in the school, were well received by the men, and proved themselves adequate competitors. In contrast to Harvard, the University of Pennsylvania medical school admitted women in 1914, Physicians and Surgeons in New York in 1917, and Yale in the same year. — *J. C. Aub.*

every single objection and making it as easy as possible to accept! Send it back to me when you have read it. Of course I have written him that I accept with joy. Only what am I to do for six months of each year in Boston. It appals me to think of it.[10]

Nevertheless, the Harvard barriers did not all come down at once, and Dr. Hamilton was not allowed to walk in the commencement procession or sit on the platform, while there were further unwritten rules pertaining to the use of the Harvard Club and tickets to the Harvard-Yale game. She handled it extremely well. She only laughed when she was told not to apply to the Harvard Club. Because of her extraordinary tact, this antagonism disappeared over the months, particularly because she had a personality that endeared her to her associates. She had a charming modest manner —honest, respectful and gracious, and she was clinically very alert. She was never reticent or retiring in her medical opinions, and always completely frank. She had a flair for making those around her speak with this same honesty. She had great courage and the ability to tell industrial men that they were endangering the health of workers. They were not able to resist her for very long and soon, without questioning, made efforts to reduce hazards and thereby victims.

Not long after she joined the Harvard faculty she went to Europe with her old friend Jane Addams, and when she came back she talked at great length about the starvation of children in Germany and Austria and urged that food be sent to them. This upset some people who were still thinking in terms of "the enemy" and Dr. Hamilton was specifically told that one of Harvard's benefactors would stop giving money if she went on talking about feeding Germans and Austrians. Alice Hamilton went to Edsall and offered to resign. But, according to her story, Edsall told her that Harvard never interfered with freedom of opinion or speech, and as for her salary, it was paid from general funds and had nothing to do with individual donors.[11]

To her relief, Dr. Hamilton found this spirit more typical of Boston than she had thought, just as Edsall did in shifting from Philadelphia ten years earlier. "I found in Boston more of the old American respect for individuals' rights, more willingness to go against the stream, than I had found in Chicago," she wrote in her autobiography. "Always there was a group of eminent men and women who could be trusted to stand up publicly for civil rights, even in behalf of people for whose views they had little or no sympathy."[12]

It was a fortunate conjunction of events that brought to a head Rockefeller interest in supporting schools of public health at the same time that the medical school had a dean with a view of public health well in advance of his time. In January 1921, inquiries came from Wickliffe Rose, director of the International Health Board, which had already assisted the Johns Hopkins public health school, and Edsall responded with particular enthusiasm. As he surveyed the field, an ideal school should have a strong staff in administration of public health, epidemiology, vital statistics, and immunology and general bacteriology. He agreed with Rose's view that "we need to have a group of men who would have as their dominating objective in life the development of public health rather than the development of undergraduate medical students to graduates."

He cited the already existing departments of preventive medicine, industrial medicine, tropical medicine, and comparative pathology, and the university's department of sanitary engineering which could form the nucleus of a new school. But to him a new "school" was not a matter of independence and new buildings, but of a point of view and an increased opportunity to do useful work, even as he had outlined it in his confidential speech to the faculty on becoming dean (see Chapter 15).

Continuing his letter to Rose, Edsall outlined the ways in which the field could be covered thoroughly, and the benefits of having full-scale departments with a professor at the head, a proper staff and a budget, as against developing the subjects as divisions of existing medical school departments.

The budgets of the four main departments already came to $87,430, although in some cases the support was temporary. With all the departments of the medical school pressing for money, those which were not central to medical teaching were begrudged any sizeable sums from general funds on anything but a temporary basis, as "this would mean crippling the activities of other departments." They were supposed to run on their own endowments.

Something like $72,000 was available in their budgets on a permanent basis for support of public health activities. The amount of teaching of general medical students done by the four departments might be considered to be balanced by public health contributions of pathology, bacteriology, physiology, biological chemistry, and pharmacology. Edsall thought it would take at least $85,000 to handle public health matters properly, but the Harvard Medical School was not able to extend itself thus far. In fact, as he said, "There is a widespread feeling in the Faculty that we have engaged in a number of expansions that tend to imperil the success of fundamental Medical School departments, and I believe there is soundness in this view. I think it would be a great pity to start a development with the sense among many people that it was unwise."[13]

A few further negotiations and Edsall was prepared to outline just what was already going on and what might go on at Harvard in various departments, a list of which showed how widely the medical school was already drawing on the university. This included the work in entomology of the Graduate School of Applied Biology, courses in climatology and poisonous reptiles, sanitary engineering under Professor Whipple of the engineering school, public health law by Professor Wambaugh of the law school, and a number of developments under the School of Business Administration.

Edsall continued to Rose:

> The feeling is strong in the Faculty of Medicine and among the University authorities that with existing Medical School funds we should rather contract when possible than expand and should intensify strong existing departments and strengthen others. I thoroughly share this view and have so stated openly ever since I have been Dean. I could not therefore without damaging my own usefulness and exciting discontent advise any further use of Medical School funds in new and largely separate undertakings beyond those indicated. Deeply as I was interested for example in developing Industrial Hygiene, one of my main interests in life, I recommended undertaking it only upon the understanding that it should carry itself without employing Medical School funds. That we have been forced temporarily into employing some of these funds in Industrial Hygiene appears to me to be a warning that increase in similar ways could not be undertaken without being properly felt by my colleagues to be either bad judgment or something

approaching bad faith. Several important Medical School departments are now greatly in need of more money that they justly deserve and that some similar departments elsewhere have, money that at present we can not provide.

The adjustments that I have indicated that we can make will cost the Medical School a considerable added sum . . . This degree of cost seems to me to be justifiable because of the added usefulness to the School in the projected Public Health development. I do not feel that I could safely add more and I feel disturbed at that and wish it could be avoided in order to feel secure of the entire approval of the developments, in case they come, among my colleagues and the University administration.[14]

Harvard's existing development all in all was quite considerable. In fact, Dr. Rosenau called it "more than a banquet for any student who wishes to prepare himself for public health service." His ideas were running parallel to Edsall's. "The great need, of course, is to obtain a small group whose primary interest would be teaching and investigation," he wrote. Although he knew Edsall's resistance to spreading his resources too thin, he feared that "the needs of the situation are so many and diverse that there is real danger of making the scheme too diffuse."[15]

A few more months and a tentative faculty met together: Rosenau, Whipple, Strong, Tyzzer, Roger Lee, Cecil Drinker, and Edsall. President Lowell had asked that the dean of the Harvard Medical School and the dean of the School of Public Health should be one person. A building had suddenly become available next door to the medical school: the Infants' Hospital was combining with Childrens' and planning more modern and convenient space. They were willing to sell the original hospital. New bequests and a certain amount of shifting of funds had provided an additional $25,000 a year. A considerable amount of the DeLamar money was already being used in the public health activities and more might be available. As Edsall wrote to Rose:

I have stated since the first announcement of the DeLamar will, that it was obvious that these funds should be used in such way as whenever possible to develop the teaching and research in public health and not in individual medicine alone . . . and that the atmosphere of the School should become an atmosphere in which public health was thought of everywhere throughout the School, just as much as individual health, and that such a manner of employment of the funds was the only one that I could see that was wise and in conformity with the wishes of the testator.[16]

In August 1921 President Lowell announced an initial gift of $1,785,000 from the Rockefeller Foundation for the creation of a School of Public Health at Harvard. There were to be new or extended facilities in public health administration, vital statistics, immunology, bacteriology, medical zoology, physiological hygiene, and communicable diseases. A possibility existed of $500,000 in future gifts. Close relations were to be maintained with the medical school; certain heads of departments were to be members of both faculties, and a number of laboratories and lecture rooms were to be used in common. The joint library was also an important item. Singled out for special mention in the announcement were the cooperative relations with industry, whereby students could gain practical experience in industrial hygiene.[17]

Thus the Rockefeller Foundation made the second of its major gifts in support of public health education which between 1931 and 1950 led them to spend $25 million for twenty-two public health schools and institutes in seventeen countries.[18]

At Edsall's suggestion, seconded by all the public health faculty, the endowment received at this time was named the Henry P. Walcott Fund of Harvard University, honoring one of the leading figures in public health in the state and the nation.[19] Dr. Walcott had been a leader in the fight for public health legislation in Massachusetts and for a clean and plentiful water supply for Boston. In 1881 he became Health Officer of the State Board of Health, which he served over forty years. When diphtheria antitoxin was discovered abroad, he led the state and Harvard Medical School to start producing it, and in 1895 Massachusetts was the first state to distribute it. At his death in 1933, the *New England Journal of Medicine* said, "To his leadership more than any other has been due the great improvement in public health and welfare which has taken place during the last half century."[20]

Edsall knew Walcott at the MGH where he was president from 1910 on, but he knew him in other respects as well. For most of his life Walcott embodied Public Health in the state; legislators consulted him as to framing public health laws, and public servants as to how to administer them. He served as a fellow of Harvard University for thirty-seven years, and on his retirement from the Corporation in 1927, Dr. Frederick C. Shattuck said of him, "Dr. Walcott was born wise and his wisdom has steadily grown through the experience and responsibility of active professional practise, followed by long service on the State Board of Health, the Water and Sewerage Board, the Massachusetts General and Cambridge hospitals . . . In any prolonged absences of President Eliot, Dr. Walcott was acting president . . . Dr. Walcott is the greatest living public servant of the State."[21]

Announcing the new school in his dean's report of 1920–21, Edsall said:

> Let one reflect upon the changes that have occurred in the past few decades in the relations of the public health organization to the individual practitioner of medicine, in the extent of the province of public health, and in its powers of accomplishment. The question will then at once arise whether an extremely powerful influence can continue long to be exerted by any institution that teaches medicine unless it makes broad provision for the changes that have already occurred and for those that are to come . . .
>
> It is a vastly different picture from a quarter century ago. The public health organization contributes through the physician to the healing of the sick much that is of essential importance in diagnosis and treatment and in increasing degree it contributes even in direct ministration to individuals . . .
>
> The chief thing needed . . . is to make the public health aspects of the physician's knowledge and of his activities more visible and living . . . most teachers of medicine are trained almost exclusively with a view to the individual sick. Adequate presentation of the double viewpoint will not come until it permeates the minds of those who conduct almost all the various courses . . .
>
> The development needed can come about most promptly and effectively by the intimate correlation that is projected between the School of Medicine and the School of Public Health and the consequent and almost inevitable infiltration . . . If this is needed now more than it is done, it can scarcely be doubted . . . that in a few years more it will be essential and demanded.[22]

With its endowment announced in the fall of 1921 and teaching of students scheduled to begin one year later, the labor of creating the actual school — its entrance requirements, its departments, its schedule, its degrees — was turned over to a three-man committee consisting of Drs. Milton Rosenau, Cecil Drinker, and Roger Lee, with Dr. Lee as chairman. It was hoped to admit women on the same terms as men, and plans were made for public health nursing, but the Corporation proved immovable on the subject. Women could attend as special students, or they could work for a Ph.D. in hygiene through Radcliffe. They were not admitted on an equal basis until after Cecil Drinker became dean of the school.

The School of Public Health duly opened in the fall of 1922, but the semester was scarcely under way when Edsall went abroad to make his survey of British medical education for the Rockefeller Foundation (see Chapter 20), leaving Roger Lee as acting dean for the first semester.*

Not only did the SPH take over the Harvard-MIT students in full career, it opened with a number of foreign students of assorted backgrounds (many sent on fellowships granted by the Rockefeller Foundation) who were slated for important posts in public health in their own countries on their return. The school's schedule had therefore to be broad and flexible, and a student's course was set up to fit his particular needs. Such foreign students, who had often been working in medicine and public health before their arrival, were a challenge, for they knew with great precision what they wanted from their teachers.

At the end of the first year Edsall reported: "The Faculty . . . made no schedule of hours or courses or credits. It particularizes with each man, requiring that he shall show evidence of having, or shall obtain, sufficient elementary knowledge of a very limited number of fundamentally important subjects, but arranging his work otherwise to fit as well as possible his special needs and the particular object he has in view. A very large degree of intellectual

* Because of this absence, Roger Lee has sometimes been called the "first dean" of the School of Public Health, which overlooks the fact that Edsall, the only official dean, had been functioning since the spring of 1921 when President Lowell requested that the two schools be headed by the same man. Roger Lee had a share in bringing the plans to birth in an actual school, particularly through his work on the Organization Committee.

Edsall's official tenure of the office began on September 1, 1922, although by an oversight this was not voted by the Corporation until May 24, 1926.

independence with the beginnings of real progress in some defined line is the object aimed at with each student." [23]

Until the depression caused a cutback in funds, the department of industrial medicine was one of the most flourishing. As Edsall wrote:

> [For] about fifteen years, I think it fair to say that there has been no place where the training of industrial physicians, research in industrial medicine and hygiene, and the acquainting to some degree of all medical students with the matter has been better and more continuously done . . . There has been a good deal of contact with a whole series of important industries and their executives and they have often turned to the School for aid in their problems and have repeatedly given large sums of money for the study of them by the staff and have often consulted the staff and the Journal of Industrial Hygiene. The number of physicians who have after training there taken more or less important posts in industry is considerable and they are wide spread.[24]

While most subscribers to the Industrial Hygiene Fund were New England firms, the list included such giants as the United States Steel Corp. The Harvard doctors did consulting work on the health problems of a multitude of firms. The relationships were fruitful all around, and some of the country's leading magnates learned to familiarize themselves with a new point of view. One can imagine the jar with which the head of one large concern read the following missive from the dean about having medical students in one of his plants:

> We discussed the question of sending students to ——, and of course it is a great privilege for us to consider that we may have opportunity to do this. I regret that we did not give you a more definite picture of the sort of work we felt our students could now undertake in your plant. You are used to dealing with men from Technology, and are able to place them on technical work which meets the best standards of technical procedure in the engineering field. In our case, however, the field is a different one. Would it be good policy for us to send men, in whom we are anxious to instill ideas of careful thorough work, to learn the technique of industrial physical examinations where our own staff have seen these examinations being made at the rate of eighty in less than two hours? If we sent men who are going through the regular course to Doctor ——, they would undoubtedly obtain benefit and in-

spiration from watching him cope, practically single-handed, with your problems. They would not, however, obtain the sort of idea the University knows they should have, relative to the real possibilities and organization of industrial medicine and surgery.

In regard to research men, the case is different. —— presents a wealth of unsolved problems, and if it is possible to arrange for such students, we shall be very glad to try the experiment of sending them. Indeed, we shall regard it as a real opportunity to be able to help them as well as you in this way.

Sometimes such projects meant going into a plant, examining the men, checking the processes, analyzing air and other sources of exposure—often a fascinating detective problem. In other cases, we at Harvard were able to do some basic research, as for instance in the manganese mills and in the lead studies.

The lead research originated when Dr. Alice Hamilton met the president of the association of white lead manufacturers, who was seriously concerned at the dangers of the trade. He pricked up his ears at hearing about the manganese studies and the work in the Ludlow mills, and told her that he was interested in getting at some of the facts in lead poisoning. As Dr. Hamilton wrote to Edsall, "he understands that whatever work Harvard undertakes will be done in the spirit of inquiry, not of propaganda for either side."[25]

As a result the National Lead Institute generously supported some basic research in lead, which as it turned out was unexpectedly fruitful not only in prevention and treatment of lead poisoning but in understanding aspects of calcium metabolism and related problems. I was put in charge of this research.

The president of the National Lead Company responded to one of the early reports of these studies:

I quite agree that it is already being demonstrated that scientific investigation of lead poisoning, even though commenced for the sole purpose of acquiring knowledge, is likely to result in great advantage to everyone whose business brings him into contact with lead. The statement that lead absorbed through the lungs is more injurious and dangerous than lead absorbed through the stomach is very interesting. We seem to have established the fact by experience, that if we keep lead out of the mouth and nose of our employees we will be free from lead poisoning — even though their hands are continually in contact with lead (both dry and wet) during working hours. We have long realized the necessity of enforcing the rule of wearing respirators in dusty places, as well as washing the hands and face carefully before eating.

We have been influential in procuring a sandpaper which will not be unduly expensive and can be used wet. We hope that it will be adopted by painters generally at an early date. Your finding that lead is relatively more dangerous when absorbed through the lungs will assist us in inducing painters to adopt wet sandpapering.[26]

We continued working with this problem until 1928. Much of our early work was done in the physiology laboratories at the school. Later we were working with patients, and the Massachusetts General Hospital gave me a little four-bed section in Ward B, their first research ward, which grew into their well-known Ward 4 when the extension was finished on the Bulfinch Building in 1925. Efforts to understand the chemistry, physiology, and clinical aspects of lead poisoning resulted in establishing how lead enters the body and why the air route was the most dangerous, how lead is handled in the body and stored in the bones, and how it is possible to extract it from the bones and in fact de-lead a person so that he is not subject to acute attacks in later years. We also found a quick and effective treatment for lead colic.[27]

As always, what concerned Edsall was the opportunity to do good work. And feeling as he did about the importance of the public health point of view, the creation of the new school was definitely something to be proud of, despite the sacrifices it entailed. As long as he was dean of the two institutions, he ran them in a unified manner, using his grants with his usual meticulous regard for the wishes of the donor but crossing the lines between schools with great freedom. There was also a unified point of view as to the teaching and research resources available, which enriched both schools, if it was at times the despair of comptrollers. In making sure that the school would have adequate Harvard support, the Rockefeller Foundation had required certain funds to be earmarked or kept in reserve which had a limiting effect on the Harvard Medical School for some years. As Edsall wrote to Abraham Flexner in 1922:

The Rockefeller Foundation was willing to aid us in adding to what we had done in public health lines so that we might have a real

School of Public Health only on condition that we (a) add a very considerable amount to what we were already doing and (b) guarantee to maintain ourselves, if necessary, activities that demand a large sum yearly and are now maintained on temporary subscriptions. Sums that would have sufficed in a way that would appear to me adequate, though restricted, in meeting the needs above mentioned, had to be set aside for (a) or kept from absorption into permanent budgets in order to be honestly able to meet highly possible decreases in subscriptions to (b).

The wisdom of undertaking the public health developments under such circumstances may seem questionable. It seemed so to me but it appeared on the whole desirable to risk difficulties in order to get live public health training and research started, and the public need for this seemed to justify it also.[28]

An interesting description of Edsall's contribution to public health developments at Harvard was written by Cecil Drinker, who followed Edsall as dean of that school. This was written in 1945, from a perspective of ten years in office.

It might be said that lacking Dr. Edsall there would have been no Harvard School. Certainly the establishment of the School would have been greatly delayed . . .

First, and most important for subsequent events, he realized that public health in all its phases must be a field distinguished by productive scholarship if it was to have a real place in a great university . . .

While it is true that the gift from the Foundation stemmed from interests far beyond any individual, I am confident that Harvard University owed the prompt action of the Foundation to Dr. Edsall's position as a scholar in medicine and to his liberal and inquisitive interest in all that might advance the public welfare . . .

A quarter of a century has passed in the educational experiment, we, at Harvard, owe so entirely to Dr. Edsall. A new policy and new idea may guide us. But few who saw the start of health education can quarrel with Dr. Edsall's first conception and with his conviction that public health education must depend on close integration with medicine and engineering in order to succeed . . .

A graduate school in a university . . . should attempt the nourishment of leaders, should provide opportunities for advances in our subject through the production of scholars and scholarly work . . .

[Edsall] was a human, interested man, not the abstraction in administration he may have seemed.[29]

CHAPTER 19

The Deanship Becomes Full Time

In his first five years as dean, working at the job part time, Edsall initiated many of his most important policies, revitalized the school's financial circulation, and generally put his stamp on the developments of two decades. It was a tremendously productive period, but the load grew increasingly difficult to carry as he went on. He was professor of medicine, chief of service at the Mass. General and head of its medical research, dean of the Harvard Medical School, and, from 1921 on, responsible for the creation and administration of the School of Public Health.

Indeed he had not been dean six months when he realized that it would require a major revision of his own schedule. Either he would have to reduce his hospital responsibilities, or else give up his outside practice and receive a salary from the hospital, he wrote to the director of the hospital, Dr. Frederic Washburn, in May of 1919.

> The important thing to be considered [he wrote] is not my own affairs but the future of the Hospital. We have here at present very much the strongest and most productive group of men in Medicine to be found anywhere, I believe, and there is no reason why similar conditions should not rapidly develop in connection with the other services. The continuance and general development of such conditions is, I believe, the most important of all matters in providing for the maintenance and increase of the Hospital standing and leadership. But this demands unremitting attention and sympathy or the men will become dissatisfied and the whole organization will break down. Competition is extremely strong and with the large developments planned elsewhere it will be still stronger for some time to come. There is a need for wise and patient attention on the part of some person or persons who can give much time and thought to that.[1]

The MGH had recently appointed a committee on scientific and educational activities, with Edsall as chairman. Edsall considered that this committee might be the instrument to carry on the development of research, "if the members are carefully chosen with sole regard to their suitability, whatever their hospital rank, and the chairman is particularly carefully chosen because best qualified through training and judgment to develop that work." For this important position he proposed to Dr. Washburn the name of Dr. Roger I. Lee, recently returned from the war and then visiting physician at the MGH.

"My suggestion is . . ." wrote Edsall, "that we give every opportunity to Lee to develop that work and that, if it would further his success in it, an extra place be made on the Executive Committee and that he be put in it as the recognized guiding head of the various clinical scientific activities of the Hospital, if his other duties permit him to undertake this."

Edsall knew that he would regret the "rapid decrease of my close relations and influence with the large group of young men here," which had been "my greatest pleasure," but he felt his responsibilities as dean could not be slighted. He told Washburn how he had made proposals to the university of "rather broad and radical changes in policy," involving increased expenses, and had received wholehearted encouragement.

> Everything was agreed to at once [he wrote] and President Lowell has indeed gone beyond this since and has informed me that he considers me responsible for the wise expenditure of the Medical School funds . . . the School budget not improbably will soon approach half a million a year . . .
>
> I have a keen appreciation of the responsibility involved in making careful and sympathetic provision for their scientific and teaching activities and a very grateful appreciation of the cordial manner in which the University authorities have met my suggestions and have even gone beyond what I expected of them. It seems obvious that I must so arrange my other work that I can do as well as I am able in that, even though it reduces my attention to things that I have grown much attached to and that also offer large opportunities.[2]

Since he believed the trustees had no intention of putting him on salary, he realized that he would probably have to increase his practice in order to make ends meet. But to his surprise, the trustees voted him a salary as head of the medical research work at the hospital, and kept him as their medical chief for four more years.[3] His suggestion for putting Roger Lee in charge of research in his place had not been accepted.

We can see now that Edsall was trying to carry two full-time jobs and even for a man with his capacity the situation often seemed impossible. Early in 1920, he was writing to Christian:

> Repeatedly this winter with committees that the Dean must sit on I have had every afternoon given over to committee meetings, often two or three committees in one day . . . Attention to these and to other School matters has forced me in the first place to give up almost all my consulting practice (which I am still free to do) and it has even come to be the case that my work at the Hospital suffers seriously . . . As I see it, no one can under the present system do the work of the Dean in a way that is at all adequate and do anything else well, nor can he conduct the Dean's office in a way that will make it at all easy to avoid serious difficulties with sensitive or suspicious men though it is of course questionable whether the latter will ever be possible.

So far it had not occurred to him that the deanship should be a full-time job or that he might accept such a position. Said he to Christian in the same letter:

> I question whether any desirable man would want that job. Any institution as complex as this, with as large a budget and in the midst of important developments, can not be well managed however without a simple and easily worked organization. Frothingham remarked to me recently that his experience led him to feel that it is the most extraordinarily complex and cumbersome organization he had ever seen. I think he is about right.[4]

At this point his solution was to resign as dean, and Dr. Christian wrote him a warm and persuasive letter urging him to stay on. In the first place, he pointed out how well Edsall got on with President Lowell and with Dr. Walcott (president of the corporation of the MGH). Then he pointed out that there was nobody else. "There seems no desirable successor except Cannon & Peabody & the former won't and the latter shouldn't."

Christian went on, "The dean of the wrong sort can do a deal of damage that can't be repaired by the individual members. As Dean you are of more value to the School; as not-Dean you are of more value to the Dept. of Medicine & the M.G.H. The former is the bigger & the inclusive. So as I see it you should hang on." He further apologized for his part in creating a "sense of suspicion & antagonism," and went on, "There are a lot of us who believe in you & we ought to do more to help your work — let us make another try." [5]

With the addition of the School of Public Health to his responsibilities, Edsall was sure that something had to give, and in April 1922 he went over his situation in a six-page letter to his friend Dr. Locke.

I am pretty thoroughly convinced that I can not long go on doing what I am doing now without doing something badly and without reaching my limit nervously and physically. I just about sail along without taking on water most of the time and then some concentration of things or some unusual anxiety nearly capsizes me and then it takes weeks to get back where I can do things with any vigor or zest. In the last few months my circulatory apparatus has been very irritable and while I think it is simply functional strain it is obviously undue strain. I think I shall have to give up either the M.G.H. & the Professorship of Medicine or the Deanship. Much the easier life would be to give up the Deanship. I can not however escape the strong probability that I have very little further to contribute to the Professor job except to steady it and hold it together, both because things move so fast in medicine that it is a hard job for anyone to act as a leader at my age * and because in my case executive work has taken me from progressive contributory activities very greatly during the last ten & more years and almost wholly in the last four years, and the proper younger man would do it better. On the other hand, while it is trying work and in large part very boresome, the Dean's work provides opportunities that a man properly trained & with proper qualities can do better at my age than a younger man if he keeps himself in touch with the changing pulse of things and I feel sure that medicine (in the general sense) and medical education in particular needs . . . some continued line of guidance and development to give it a soundness and stability that it has nowhere reached in this country . . . If that can be contributed to largely it seems to me a larger service than one would render in charge of a department . . .

The Dean's work, looked at as I see it is a full man's job if it is to be well done in a subject that is developing so rapidly. The amount of

* He was fifty-two.

time that I have to give to University Presidents & Deans who come for rather extended discussions of policies and of personnel has grown very serious in the past years, and yet all such things are very important in placing our younger men elsewhere and in increasing the School's prestige in other ways; and the amount of time I spend with hospital managers is similarly serious, and yet obviously I can't put them off or slight them. And the very internal policies and activities of the School don't get done or get done in a very haphazard way unless a good deal of central thought and pushing is given to them.

The questions then are – should I personally do well to do the one or the other, and, if it should be the Deanship, how can the M.G.H. be arranged so that it will progress and not slump? There is serious danger there – more serious than anywhere in relation with the School – of a relapse into the unprogressive old methods unless a strong hand can be kept on things and I do not quite see the way out of that.

Further should I assume a full-time Deanship there would have to be a very large increase in the salary. However much I might be willing to do what is the right thing for the School, I have already cut out any opportunity to lay aside anything of consequence for my family and am now only a very little more than able to meet my obligations year by year in caring for the family, educating the boys and providing aid for relatives that I must continue or bring hardship upon them of a degree that I can not contemplate. So I can not plan any noteworthy further financial sacrifice. The Public Health School could properly make up a good deal of the difference needed, but not all, and the total needed, while what they are paying or offering at Columbia & a couple of other Medical Schools is so much more than they are accustomed to here in any Deanship that it would embarrass me excessively to propose it even though it means the Deanship of both the Medical School and the School of Public Health . . .

I anticipate being away – probably abroad – the coming half-year, so I can stick it out for next year, but I doubt the wisdom & even the safety of going beyond that.[6]

This letter shows Edsall's desire to simplify his life and arrive at an economic arrangement to give him an adequate income for his growing family. The problem of keeping the medical school on an even keel and keeping the MGH, his first love, improving, both absorbed an increasing amount of time and tact. He had the affection and trust of the staff at the MGH. The situation at the medical school was more complicated, and many men distrusted his motives and clung to their prerogatives without thought of the needed changes and improvements which might occur to the school as a whole.

One of the situations that led Edsall to become a full-time dean was the problem which arose over the professorship of pediatrics. The previous professor was characteristically a straight clinician interested only in the kind of food to give to babies and very much involved with the problem of calories. He did little investigation on his own. When the time arose to choose his successor, Edsall reached out for Oscar Schloss in New York.

Schloss, a brilliant clinician, was also completely dedicated to research. There is little doubt that it was this latter characteristic that made him beguiling to Edsall. At any rate when the offer was made, Schloss accepted and came on to Boston. From the beginning Schloss had a difficult time. A Jew and a very shy person with a complicated personality, Schloss found it difficult to adapt to the Boston environment. And no one made things easy for him. Schloss, for example, came from a very poor family, and in addition to supporting his own family he supported his parents. In New York he was able to do this with some ease because he had an enormous practice. He should have been able to do this in Boston as well if his income as professor had been supplemented by consultations, but nothing came his way because the previous professor of pediatrics had arranged it so no one would send him a consultation. It was an ironic situation because there is no doubt that Schloss was one of the best clinicians to reside in the Boston area. Then, although Schloss wanted to do research, obstacles were put in his way at the hospital. When the situation continued to deteriorate, Schloss decided to go back to New York to his old professorship which fortunately had been kept open for him. Edsall tried hard to persuade Schloss to stay but it was then too late.

In the end there were many in Boston who blamed Edsall for Schloss's departure, claiming that if he hadn't been too busy with the MGH and had devoted more time to Schloss he would have been able to keep him in Boston. Some people made a good deal of this incident, hoping that it would drive Edsall into resigning the deanship, but it had just the opposite effect.

As one would expect, Edsall seemed to have no real prejudices

— certainly socially he had none of those so prevalent in those years. At the Massachusetts General Hospital a single Jewish medical house officer had been appointed soon after Edsall arrived there, this at a time when new applicants for New York hospitals had to state their place of birth and religion. Such applicants were not allowed to take the examination if their answers on the application were unsatisfactory. Edsall could not countenance this and applications for examinations were promptly open without prejudice at the MGH, with the result that appointments were soon made without discrimination in most hospitals throughout the East, and this particular need for establishing sectarian special hospitals disappeared.

In general there was no longer social or religious bias in hospital appointments and the same has been true of appointments to faculties. Neither were appointments to the faculty at the Harvard Medical School governed by strong social influences when Edsall settled in the dean's office and was strongly supported by President Lowell.

The battle over the professorship of pediatrics was a crucial one, for the resignation of Dr. Schloss and his return to Cornell Medical School in New York established the need for a respectable reception for high appointments at the Harvard Medical School. It was thought that this unfortunate situation must not be repeated and that the administration must make sure that similar misfortunes would not occur again.

Once Edsall had become dean not only of the Harvard Medical School but the School of Public Health, it was clear that the post would become a full-time job for someone before very long. Edsall was not sure he would be the man, but he began to look for the practical means of supporting the post, and broached the matter to Abraham Flexner and the General Education Board. Flexner had asked him whether the medical school would be "satisfied" for a while, after one more major gift then under discussion, and Edsall had responded by showing urgent needs in a number of directions. His last point was the deanship.

Beyond this the only item of importance that I see at present is money to increase the salary of the Dean to nearly or quite a full-time salary so that he might relinquish other serious responsibilities . . . I am not *quite* sure how far this should go, but, whereas my impression was strong a few years ago that this would be unwise, I am thoroughly convinced now that what I have said above *is* wise and that much of the instability in medical education could have been avoided if . . . keeping the whole rather than its parts before the faculty had been rather generally stamped as an important thing . . .

I have not discussed this with the University authorities or the Faculty and they might not agree with my views. The effect of the present system of having two important positions held by one man is to have both ill done, and the trying work that yields least credit is the administrative so it gets less attention than its central importance demands.[7]

It had taken him a long time to come to that conclusion, and even now he knew he had a contribution to make either as professor or as dean. He did not seek an administrative post for any purpose of self-aggrandizement. He wrote to his friend Locke, "I do not want to be Dean unless the Faculty as well as the President think I should be . . . It would be a very unsatisfactory thing to give up other things unless it was thought I should be sufficiently *more* useful as Dean to make it worth while & my duty to do it."[8]

"The routine duties of the Dean would not interest me as an occupation," he wrote to President Lowell on April 11, 1923. "On the contrary they are merely tolerable as a necessary accompaniment of the interesting opportunities for developments."

He wanted to make sure before committing himself that he had the necessary agreement of the university to his general policies, which had already begun to take shape in the preceding five years. He wrote further to President Lowell:

In most important ways there seems to be agreement. Were that not the case I suppose this step would not be under consideration and as I stated to you I have had most grateful relations with you. In regard to educational aims I have not had any thought of a likelihood of any important differences of opinion.

I do not feel so clear about another aspect of the work, however, that might get embarrassing, namely, I do not feel sure that you and the Corporation consider me wise in having allowed the general costs at the School to become what they are and in having increased individual budgets of departments to the extent that I have; whereas I feel that it is well nigh certain that had they not been thus increased we should have had a number of disastrous losses, and I believe also that the costs are actually economically low now as compared with what we are competing with.

I believe that the conditions in Medicine have become much more changed, financially, within ten years than is appreciated by the Corporation or indeed by any except a few persons who have carefully looked into it in some detail; more changed I think than in any other branch of education. This is true not only in those places that are known to have received large amounts of money. It is true also in many other places including State universities. Our departmental budgets are, with the possible exception of two, lower than budgets that I am familiar with for the same sorts of departments in a considerable number of other schools, and in some instances they are very much lower indeed. This is true of those of our department budgets that have been decidedly increased in recent years, and still more true of those that have not.

There is a tremendous competition in Medicine in securing both senior and junior personnel. Even since I have been Dean there have been no less than nineteen instances in which men in charge of our essential departments have had opportunity to go elsewhere, usually to very attractive places, and with budgets decidedly larger than those they now have here, sometimes very much larger . . . The decisions of the men to stay with us have sometimes been made largely because they liked the general situation here, but nearly always there has been a large element of real loyalty in the decision and they deliberately made what they felt were sacrifices and some of them thought that I was pretty tight with the purse-strings as compared with what other institutions are doing . . .

In four instances in which we have asked men to come here as professors and they have declined I feel assured that an important deciding factor was what they considered the inadequacy of the budgets offered. This has been true repeatedly of junior men also.

The total costs of our school as compared with some others are significant likewise. Our total costs per student this year, last year, and prospectively next year . . . are about $1170 . . . In the three schools with which we chiefly compete the costs per student are from over $1700 to over $2,000 . . . In point of fact however these figures do not represent the real situation for those schools offer less varied opportunities to the student than we do. We have departments that they do not have, such as Comparative Pathology, Industrial Medicine, Tropical Medicine &c. These unusual departments cost us a very large sum each year all told. If in order to get a direct comparison with what the other schools mentioned offer we eliminate our unusual departments from the calculation our yearly costs per student would be about $870. This very large difference from their $1700 to $2,000 is explained by the very large budgets they provide for their essential departments . . .

A number of our departments now have amounts that are in my opinion fairly adequate, even though some similar departments elsewhere have decidedly more. There are however some of the most important that can not now do effective work, and especially can not get junior men when desired from elsewhere. Several of them must within a few years have considerably more. For example if I leave the medical department at the M.G.H. there will be available for my successor all told from the School and hospital together about $22,000. Nearly half of this is provided by the hospital. One of my earlier assistants, now a professor of Medicine has $115,000, another $75,000, another $70,000. Several offers have been made to my men of budgets of $40,000 to $50,000 . . .

It has been exceedingly difficult for me to get along on that amount recently, however, and whereas I have several times wanted to get new young men from elsewhere or bring back a man who had been there I have been unable to do it . . .

Medicine at the Brigham is nearly as limited as at the M.G.H. as is Surgery at the Brigham. Surgery at the M.G.H. and at the City, and Obstetrics have hardly what can be called a budget at all . . . Pathology with us does more work (for our affiliated hospitals) than does any pathological department elsewhere that I know of, but our department of Pathology has the smallest budget that I am familiar with in a leading school . . .

What will remain from the DeLamar income after things that we are committed to are met will not be adequate and it will be necessary to have sympathetic cooperation in securing more funds. I am ready to do all that I can in this, but I have recently found myself blocked several times in trying to secure funds for special purposes and have found a belief that the Harvard Corporation considers that the Medical School will get along financially without special effort . . .

I do not of course think that money is the only determining factor in success, but when it is necessary it goes far to determine success. The money that has come to the Medical School in recent years has been chiefly without effort on the part of the University and it may readily prove in the next few years that whoever is Dean will be unable to carry out proper developments without active aid in securing money.

I have a strong repugnance for extravagant methods, particularly with trust funds, and it would be very trying to be Dean if I appeared to my superiors to be extravagant. But what I have pictured is, as I see it, what the situation is likely to demand if the opportunity we have is to be effectively and successfully met . . .

If the sketch that I have given seems a reasonable one and the other matters that I mentioned in the beginning are such that I could feel safe in undertaking it, I feel that I should probably do more useful work in the Deanship alone than by going back simply to the work at the hospital. If the sketch seems unreasonable or unwise in its general features I believe that I should be more useful at the hospital, for I have given the matter much thought for a number of years and the statement I have made is a mature judgment.[9]

It was a very difficult decision. As Edsall told Dr. Washburn, "I have never approached any question with so much reluctance. I have given very great thought to it and have consulted a number of friends whose opinions seemed valuable, not only those here, who might be biased, but also some elsewhere. I have met a very general feeling that my chief future opportunity for usefulness lies in the Deanship. I think myself that this is true."[10]

The response from the university authorities left no doubt as to his full support from that quarter. Lowell had written him, "I think it is absolutely essential that you should remain Dean of the School; that your place there would be far more difficult to fill than the one at the Hospital; and that, in view of the great development of our Medical School, it is much more important."[11] "I can have no grounds for indecision on that score," said Edsall to Washburn.[12]

One thing that weighed with Edsall was the fact that he had been growing away from active research in the preceding years. He told Washburn:

I am sure that I am less suited now to aid such an important clinic and to keep it active and progressive than I was a few years ago and I shall be still less fit for it in a few years more. A younger man, on the other hand, who had maturity and capacity would be better able now to handle present medical knowledge and progress, and he would become more valuable to the Hospital for years to come. So much of my time has necessarily been spent in the last ten or twelve years in organizing administrative work that I have not kept step with the rather dizzy progress of Medicine to the extent that is desirable . . .

I have become so affectionately attached to the Hospital itself, I have had such kind treatment from the trustees, and I have had such happy relations with and such loyalty from the men with whom I have worked here that it is a great wrench to contemplate leaving, more so than anything that I have had to do. No less than any of the things mentioned is the sense of loss in giving up the close association with you personally. Your counsel and your ready support in all things that could be shown to be useful and wise have been the largest individual item among the many things that have made the Hospital appear to me to have an actual as well as a potential attraction and power for good that are surpassed nowhere.[13]

Dr. Washburn replied, "It is easy to say too much under the circumstances. I will content myself with the statement that in my opinion, it is the most serious loss which the hospital could sustain. I shall feel keenly the personal loss of the stimulus of close contact with your high ideals and wise thinking." [14]

Edsall gave up his courses in the department of medicine, and his headship of the medical services of the Massachusetts General Hospital. Dr. James Howard Means succeeded him as Chief of Medical Services of the hospital and as Jackson Professor of Clinical Medicine. Edsall now shifted his base from the office in the Bulfinch Building to the dean's office in the administration building of the Harvard Medical School. As a doctor he no longer had an office anywhere and saw no more patients. He became the first full-time dean of the Harvard Medical School and of the School of Public Health.

CHAPTER 20

Advances in the Medical School, 1923–1928

WITH EDSALL settled full time in the dean's office, the time of crisis had just about passed. Not that there was any lack of effort and excitement. Large sums came in for the development of psychiatry, neurology, and ophthalmology. The new Lying-In Hospital rose at the far end of the medical school's green quadrangle, across Longwood Avenue. On the opposite corner was built Vanderbilt Hall, Harvard's first dormitory for medical students. The Rockefeller Foundation endowed the School of Public Health with another million dollars.

In Edsall's first ten years the medical school endowment rose from $4,371,000 to nearly $13,660,000, while that of the School of Public Health grew from nothing to $3,415,000. The custom of yearly gifts from benefactors of the two schools had grown to the point where these averaged the equivalent of income on another $3 million. Cooperating hospitals contributed another $100,000 a year in the form of funds for salaries and for research.

The budget of the Harvard Medical School at the beginning of Edsall's term of office had been about $270,000, and the school had been running at a deficit. In ten years' time the deficit had been wiped out and the budget had risen to $810,000. Counting gifts and income for both the schools, the medical and public health expenditures were over a million dollars a year.[1]

A great deal of this golden endowment came from the Rockefeller General Education Board and the Rockefeller Foundation. Without their support the Harvard schools of medicine and public health could not have had such a flowering. Earlier, while the General Education Board and the full-time question held the center of the stage, Harvard received relatively little, the major sums in support of full-time academic medicine going to Hopkins, Washington University, Yale, and Chicago.[2] In Edsall's earliest days as dean, he had a long, friendly, and frank discussion with Abraham

Flexner, by letter, on the knotty question of full time. Flexner, wholehearted in support of full time in clinical medicine, felt that the necessary reforms could not be made without a break with the past. Edsall described the more liberal point of view of the MGH trustees:

> It is distasteful to them to lay down laws regarding men's fees and activities and they have preferred to say "do what work you think is desirable" in the office and private ward and they do not care anything about the fees except that they are not exorbitant. I think a large proportion of people everywhere, including Baltimore, think that is the way in which things can be simply and safely adjusted . . . It is less a question of receiving the money than the question of dignity and justice, but it has some real importance. Personally I think that, having chosen men with purposes and ideals such as Longcope has, and there are a considerable and growing number of them available now, there would be very few who would care to have the getting of money trespass upon their intellectual activities.[3]

There was another aspect to this which came to have more importance to Edsall as the years passed. He wrote to Flexner:

> I am thoroughly convinced that we can not grow successive crops of good clinical teachers and investigators without actually seeing to it that they do, as they develop, get contact with the differing types of disorders and the methods of therapeutics that are used with the non-hospital class of the community. I think this more and more forcibly as I grow more convinced of the need of a full-time plan. In fact I know from personal experience that fine hospital internes and young instructors who have had only hospital experience with ward cases have so very distorted an idea of what good medical service to the community is that were I to choose between continuing the present conditions everywhere on the one hand or on the other hand having medical education entirely dominated by a group of men who had been brought up strictly in hospital wards, I know I should enjoy the latter more in my own work but I think I should feel that the other was right for the public.[4]

But the vision of ideal medical education which had enabled Mr. Flexner in 1910 to weigh American medical schools and find them wanting, much to the benefit of those which survived, led him to hold fast to a more ideal arrangement. He replied to Edsall:

> The full-time scheme is, in my judgment, experimental — a state-

ment which you also make. Instead, however, of giving the university the right to prohibit fees, after a more or less prolonged trial on the basis you suggest, it seems to me that we are more likely to work out the problems involved, if we begin with a definite limitation . . . I believe that the clinical men at the Rockefeller Hospital and the Johns Hopkins, who are most productive, feel that the full-time scheme, as operated in these two places, gives them a degree of protection and independence that they could not otherwise procure. The points involved cannot, I think, be settled until we have a group of men who have grown up as Palmer and Dochez are growing up under the full-time scheme . . .

I should like to see the scheme, in its most rigid form, tried out during the next ten or fifteen years in a half dozen institutions and then ascertain whether it needs modification . . . I suspect that, if the funds are adequate to give decent salaries to men working whether as chiefs or as assistants on the full-time plan, the outcome will be favorable to the more rigid conception.[5]

In rebuttal, Edsall stated, "In all the experimental work that I have had to do and have observed one of the most important elements is always flexibility rather than absolute rigidity."[6] He cited the mutual benefit derived by men of various types working together as he had seen them do at the MGH, and said:

I am hopeful that a plan such as I have outlined above may meet this; namely, that the full-time men would be kept free from the distractions of practice and the fee system but they would be kept in considerable touch with it by the very limited number of cases that they would see (without receiving the fees for them) together with their constant contact with the group of men whose interests still remain almost entirely in the hospital and research work but who do need to adjust themselves in their work to the conditions of private practice.[7]

By 1924 he could look back to his experience with various kinds of arrangements at the MGH, experience which confirmed him in his adherence to the less limited system. "Acceptance of complete personal responsibility in the usual and natural way with a few private patients in a hospital is much more effective in achieving the academic purpose in seeing such patients, than is the method that makes it purely a part of the duties to an institutional organization," he said to his colleagues in the Massachusetts Medical Society in 1924. "Skill in practise . . . is retained not by practise only, but by undivided and keen responsibility in practise." He went on to describe the various methods as they had affected him, which helps to explain why he retired from practice and went fully into administration at the time he did:

> Circumstances, not regulations, have led me through all stages of experience in this. I for some time did general practise, then for a long time only consulting practise, then a distinctly restricted consulting practise, then none. I believe, I trust with reason, that I became better as a consultant as I went on, but I am sure I became coincidently less and less skillful in the duties of family practise. Then in the several years in which other duties grew so pressing that consulting work, though continued, became in considerable degree a task rather than a highly interesting and alluring thing, I could clearly see that my skill, such as it was, decreased, not simply as a consultant, but even in teaching and research. I imagine that this is usual.[8]

From the time that he became dean, Edsall went quietly ahead smelting out what became known as the "Harvard system" of full time. Having tried his hand at two revolutions, he was content to follow an evolutionary course in carrying this policy forward, making changes department by department "chiefly when chairs became vacant and it became feasible to do so without conflict or unfairness."[9]

In 1924 the arrangements had reached a definite enough form to be committed to paper, and were accepted by the President and Fellows of Harvard College in November 1924:

> It is the understanding that when clinical teachers are given larger salaries because they relinquish active practice and give "full-time" to their Medical School and Hospital work, they are still at liberty to see and to receive reasonable fees from a small number of patients on the Private Wards of the hospitals where they serve, — the amount of this work to be so limited that, while it keeps them in touch with the relations of such work to their academic work, it in no way interferes with their duties in teaching, in administration and in the conduct of investigation. For exceptional reasons that may develop in particular instances, consultations may be held outside the Hospital.
>
> Since arrangements regarding the sources of the salaries differ with the several hospitals, the salary of a clinical teacher is held to be the

amount he receives as a yearly stipend from the Medical School and the Hospital together.[10]

By the spring of 1928, looking back on the accomplishments of ten years, Edsall permitted himself a cautious optimism on this achievement, and wrote:

"A great deal of the betterment in the tone of the School and in the interest of the students in their work has been due to the extensive establishment of our system of modified full time in a large group of the clinical chairs." He was careful to point out, however, that the Harvard Medical School had "always maintained in each department a group of clinicians in active practice, as an essential part of the proper training of students."[11]

By 1928 only three of the clinical chairs were still operating under the old method. Eight had become formally full-time chairs, including two professorships in medicine, those of James H. Means at the MGH and Francis W. Peabody at the Boston City Hospital; two in surgery, those of Edward P. Richardson at the MGH and David Cheever at the Brigham; C. Macfie Campbell in psychiatry; Kenneth D. Blackfan in pediatrics; Harris P. Mosher in laryngology; Stanley Cobb in neurology. Edsall referred to Christian and Cushing at the Brigham when he said, "the remaining chairs of Medicine and Surgery, while not definitely established on that basis, are conducted by the incumbents in large degree in the same way."[12] At least in the early days, their system had been closer to half- than to full-time.*

Much of the support for these developments as they were made, step by step, came from the Rockefeller boards, and Edsall was able to win such a measure of support in spite of Harvard's refusal to adopt the Rockefeller style of doing things. In one instance, Edsall made sure a newly-established professor was to have the privilege

* In 1923 Edsall and Christian had an enlightening exchange of letters when Christian claimed that he and Cushing had been Harvard's first full-time clinical professors, though not formally recognized as such by the university. Christian was understandably indignant when he asked his salary be raised from $2500 to $5000 on the part of the school (he had another $5000 from the hospital) in line with the other top full-time men, only to be told by the university that he was not a full-time man and if he wished to increase his income he could take on more outside practice. He felt that for ten years he had been carrying out their full-time scheme in every detail – except salary – and Cushing had done the same.

of seeing private patients although he knew it would scarcely be used. "I arranged the 'privilege,'" he wrote to Christian, "because I was dealing with the General Education Board and I did not want them to think we were bending the knee to their full-time plan."[13]

Practical experience eventually settled this vexed question. In his last Dean's Report, Edsall wrote:

> The dispute, which was long a hot one, as to whether it must be absolute full-time or near full-time, has long since lost interest with recognition of the fact that a clear acceptance of the objective is much more important than the details.
>
> The belief was expressed by even some of the ablest and most flexible clinicians that it would make "exotically" trained doctors. This might well have proved correct had the extreme view been followed and only full-time clinical teachers employed, but that view has now essentially disappeared . . . The devotion, however, on the part of the full-time staff of their chief effort and thought and time to the academic work has altogether transformed the character of the teaching and its acceptability to superior students, as well as the extent and effectiveness of the research accomplished in clinical departments . . .
>
> The cost of this development has been very great. It has been impossible to extend it to all branches, including the definite specialties, in most places . . .[14]

While Harvard had been working out this more liberal way of doing things, Washington University had been less happy with the strict full-time system. Dr. Dock wrote to Christian in 1922 after resigning from the department of medicine, that running his department on its 1916 budget had become increasingly difficult under the full-time plan.[15] Christian replied:

> It has always seemed to me very questionable as to whether the plan put in effect there was workable with any reasonable amount of money, or to put it another way, whether it was not an uneconomic way to accomplish a given job. My own feeling has always been that a different plan with more of the men on part time basis and the whole time men rather limited to the younger men in training produced more of a result with a given expenditure of money, even up to a very large sum, than the plan by which the bulk of the staff was on a whole time basis. I feel that it is better for the Chief to be on a part time basis and have outside remunerative contacts, though of course not too much private work.[16]

Raymond B. Fosdick, writing in 1962, noted that only two medical schools, Hopkins and the University of Chicago, had maintained the full-time plan as Flexner conceived it.[17]

The change in the relations with the Rockefeller came soon after Edsall was appointed as dean. In 1920 the Division of Medical Education of the Rockefeller Foundation had a new head, Richard Pearce, who had been Edsall's colleague in Philadelphia and was always a warm and understanding friend. Alan Gregg was Pearce's assistant director, becoming director when Pearce died in 1930. In 1923 Wickliffe Rose became president of the General Education Board. This ushered in an era in which Harvard was one of the major beneficiaries of Rockefeller grants, along with Columbia, Cornell, Tulane, Western Reserve, and Rochester.[18]

There was the grant for the Harvard School of Public Health, $1,785,000, followed later by another $1 million, $300,000 for the Lying-In Hospital, $100,000 for new laboratories at the MGH, $350,000 for psychiatry and neurology in 1923 and $350,000 for neurology in 1925, $40,500 to support the experiments with the tutorial system, $175,000 for ophthalmology, $22,500 for physiology, and other major sums.[19] By 1927 the grants from the Foundation had reached $3.5 million dollars,[20] while the General Education Board between 1914 and 1960 appropriated $1,393,000 for Harvard.[21]

Among other notable gifts was the one made in 1930 by the Rockefeller Foundation of $175,000 for research in physiology under Dr. Cannon, and for physical chemistry. This was a new departure for the board, as Edsall wrote to Cannon when he had the happy duty of announcing it to him. "This is the first time that the policy of the Foundation to make definite support of a leader has been brought before the Board. There have been repeated instances before of support of particular investigations under distinguished men, but the proposal in this case was to make a definite example of the policy to support a man who had made himself a leader . . . I am sure you would have been deeply gratified to have heard the comments made when this proposition was brought up." [22]

Dr. Cannon's wife wrote of the results of this gift:

If quantity of production is any criterion the Rockefeller Foundation ought to feel repaid for its gift, by the output of the laboratory in the years 1930–42, which consisted of eleven large volumes of collected papers by members of the Department.

The laboratory, at last adequately supplied with funds . . . was furiously busy from that time on. Studies of the sympathetic nervous system, the chemical mediation of nerve impulses, the discovery of sympathin, further divided into Sympathin E and Sympathin I, the significance of the emotional level, traumatic shock, unsolved problems in Gastro-Enterology, effects of sympathin and adrenine on the iris, the adrenals in hypertension, the problem of ageing, homeostasis in senescence, effect of emotions on digestion, the adrenal medulla, cortical responses to electric stimulation, and many more, resulted in almost three hundred papers by W.B.C.[23]

In 1922 the Rockefeller Foundation sent Edsall abroad to study medical education in Great Britain in comparison with our own, and a long report, full of sagacious comments on both systems, went to them on his return.

He paid particular attention to the quality of the honors courses in Oxford and Cambridge, with their use of the tutorial method, and their significance in medical education. He told the Association of American Medical Colleges:

One thing which impressed me very much indeed was that in every medical school I visited, some students stood out a great deal in their general mental poise and their attack upon subjects and their whole attitude toward the work, quite differently from the rest of the students. Whenever I inquired into them, I found they were always men who had been through some of the great universities, and in a great majority of the cases they were men who had gone to Oxford or Cambridge. It seemed that there was one thing which might readily explain this, and that was the influence of the kind of training that they had had with tutors in those two institutions . . .

In Great Britain this method is used in Oxford and Cambridge in connection with that course which men take there preparatory to taking medicine, that is the Honors School of Physiology (which includes bio-chemistry, and they usually get their anatomy there also). The men there seemed to me to get in that time a degree of comprehension and poise in the subject that is quite unusual in students of that age in this country. It seemed to me also that since it is not used at any further stage in the medical course in Great Britain, that I could actually see these results falling away from many of these men during their clinical training in Great Britain. Now, if it is of any use at all, (and it seems to me likely to be of a great deal of use) the time in medicine to develop the best there is in most of them, is more in the clinical departments than in any other. I don't think it has ever been anywhere used throughout the whole course and I am particularly anxious to see that it is done in the clinical branches.[24]

"I was greatly struck," wrote Abraham Flexner to his friend President Eliot after reading Edsall's report, "with the thoroughness and fairness of his observations, and with his desire to utilize hints derived from England for the improvement of education in this country. About the same time I had an interview with another American medical professor who had been in England a few months and who reported to me—to quote his exact words—'I didn't see a thing that was good.' There are none so blind as those who won't see!"[25]

In 1923 Edsall became a member of the Rockefeller International Health Board. In 1926 they sent him to China to the Peking Union Medical College (as told in Chapter 14). In 1927 he was appointed to the board of trustees of the Rockefeller Foundation, made a member of the divisional committee of the Division of Medical Education and of the executive committee of the Foundation.

In 1928 Edsall was chairman of the special committee in the Medical Education division, which was to consider a new policy in that field. He wrote the report which led the Foundation to shift its emphasis from raising the general level of medical teaching—a goal that had largely been accomplished—to creating optimum conditions for the best and most promising researchers (for example, the grant to Dr. Cannon a few years later).

Some of Edsall's fellow trustees jokingly remarked to him that they had made him a member of the board because they couldn't afford to have him off it.[26]

There is a story that once when Edsall was outlining to Pearce all the things he thought the Foundation should be doing (not in Harvard terms, of course), Pearce responded rather testily, "After all, Edsall, you've got to remember we only have two hundred and seventy-five million dollars."[27]

Alan Gregg told me once that Pearce relied on Edsall very much as a "trustee-adviser and critic." Gregg himself regarded Edsall as one of the best-informed and most courageous trustees.[28]

Frequently on his trips to New York for one or another of the Rockefeller meetings Edsall would stay with his close friends Carl Binger and his wife Chloe. The Bingers had provided a special couch that would fit him. In an ordinary bed he had to sleep folded up or catty-cornered and spent a miserable night as a result. Dr. Binger recalls how Edsall always wore Irish homespuns, never took a taxi, and never let a porter carry his bag. It was good that he had intimate friends to stay with, for he was often troubled about something he wanted to accomplish in dealing with all the powers in New York. The Bingers listened well and their advice was good.

During this time of development in the twenties, the life of the medical students was made easier and more fruitful. One of the most important advances was the building of their own dormitory, made possible in large part by the generosity and enthusiasm of the alumni and the leadership of Drs. Joslin and Rackemann, and finally by the generous gift of William K. Vanderbilt. Until the completion of the first phase of Vanderbilt Hall, medical students slept where they could find a rooming house, and ate where they could pick up a meal, often trying to stretch a tight budget in a very irregular fashion. The dormitory provided not only rooms and food, but opportunity to exchange ideas and to exercise. Medical students tend to be an ardent and hard-working group and their proximity to each other was a stimulating thing. All in all it was a healthier and better life.

In the twenties medical students were not famous for healthy living. Many of them abandoned regular exercise, stinted themselves on sleep, and ate in a nutritionally scandalous fashion. Meanwhile they caught one disease after another as they studied it—in imagination. A yearly physical examination had not yet come to be accepted by the medical profession. Many doctors thought it a waste of time to examine a healthy patient. A change was coming in this field, however. The President's Committee of Fifty on College Hygiene deplored the fact that only four medical colleges required regular health examinations of students. Dr. Roger Lee kept up a running battle on behalf of regular health examinations, and gradually Harvard moved toward setting up better health protection. It was shocking to pick some student in class to demonstrate a point in X-ray or fluoroscopy, only to find oneself confronted with a plain case of unsuspected tuberculosis.

In 1926 at last the whole student body of the medical school was examined at the Deaconess Hospital, and Dr. Lee reported to Edsall, "I think that the staff of the Deaconess Hospital were rather amazed at the rather poor general condition of the Medical Students and the heedless neglect of ordinary rules of health in many instances." [29] Fifteen per cent were referred to specialists with conditions to be followed up.

In 1927 Dr. Reginald Fitz became physician to medical students, and set up a system in which the third- and fourth-year men helped with the examinations, thus learning what a proper health examination should be. One year he reported to Edsall:

"This year all the boys seem to think that they need to have their eyes examined . . . Next year it may be their tonsils will get on their minds and they will want to see a Throat Man, or the fact that they are getting bald so that they will want to have special dermatological advice. I had forgotten what peculiar medical opinions medical students pick up. It is important for their peace of mind that they should receive the best possible guidance." [30]

After Edsall's time medical students came under the general university health plan.

During this same period the tuition fees were advanced. Knowing how this would add to the strain of paying for a medical education, Edsall provided that none of the increase was to be used in the medical school until the money had built up a considerable student loan fund. It was expected that this would take several years, but Dr. Frederick Shattuck stepped in to provide a large part of the $150,000 needed. This made it possible to aid a large number of students every year.[31] Shattuck also saw to it that this was called the David L. Edsall Revolving Loan Fund, and stated, "It is a pleasure to recognize, even inadequately, the service which Dr. Edsall has rendered the Medical School in the sixteen years of his connection with it, ten of them as Dean." [32]

Edsall was never the type of man to become a well-loved "Papa Edsall" to the medical students. For their day-to-day problems the students went to Dr. Hale, or to the School's secretary, Miss Louisa Richardson, and many of them thought of Edsall as a rather distant, even forbidding figure. But Edsall had a sympathetic understanding of their problems and fought for their welfare. At one point Dr. Christian complained about the way students were allowed to make changes in their schedules, and, for administration reasons, he asked for a three-week deadline on changes in third- and fourth-year assignments. Edsall replied, "I have been heartily opposed to rigid rules of any kind in the Medical School because we are dealing with a group of fairly mature and very intelligent men; and when they have real reasons for making changes, I believe that we ought to try to cooperate with them rather than treating them like school-boys." [33]

When he did see a student, Edsall had time to consider his problem. One of them told me how impressed he was, on being sent to the dean's office for something, to find Edsall reading a philosophical treatise and ready to talk to him with no sense of hurry.

This readiness to take the necessary time for important things, and not let time for consideration be nibbled away by hurry, was always one of Edsall's great gifts. He was a fascinating person to go to for advice and had a flair for straightening out an assistant or associate. I remember how he helped me when I was trying to decide whether to leave Harvard. I went out to the Middle West to look over a professorship of physiology after I had been in the department of physiology with Dr. Walter Cannon for three or four years. The position was never really offered to me officially, but I went out to look them over and to be looked over.

It was the middle of the summer vacation. The building appeared enormous and tomb-like, there were no assistants around, and the professor who was retiring had obviously petered out. There was no interest or enthusiasm in this department, nor any activity. I enjoyed my visit because of friends, vigorous men who were in other departments. Although the whole large building was devoted to physiology, there was a small amount of rather archaic equipment and practically no personnel to start with, and a large class was planned for, as one would expect in a state medical school. The whole department would have to be staffed and organized, with no inheritance from the past.

I devoted several days to this visit, and on my way home on the train I could not understand why I was so utterly depressed at such a splendid opportunity. So I went to Edsall. He had a wonderful way of sitting you down in a comfortable chair beside him and listening attentively. I knew he would listen to my problem seriously. We talked for several hours. He started by using some well-chosen phrases which started me talking and thinking about what this meant to my future. And before long, I was telling him what I wanted to do with my life and why I did not want to take this job. It became clear that though I loved physiology and really wanted to devote my life to research, I could not bear the idea of dissociating myself from sick people and clinical medicine. So we decided that this job was not for me and that I had to find a niche which would allow me to see patients and teach clinical medicine. Edsall made this very easy for me and invited me back to the MGH to make ward rounds and teach young doctors what we knew about physiology and disease. So at about thirty years of age, Cecil Drinker and I began visiting on the wards at the MGH — this was a departure for the MGH, for the official visiting men were older, over forty, and had devoted many years to an apprenticeship in the Outpatient Department.

I was lucky. To visit on the wards when my first house officer was older than I, was hard on me, but my next two house officers were superb, and we had a wonderful time talking about the physiology of disease. These two were Joe Stokes, now one of the outstanding pediatricians in America, and Jack Starr, who became the well-known professor of pharmacology, both at the University of Pennsylvania.*

When Edsall could advise young people about their work he was particularly happy. He often felt lonesome; he said people were

* On my first ward round I had a trick played on me. The men in the physiological laboratory, always practical jokers, filled my stethoscope with wet paper so that when I tried to listen to a heart I couldn't hear anything. I thought I might really have been away from clinical medicine too long, the way some people had warned me — until I got back to the laboratory and found out what they had done. I then spent the rest of that round talking about *why* things were happening, why did this case of pneumonia have such rapid breathing, what auricular fibrillation did to the circulation, various things that I had picked up in physiology. They were interested, as it suddenly opened a new approach to medicine. *J. C. Aub.*

prone to call on him only to ask for something — they did not come to visit him because of friendship or to pass the time of day. This was not true universally, of course. For instance, Locke dropped in frequently and was always warmly received, for they had a very close friendship. But certainly most people came to the dean's office to ask favors, something for themselves or their department.

Often in later years if Edsall was driving to Atlantic City for a medical meeting, he would invite me to keep him company. These drives were a particular pleasure for me, though Edsall's main insistence was on being punctual, a characteristic I always found most difficult!

In 1927 the Harvard Medical School developed teaching affiliations with another important Boston hospital. The Beth Israel was completing its move to the Harvard medical area, where it began seeing outpatients in the spring of 1926. The first joint Harvard-Beth Israel appointment was that of Dr. Monroe I. Schlesinger as Pathologist-in-Chief. The new hospital complex was dedicated in August of 1928, with Dr. Harry Linenthal, Edsall's colleague in the days of the Industrial Disease Clinic at the MGH, as physician-in-chief. Also in 1928, Harvard helped them to organize a medical research department, which was headed by Dr. Hermann L. Blumgart. Through the years since, many staff members have also been teachers in the medical school.[34]

When he sat down to write his Dean's Report in the spring of 1928, Edsall could look back over ten years of solid accomplishment. The great DeLamar bequest and other important gifts had brightened the financial picture. "These figures as to money are, of course, interesting chiefly in relation to what they have made it possible for the Faculties of the two Schools to do," Edsall wrote.

"The most important effect of all has been to change the atmosphere from one of uncertainty and apprehension and dissatisfaction on the part of many of the staff, to one of stability and confidence and reasonable contentment. There has, in the first place, been made possible the provision for budgets for the individual departments that make feasible reasonable salaries as against quite inadequate recompense, and reasonable provision for the cost of teaching and investigation. All this has involved nearly the whole of the organization."[35]

The faculty had been preserved against disintegrating forces. The plant, allowed to run down because of financial stringency, had been rehabilitated to the tune of $200,000. The School of Public Health had come into being, and there were new departments and new work undertaken in old departments. There were new relations with hospitals. There had been phenomenal growth in research. As to this aspect, Edsall wrote further in his Report:

Just before the disorganization caused by the war there were only about twenty of the senior personnel who were really trained for, and seriously engaged in, contributing to the progress of knowledge and devoting their major time to this and teaching, and only about the same number of younger men of that type on the staff, with almost none from abroad or from elsewhere in this country who were voluntarily getting training of that sort here. Contrasted with this, there were . . . two years ago, just under sixty senior men doing very serious investigation as well as devoting most, usually all, of their time to the School activities. There were more than that number of younger personnel on the actual School staffs doing research work of admirable type. There were twenty-nine Fellows here sent by, and the salaries paid by, organizations in various parts of the world, and there were others from elsewhere working here voluntarily. All told there were over 150 persons engaged in serious research and they represented all parts of this country and twenty-five other countries.

In summary he added:

This report has dealt almost solely with progress, purposely in order to summarize what has been . . . a critical period. It might readily give the impression that there is at these Schools some sense of completeness and of relaxation, but it should not.

Surveying the decade, it seems clear that the apprehension that surrounded the beginning of it is gone, so far as fear of immediate deterioration is concerned and that there has been progress. It is, however, never time for complacency or for rest from progress . . . Complacency or easy-going methods or lack of skill in the choice of new personnel or other such matters have, within a comparatively few years, transformed institutions from an admirable state into one of greatly lessened effectiveness and prestige. Unremitting care, especially in the choice of personnel, both senior and junior, and constant effort to take advantage of the new opportunities for service that now offer . . . are necessary if the Schools are to be in future of high credit to the

University. And there are many things still in unsatisfactory state, some of major importance . . .[36]

Nevertheless, his mood was one of confidence. Despite problems and factions, Harvard was a good place to work things out. Edsall wrote once to Zinsser:

"I have seen something of three or four places actively, myself, and I have never found one that wasn't associated with some difficulties and problems, but the thing that has always impressed me since I have been here, is that there is a consistent and loyal progress toward the solution of them. I have long since come to the conclusion that if there were no problems and no difficulties, that most of the fun would be gone." [37]

CHAPTER 21

Second Spring

For a number of years Edsall's marriage with Elizabeth Pendleton Kennedy had become more and more difficult to sustain. He made several attempts to persuade her to arrange for a divorce, but she refused. Finally he went to Nevada in 1929 and obtained it himself. Pen told her wrongs far and wide in Boston and Cambridge, saying that Edsall had only married her to bring up his children for him and now that they were grown he had thrown her aside. Eventually she went back to Millwood which was made over to her. Edsall's side of it was not heard, but his friends felt that he had suffered enough from his wife, and that now his children were of an age to leave home she was driving them away from him. It was this that caused him to take such an uncharacteristic step, one which in those more strait-laced days was even a dangerous one for a man in his position.

After years of increasing unhappiness in his personal life came a break in the clouds. In room 101 of the administration building of the Harvard Medical School sat the dean himself, and the assistant dean, Worth Hale, and a very remarkable woman, the secretary of the medical school (an official not to be confused with a stenographer) — Louisa Richardson. People who knew her said that she really ran the medical school, as far as keeping the whole complicated machine on the road. Proper records began to be kept, departments knew what their budgets were and how much they had spent of them, a friendly, calming, statesmanlike influence was at work in the day-to-day affairs of the school. Edsall found the support of a mind as broad-gauge as his own, and, increasingly, a sympathetic personal understanding.

Louisa Richardson came of a well-known Boston family, though she did not call President Lowell "Cousin Lawrence" as her predecessor, Elizabeth Putnam, had done. Her father was John Richardson and her mother Louisa Storrow Cabot. Born in 1884, she graduated from Radcliffe in the class of 1906, after going in for dramatics, heading the college literary magazine, and captaining her class and college basketball teams. After college she worked for the Associated Charities of Boston, handling probation work with girls and with families. During the war she worked for the Red Cross, and when that was over spent a year in the Medical Social Service at the Massachusetts General Hospital. At that point her cousin, Elizabeth Putnam, the Harvard Medical School's first woman secretary, was retiring to get married, and she was responsible for Louisa's start at the medical school.

In the midst of all this, Louisa also found time to study for three years at the Boston Art Museum school, and her hobby of portraiture continued the rest of her life. An excellent portrait of Dr. James Putnam, painted by her, hangs in the staff room of the Mass. General Hospital.

She was a charming person, with warmth and understanding for young people, a quality which comes out clearly in the little note she wrote for the fiftieth anniversary of her class in 1956:

> One of the chief things that stands out among the ups and downs of the last 50 years is the great increase in opportunities for women, and their great freedom in so many ways. Gone are the days when the undergraduates 'must wear gloves in the Square'; must *not*, in a dramatic production wear men's trousers; must *not* loiter around Harvard Yard. But there is a whole new set of Musts. She must know something of what is going on in the world; she must try to understand the people around her so that there may be peace on earth; and she must pick up her own clothes, otherwise they will stay permanently on the floor. I hold, as I am sure you do, that all these changes are very salutary, very developing, and make for a more useful and happier lot of people. And have we not seen in the younger generation a capacity and a readiness that are astonishing and admirable; a determination to think out what part they really want to play in the great arena before them, and then to go ahead and do it? It is all most encouraging.[1]

She was then 72.

Edsall and Louisa shared many interests, among them a love of music and a fondness for real walking. After his divorce, Edsall used to take Louisa walking on Saturday afternoons, carefully chaperoned of course by a party of friends. But both he and Louisa were very long-legged and would be off up one of the Blue Hills at a pace few of their panting friends could match. Mary Lee remembers how the party would catch up with them at the top of the hill, sitting among the blueberry bushes admiring the view, whereupon Edsall would clear his throat and say, "Well, Miss Richardson, shall we push on?" and off they'd go up the next hill. The Blue Hills south of Boston were a favorite summer stamping ground. In winter they would often go with a party to Fitzwilliam, near Mount Monadnock, and explore the woods on snowshoes.

Some excerpts from the little notes Edsall wrote to Louisa during the winter before they were married are revealing of the pleasures and concerns they shared.

Dec. 30, 1929

I have sat here happily listening to Beethoven, Bach, Cesar Franck, Brahms, Wagner et al. Now I have most of the things that I *deeply* crave, except some of the cantata music and some Tschaikowsky. And then there are a thousand others always! But I do see ahead, when we get settled, a real extravagance, namely an electric instrument, for they are ever so much better, nearly like the real thing. Mr. —— can get me quite a rake-off I believe, and I will get a cheap car next time I have to have one and balance my extravagance that way. I am a little like Geoff when I get "set" on the trail of Music. All my boys are devoted to fine music . . . At any rate it is a great resource when shut in alone.

Dec. 31, 1929

Mr. Kent * died last night, I believe of a stroke. I am not surprised. Ten years ago or thereabouts, he was in the Brigham under Francis Peabody who said he had so bad a heart that he thought he would never get about even then. I feel much grieved. I was fond of Kent and appreciated his stern devotion to the School greatly. It will be difficult to get as valuable a man. I ordered a wreath sent for the Faculty.

Jan. 1, 1930

The day began bright and cheerful and I likewise, but I got rather shot to pieces early by a telegram from Mrs. Dunham saying John's hand and arm had got red and swollen and he had gone in to the Presbyterian and they had kept him there . . . I went to the School

* David F. Kent was Superintendent of Buildings at the Harvard Medical School from the time the new buildings opened. His two daughters both married doctors. The 1929 Aesculapiad printed his picture as a salute to his twenty-three years of service.

thinking a telegram might come to me there and not get to me on a holiday since Kent is not there to see to it. I found Miss Kent on the job, brave and composed and find she was here yesterday. I got back to . . . lunch and was called to a "New York wants you." After much delay and feeling sure it meant serious trouble I found it was Dr. Auchincloss very considerately telling me that all John's flare-up had disappeared after a day of active treatment . . . it was *very* kind of Dr. Auchincloss to do that, but after Mrs. Dunham's telegram I felt, while trying to get connected, somewhat as if I were before a firing line waiting for the guns to crack.

Jan. 2, 1930

I went to Kent's funeral this morning and found Hale and Drinker there. The service (the first Catholic funeral I was ever at) was very impressive and much of the music of the mass very fine, and the organ played the Dead March in Saul as they brought the coffin in, with a detachment of marines and color bearers carrying the Stars and Stripes and the Union Jack, the coffin draped in another flag. As the service ended a bugler played that lovely plaintive "Taps" from a recess in the back of the Church and they marched out to the funeral march from Beethoven's "Eroica" Symphony. A beautiful appeal to the emotions must soften a tragic time. I believe one of the great sources of the strength of the Catholic Church is that they never neglect or belittle the things that make religious exercises seem solemnly beautiful. To sit in a great Cathedral amidst its beauty and hear a wonderful service finely done makes me feel pious for days after, and I still remember with much feeling all the effect of many services I heard years ago in England and on the Continent, the finest of all a Christmas Eve service in the Cathedral in Vienna in 1894, with two cardinals, many priests, an orchestra with the organ and singers from the Opera. As we walked out after an hour spellbound in the unheated church, shivering with cold my companion said "If I ever join any Church it will be the Catholic, it gives you most of what you want." — and he belonged to a Catholic family but had never become a member because his reason would not let him accept what they demand.

I have written Hunnewell requesting permission to continue Mr. Kent's salary, just like a professor, for two months, and then I will see that Mrs. Kent gets some money from the Skillen Fund for the rest of her life. She looks very old and frail. Kent had served the School twenty-five years.

Jan. 4, 1930

I have been out this afternoon with Geoff for a brisk walk of an hour and a half in the Blue Hills . . . we came out on the road under Chickatawbut and then got lost on our way back so that we had to retrace our steps for some distance. It has been a beautiful day, the first since you left, most of them having been very wet and warm . . .

Jan. 8, 1930

I had a letter from Day, of the R.F., today telling me my last letter had given him all he wanted and that he and Pearce are "enthusiastic" over the industrial work of several kinds here, are discussing it with the officers and "hope that they will make it a substantial grant." That means, I think, that they will and it is likely to be $150,000. or more for ten years, and more to come in future, I think. Now, if we can get psychiatry cared for in a year or so, I think all the major money needs will be met, for some time to come, and attention can be devoted to polishing up what exists and steadying it.[2]

On May 3, 1930, Louisa Richardson was married to David Edsall very quietly in the garden of her mother's home in Newton. From May to September the two went abroad on their wedding trip, happily combining visits to eminent scientific men with sightseeing around Europe by car with Richard and Geoffrey. On their return they expected to settle in the newly enlarged dormitory of the medical school, Vanderbilt Hall, where an apartment had been built for the dean. They hoped to have a good time there, entertaining the students and making it a homey situation. But things were not to work out so smoothly.

Edsall's opponents, whether really indignant at the dean's behavior, or scenting an opportunity to force him to resign, set out to make trouble. They said the students were upset and indignant at the idea of having the dean and his new wife in their midst. This was not so. The man who was their spokesman was Cecil Drinker. He called a group of faculty members together and wanted them to tell Edsall that he ought not to live in the dormitory. This committee was like putty in Cecil Drinker's hands. Zinsser was even persuaded that this was the thing to do. Finally they seemed determined to write a letter to Edsall signed by all of us who were there saying that it seemed wiser to tell him this, that we didn't want to hurt his feelings, but that it was unwise for him to live in the dormitory with his new wife. I was very aroused by this. I couldn't let such a thing go by particularly when Edsall was on his honeymoon. I said it was a disloyal thing to do, and I remember Minot and Means agreed, and we managed to arrange that somebody else who was much less official would write and say that it was wiser not to live in the dormitory. Bliss Perry wrote a tactful letter and the whole problem reached them in a much less painful way. In July Louisa wrote to her mother:

"I continue to be reconciled to not living in the 'Masters House.' D. had a nice letter from Mr. Lowell—and he, (Mr. L.) is arranging to have it fixed up for students rooms—so that's the way that is. Tell Charlotte and Jack.* He said *he* had heard that it was the *students* who did not want to have the Dean living there and keeping an eye on them."[3]

In Paris they spent considerable time snarled in the red tape over drivers' licenses, and Louisa sent post haste to her mother for a new birth certificate and marriage certificate.

In Stockholm, Edsall wrote to his new mother-in-law, they had an amusing time. Edsall had been seeing men throughout Scandinavia with the idea of possibly getting the best to come to Boston, and in Stockholm he asked the hotel porter to call up Professor Dalén, the professor of eye diseases, and say he wanted to see him. Edsall wrote:

I was told he was on the phone so I started talking and then another man said Dr. Dalén could not speak English but spoke German. We talked German a few minutes and he could not understand what I was after so he came to the hotel and walked right into the room where Louisa was napping on her bed. He was very cross-eyed, which I thought odd in a professor of ophthalmology! After some minutes of German labored explanation on my part his face brightened and he said I had the wrong man — both have the same name . . . but the portier had got the one who simply has a "bath-cure." I got the right one later and had a nice time with him.[4]

In Copenhagen, as Edsall recounted for his mother-in-law, they had a delightful time with two other men he had hoped to meet:

Prof. Krogh, one of the most distinguished physiologists in the world and Dr. Madsen who is at the top notch in public health work. The latter asked us to lunch at once and showed us his laboratories

* Her sister and brother.

which are perhaps best known of any of their sort in the world. He is a rarely charming and cultivated man and his wife, almost 45, one of the handsomest and most gracious women I ever met. Next day he took me for a drive and tea out opposite the island where Tycho Brahe (in "Watchers of the Sky") did his astronomical work, and on the road to Hamlet's castle. Meanwhile he told me in a fascinating way: what I had never understood before – why Denmark is preeminent in many ways in public health. He attributes it to a very interesting foundation – that for about 100 years they have given adult continuation courses to the farmers (it is overwhelmingly an agricultural country and the finest dairying country in the world) and the courses are in history, literature, art etc. So the farmers become more educated and intelligent in a general way, and want intelligent care of health. And then he says bacteriology clinched it for it was easy to show them the effect of bad bacteria on milk and dairy products and thus on their pocket books, and hence their influence on human health. So their infectious diseases, with ably managed health administration, have been vastly reduced and in most of them is the best record anywhere, and their obstetric mortality is *the* best and only about one third the average in the U.S.A. So instead of being a *system* that does it it is a long education of the people and cooperation on their part.

Krogh and his wife gave me tea when I called there, and when he found Louisa was with me he came next day and spent the afternoon in showing us the magnificent ethnological collection from the stone, bronze, and iron ages in the National Museum (in which he is much interested) and he and Mrs. Krogh gave us a big dinner party with four or five distinguished medical men and their wives and afterward he showed us his laboratories. I was glad to have Louisa meet these people so pleasantly. They are two of the most important scientific men in Europe.[5]

In Munich the chief attraction was the Institute for Psychiatric Research, "which I had much wanted to see as the place that beyond all others works on problems in Psychiatry in ways that really promise *some day* to give sound progress of knowledge."[6]

Once Geoff and Richard had sailed for home after some rather strenuous sightseeing, the parents went off to Beddgelert, a little place in Wales Edsall had remembered from twenty years earlier, Here they stayed at the Royal Goat Hotel, walked several hours every day, napped, devoured large and delicious teas, and generally relaxed.

"I have been having more rest and relaxation than for years," Edsall wrote to Locke. "It is the most charming place for walking I ever saw, I think . . . We have not attempted the harder mountains but have clambered about daily for two to four hours, rewarded every few minutes by a new lovely view."

At sixty the mountaineer was still happy afoot. He added in a postscript to Locke: "Don't think I am not in good condition — I only want to continue the good work. After years of 4 or 5 hours of uneasy sleep, I am putting in recently 6 to 8 a night and an hour or so during the day & Louisa terms me now 'the eighth sleeper,' and I hurry up these young mountains, that are pretty steep, for an hour or two with much enjoyment & stop then only because of a cautious disposition."[7]

They cut their trip short because of some important fall meetings of the Rockefeller Foundation and various private concerns, but Edsall had no particular wish to get back to work. That fall they went to Greensboro, Vermont, and to the North Carolina mountains. After they came back to Boston, they lived for a time with Louisa's mother in Chestnut Hill, an elegant old part of Newton. The house's Victorian porches gave a burglar a chance to creep in one night, and Mrs. Richardson woke to see him going through her bureau. With that moral courage so characteristic of the Boston tradition, she wasted no time in leading him to see the error of his ways. "Young man," she said, "have you a mother?" But the burglar did not wait to hear more. He gagged her with a stocking and made his escape.[8]

Milton was still Edsall's favorite town in the Boston suburbs, and 1931 saw them settled in the Alexander Forbes house for the winter. Summers they spent in Greensboro. Greensboro had been a favorite summer retreat for the family ever since 1923 when a visit to the Bliss Perrys had revealed the beauties of northern Vermont. In the fall of 1927 Edsall decided that he wanted a house of his own there and he and John went looking for a house to buy. High on the side of a rugged hill overlooking Caspian Lake they came to a farm with a house and 110 acres of land. The barn had just burned down and the farmer had decided to sell the place rather than rebuild the barn.[9] Thus was Edsall introduced to the never-ending pleasure of remodeling an old house to suit his desires. They saw a great deal of the Perrys, who had been summering in Greensboro since 1896, and whose house was only some three miles away in the village. Now the Perrys, the L. J. Hendersons, and the Edsalls were all residents of Orleans County, on Vermont's northern border; Henderson's place was on Lake Seymour, away to the northeast in Morgan.

CHAPTER 22

Renewed Crises, 1928–1933

In 1928 it seemed as if the Harvard Medical School would progress steadily onward and upward, just as the country itself was spiraling onward and upward and not yet beginning to feel giddy. In the medical school it was apparent that even Edsall's emergency measures at the beginning of his administration had been well taken with a long view in mind, and an enlarged and strengthened medical school had come into being.

Nevertheless conservative Harvard was continuously faced with competition from its rival medical schools, where on the whole money flowed a little more freely than it did in Cambridge. In the spring of 1929 Lowell felt some concern that the medical school salaries were overtopping the Harvard maximums. It was Harvard funds he was particularly concerned about, and of course where men were paid from a number of different sources, the Harvard portion of their salaries might appear very inequitable. Edsall pointed out that one had to be concerned with a staff member's total salary, which might come in part from the medical school and in part from the hospital where he did his clinical work. Once the principle had been established of counting the whole of a full-time teacher's salary, it was possible to maintain a rough kind of justice between one and another. However, not Harvard, but competition from rival medical schools determined the level of salaries, and, said Edsall, "The conditions in other Medical Schools are extremely pressing."[1]

In writing to President Lowell in May 1929, Edsall listed seven full professors with salaries of $10, $11, or $12,000 while the Harvard ceiling was supposedly $10,000; he listed lower ranking salaries also out of line with the university. "My effort has been to meet the situation as regards competition from elsewhere, and the competition against having men go into practice with as little cost as possible, and as slowly as possible," Edsall told Lowell. "In *all cases*, we pay less to any of our Professors, Associate Professors or Assistant Professors than a considerable number of others of the most important places are now paying."[2]

Then Edsall looked back over the rapid rise in salaries since he had been dean:

> You will remember that some years ago offers of $15,000. a year came from the Mayo Clinic to two of our men,* and there was a good deal of doubt as to whether we ought to meet it by a raise in their salaries here well above the established maximum. I advised in those instances that it should not be done, particularly because the Mayo Clinic is not a strictly academic organization, and I felt it would be hopeless to attempt to compete with extra-academic salaries, and that we should have to take our chance of the loss of those men, grave as that would have been. They did not go, chiefly because it was not an academic situation there. Those salaries seemed quite extraordinary then. In the interval there has been an enormous change in the actual academic situation as regards salaries elsewhere than here. Some of this, I am quite sure, is wholly desirable, and I am quite sure that a scale of salaries larger than those customary in Harvard College or in any other colleges is inevitable and is desirable. As I have said to you before, however, I am far from sure that some of the rather extreme things that have been done and are being planned are desirable . . . in all cases we are giving lesser budgets to our departments than some others of the most important places are giving to the same departments, with the exception of one or two departments here for which we have received special money (like the Neurological Department), which particular departments have never been developed elsewhere except as very trivial affairs . . .[3]

The salary question was one of the most difficult problems that Edsall faced as dean and put him under continuous pressure. As causes, he cited to Lowell "the enormous lot of money that has been given for Medicine, and the perfectly extraordinary developments that have occurred in a great many places," conditions which did not prevail in other university departments. There were not enough able men to go around and as a result, "everybody is seeking the obviously valuable personnel." Harvard had been fortunate in not sustaining any crippling loss of men, but Edsall noted the difficulty of getting men to come to Harvard from elsewhere. He told Lowell:

* Cannon and Folin.

In one instance (Dr. Zinsser) we got a Professor at $2,000. less than he would have had if he had stayed where he was, but this was simply because he happens to have a comfortable personal income and preferred to be here. Several others have declined full Professorships, and mainly because of questions of salary or budget.[4]

He was even more concerned at the difficulty Drs. Minot, Means, and Christian reported, when in at least eight instances they had been unable to get "very important young men" for the department of medicine.

To indicate what is being done elsewhere, let me say that at Hopkins, having started at $10,000. salaries for Professors, they have gone up to $12,000. and now to $15,000., and are paying proportionately for younger men. At Columbia, Dr. Walter W. Palmer, as Professor of Medicine, gets $18,000., and his first man gets $12,000. and I believe the third man gets $10,000. Dr. Sturgis went from here from an Assistant Professorship to a Professorship in the University of Michigan at $15,000. with, I think, the privilege of some practice besides. Dr. Wearn is going from here to a Professorship in Cleveland, and I am told at a salary of $17,500., and I know with privileges of practice. Chicago University,* Washington University (St. Louis), to some extent the University of Pennsylvania, Cornell and other places are doing similar things . . .[5]

Edsall's advice was to put "as much back pressure" on things as possible and not attempt to compete with the tempting salaries being offered (he had heard plans for $20,000 salaries for heads of clinical departments and $25,000 for deans). However, drag his feet as he might, he could not let conservatism in salaries damage the school. "We cannot hold to a salary scale that is altogether disproportionately below this," he wrote to Lowell, "or we shall entirely disorganize what is the essential in any organization — namely, the personnel. I can state with great confidence that if it had not been for our above-mentioned increases beyond the College scale of salaries, we should have been unable to fill with creditable men most of our important clinical chairs, and should have been unable to hold almost all of our really important younger men, and I am entirely confident that that must be our guide."[6]

Edsall recognized an inflation of values due to increased money

* The University of Chicago Medical School.

available throughout the field, and ascribed a "noteworthy part" to Abraham Flexner's support of his full-time ideas "at all costs," but he felt that much of the increase was based directly on the competition of practice, and of course more money was also available there. "On the whole," he wrote, "I think our men here have been extraordinarily loyal and unselfish in their attitude toward it." [7]

The kind of standing for the Harvard Medical School which Edsall was striving to maintain, meant not only a school strong in itself but one which spread its influence across the country, and (particularly in the public health field) throughout the world. In 1929 Abraham Flexner asked him for a list of persons who had graduated from the Harvard Medical School and had attained professorial rank elsewhere. It took a while to assemble the data, but by the following April Edsall sent Flexner a list including 67 graduates and 175 who had had a substantial part of their advanced training at Harvard. Flexner responded:

"My conscience smites me when I see the labor which my request imposed upon you, but I confess that I am delighted to have the statement which you send, for it bears out the impression which I had otherwise obtained that Harvard Medical School has been a very substantial contributor to medical teaching and research, especially in recent years." [8]

The list was indeed impressive. Of the sixty-seven Harvard graduates (including a few university scientific graduates) there were thirty-five full professors in such schools as Johns Hopkins, Yale, Columbia College of Physicians and Surgeons, Michigan, Western Reserve, Chicago, and Peiping Union Medical College, as well as another thirty-two assistant and associate professors. Men who were not graduates of the Harvard Medical School but had trained at Harvard, Folin's men, Cannon's men, Wolbach's men, numbered 175, of whom 97 were full professors.[9]

Another indication of the school's standing was shown in the changing view of students looking for the best medical education. Edsall wrote:

> In consequence largely of the increase in the quality of the personnel and of the progress in teaching and research, the School within about fifteen years has quite obviously altered from one that was chiefly a local school to a school of national prestige and influence. In 1910,

65 to 70% of the men were from local sources in New England or very near by. In 1930, the figures had changed directly so that about 67% were from further away. There has been a special increase in persons from the South, the Southwest, and particularly from the Pacific Coast and the Rocky Mountain states, and the average quality of the men coming from a long distance is higher than the average quality of the men coming from near-by, which appears to be because the most able and determined men are, by and large, those who go to any distance and any expense to get what they want.[10]

One of Edsall's most forward-looking ideas was his recognition of the importance of prevention in medicine, an idea which had taken shape in his work on industrial diseases and had permeated his general outlook through the years. It was not mere face saving that led him to take the position of Professor of Preventive Medicine in St. Louis. At Harvard, he saw courses set up to teach preventive medicine in the School of Public Health, but he wanted to awaken the whole of both schools to a sharper awareness of the subject. In the late twenties, with Rockefeller assistance, he tried some new methods.

One step was the establishment of clinical exercises in preventive medicine for the third-year students, developed from the work of Dr. Waller S. Leathers when he was Professor of Preventive Medicine at Vanderbilt University in Nashville, Tennessee. For the year 1929–30, Edsall asked me to organize a series of clinics, leading off with a general one by Milton Rosenau on preventive medicine and hygiene, and running through thirty more sessions. Diphtheria was handled by Drs. Blackfan, Zinsser, and Benjamin White * from the point of view of diagnosis, treatment, prevention of spread, duty to contacts, etc. Three sessions on mental hygiene were conducted by Dr. Macfie Campbell on the psychoses, Dr. William Herman on neurosis, and Dr. William Healy on modern methods of studying personality. George Minot spoke on Prevention of Defects due to Faulty Diets. There were clinics in such things as typhoid, poliomyelitis, tuberculosis, fatigue, and industrial diseases (including a visit to an industrial plant with Alice Hamilton).

* Director of the Biologic (Serum and Vaccine) Laboratories of the State Department of Public Health.

Richard Pearce asked for a paper describing the new course for the Foundation's *Methods and Problems of Medical Education* (1930) and Edsall and I wrote it together. In closing we stressed the change in point of view which the clinics symbolized:

The clinics should include the clinical aspects of disease as well as stress prevention and treatment. In its final form the course will probably serve its greatest purpose in four aspects:

1. In infectious diseases there will be taught not only the clinical manifestations of the disease but also at the same time the prevention, public health aspect, and immunological aspects of the disease. In one afternoon, then, the diverse problems in one disease will be combined by a group of teachers.

2. The relationship of the environment to the psychology of the individual can well be taken up; these clinics were particularly successful this year and this is a field in which third-year medical students particularly need instruction.

3. It is an excellent opportunity to discuss industrial hazards and diseases as they influence the individual worker. This aspect of disease is far too little emphasized in the usual medical teaching.

4. The food deficiencies and dietary abnormalities should be seriously considered and in this group not only the usual abnormalities but also obesity, undernutrition, and vitamin and inorganic salt deficiencies should be included.

These four divisions should properly be given as an elaboration of the regular medical lectures, the diseases as a whole should be discussed, and the problem of prevention and environment stressed.[11]

Closely related to this project was the outlining by each department of the preventive aspects of its own subject. The resulting outlines were combined into the working draft of a book, *Synopsis of the Practice of Preventive Medicine*, and the whole was intended to develop into a handbook of prevention in medicine.

Edsall's hopes in this matter were shown in his preface to the book, in which he traced the growing importance and the growing possibilities of prevention throughout the medical field, and noted:

There has been in most medical schools little progress in altering the general influence of the teaching from one that has curative measures alone as the dominant object in view . . . Only a small proportion of those who are now engaged in the practice or the teaching of Medicine had in their training any systematic instruction in even the principles of real Preventive Medicine . . .

In this Medical School during the past fifteen years considerable attention has been devoted to the preventive aspects . . . Heretofore, there have been authoritative books covering the scientific and public health aspects of the subject but the practical applications have never been put systematically together . . .

We undertook to get together a brief summary of the pertinent things relating to prevention that each department felt could and should be taught in that department . . . It provides essentially a small, collective text-book of general practice of Preventive Medicine, and is intended to further the permeation of the whole curriculum with the atmosphere of Preventive Medicine in equal strength with that of curative medicine.[12]

The little book appeared in the fall of 1929, interleaved to facilitate the expected revisions, and was distributed to the heads of departments in medical schools throughout the U.S. and Canada before anybody quite realized that publication had been premature. The Harvard Medical School blushed a collective crimson. Some departments were faultlessly represented, but others who had made a rather hasty response to the original questionnaire saw their outlines appear in print without their having had the opportunity of revising the proofs, and at least one department which had made no response at all found that the chapter attributed to it made a travesty of fifteen years of medical progress. Some departments had simply been omitted.

Reactions ranged from pain ("One could hardly have compressed into four pages a more misleading impression . . .") to outrage ("Mediaeval and pernicious"), and Dr. Kenelm Winslow in his review called it "altogether too sketchy, incomplete and unsystematic . . . a collection of rather casual, random, incomplete suggestions from twenty different departments of the medical school, none of which was sufficiently interested to see that the result reflected credit upon the institution."[13]

Edsall replied in his dignified way to angry letters, thanked their authors for helpful suggestions, and went ahead with plans for an improved edition. He wrote to Christian in March, 1930,

> I am just about to send out some material about a new edition . . . have been very much astonished at the way that the editing of the previous edition was carried on, particularly in view of the fact that I was, I thought, clearly informed that every department had gone over and approved of the final form in which it was printed before printing.
>
> I felt disappointed a good deal in it myself . . .
>
> The new edition will be without a definite editor and with the responsibility thrown flatly on the departments.[14]

A committee of the administrative board, on which I served very unwillingly, worked out some plans for revision, but in the end the decision was to drop the whole thing. Cushing with restraint called it "this unfortunate book" and hoped that it would be forgotten as soon as possible.[15]

Cushing and Edsall had started out with high respect for one another when they had both been newcomers to Boston, but once Edsall became dean, their relations were almost certain to become strained. Cushing was not a man to take kindly to deans. Moreover, in this case there was an added difficulty, since Cushing felt he had every justification for considering himself the one head of the surgical department. Edsall was equally certain that the other branches of the surgical department were not subordinate to Cushing. At times they agreed—for instance in their broad view of medicine. Cushing wrote Edsall once, "I agree with all you say except that I don't like to have you feel you are any less interested in surgery than in medicine. Personally, I have been fighting for years to make it clear that there is no essential difference between surgery and medicine, even though I do not feel that the internists in general care much for this attitude." Responding to some proposals of Edsall's on developments in outpatient teaching, he replied (in the same letter), "I am very sympathetic with the program," but continued with his characteristic snap: "Don't for heaven's sake arrange to have anything of this sort done merely in medicine. It is my constant complaint here in regard to Christian's attitude that he goes ahead on his own lines without realizing that shifts of program, if for the better, can be made much more effective if they concern both the major clinical branches at the same time."[16]

But more typical was a state of open warfare, which broke out at times over very trivial matters. Locke described a time when he was in Edsall's office and Cushing came storming in and abused Edsall because some equipment in the Brigham operating room had not been properly installed. Cushing was very abusive and after he left, Locke asked Edsall why he had taken such insolence without a word of reply. To this Edsall only shrugged his shoulders.[17]

Edsall's very success in getting money created deep rifts with Cushing. When Edsall received money from the Rockefeller Foundation, he quite naturally distributed it as was his prerogative. Whenever any of these funds went to the MGH, Cushing always raised his voice in opposition and said it should go to the Brigham. It is true that the Brigham needed money at the time, but so did the other hospitals. Edsall was interested in creating a firm base for the development of clinical medicine and surgery at Harvard, not in advancing one hospital over another. Cushing always declared that the funds went to the MGH because Edsall was partial to the MGH. He was a tough antagonist and what he said could hurt.

Edsall was a very sensitive man, and I know that whenever he received a letter from Cushing during these squabbles it would depress him greatly (this was during the time when I was living in Edsall's house). Edsall would come home, start a game of solitaire, and begin mulling over Cushing's letter and then talk about his strategy in dealing with it. He was in continuous battle with Cushing, if not on the allocation of funds, then on appointments, and if not on these then on the problem of whether Harvard Medical School should have several hospitals or just the Brigham.

Two of the oldest named chairs in the medical school were the Moseley Professorship of Surgery (which was held for many years by Dr. John Collins Warren), and the Hersey Professorship of the Theory and Practice of Physic. Cushing held the one and Christian the other, although previously they had automatically gone to the Massachusetts General Hospital because of its antiquity, its large clinics, and because of its high reputation. While the Peter Bent Brigham hospital was being built, Cushing wanted these titles transferred to the Brigham permanently on the basis that it would be *the* university hospital. Dr. Edsall told me that President Lowell objected to this, for he also thought Harvard's strength lay largely in the wealth of clinical material which came from the use of many hospitals.

Cushing continued to nurse a grievance against Harvard on the score of its financial support of his work. His biographer Dr. John Fulton states that by 1931, the medical school contributed $11,600 to Cushing's budget of which $4,000. was supposed to support both the laboratory and the Arthur Tracy Cabot Fellow.[18] Fulton adds that Cushing's "research budget from the Medical School and the Hospital was not materially increased after 1912 beyond a few hundred dollars assigned for office expenses." Grants from patients helped things out, but Cushing himself paid for a good deal of the work which he thought Harvard should have supported. Says Fulton: "Cushing's annual expenditure to keep his photographic service, artists, and pathological laboratory going in the postwar years amounted to somewhat over $35,000, so that he was obliged to raise or pay from his own pocket a total of $30,000 each year. At no time did the Dean of the Harvard Medical School or the President of Harvard University raise a hand to lighten this burden." *

This was certainly the way Cushing *felt* about it, but such statements are clearly, to some extent at least, an exaggeration. We have, for instance, Edsall's statement in 1934 that "the funds relating to Surgery" had increased from about $20,000 to $150,000 during his administration. "Conditions in Clinical Medicine and in Surgery have been vastly improved," Edsall wrote, "by the persons in charge of those departments and very greatly increased amounts of money provided for this, and the numbers of accomplishments of real importance and of able persons at work have greatly increased." [19]

There is also the evidence contained in Edsall's letter to Charles P. Curtis at the Peter Bent Brigham. The date is 1929.

> In Surgery, the budget ten years ago was . . $14,243.28
> and next year the budget is . . . . 47,374.20.
> of which $25,117.58 (namely, more than half) is under Dr. Cushing's direct control, whether he uses it in the Brigham Hospital or in the laboratories of the Medical School; whereas the other two units have left $22,000. devoted to the Massachusetts General Hospital and the City Hospital aside from some sums that are temporarily assigned to the Boston City Hospital. The Massachusetts General Hospital also uses a considerable amount of its own money in Dr. Richardson's department of Surgery . . .
>
> As I said, there will have to be considerable sums added in Surgery

* From Fulton, John F., *Harvey Cushing*, 1946. Courtesy of Charles C. Thomas, Publisher, Springfield, Illinois.

at the Brigham Hospital and in the other surgical units . . . it is quite obvious that without enormously increased resources, I cannot add great sums to all these, and I cannot add disproportionate sums to any one unit without being unfair to the others . . .

I find various members of the force are constantly getting the impression that other persons are getting more money than they are.[20]

The surgical laboratory through the years remained a very sore point with Cushing and a fruitful source of misunderstandings. Cushing had done much of his brilliant early work at the Hunterian Laboratory at Johns Hopkins, and when he moved to Boston he specified that he was to have adequate space for animals. He was given a laboratory at the medical school in Building D, which while not the equal of the Hunterian seemed to satisfy him, but some evil fate caused a shuffle of school space in which Cushing was moved into quarters which never satisfied him thereafter. "I always felt it was a very small action on the part of the school to oust us from our quarters and then neglect us after we were moved," he wrote to Edsall in March 1923.[21]

In January 1925 Edsall saw an opportunity of getting an endowment for the surgical laboratory and asked Cushing for "a statement concerning the accomplishments within the last few years . . . and the arguments for the existence of such a separate laboratory . . . I have already had to meet questions as to the desirability of that rather than those associated geographically with the surgical clinics."[22]

Cushing apparently read this letter against the background of five years' accumulated wrath and leaped to the quite erroneous conclusion that now they were threatening to take his laboratory away from him. He held himself down long enough to outline the accomplishments of the Arthur Tracy Cabot Fellows, and then cut loose:

The Surgical Laboratory, too, is a place where students should be taught surgical therapeutics, in other words, where they should have their first experience in the making and repairing of operative wounds. To my great disappointment, this has been made practically impossible in the Harvard Medical School partly through lack of funds, partly through the action of the curriculum committee – in short, for the want of school support . . .

I may say that when I accepted a position here at Harvard, the only request I made was for better facilities for the care of animals in connection with the surgical laboratory . . . I little realized till I came here that it would be necessary to solicit funds from outside sources to meet ordinary running expenses. Had it not been that the A. T. Cabot Fellowship was at my disposal . . . the laboratory ere this would almost certainly have been closed . . . We are now asked what is the good of a surgical laboratory at all?

In fact, no opportunities or encouragement for surgical research exist here in any way comparable to those I enjoyed as an Associate Professor at the Johns Hopkins twelve years ago . . . It has been a constant struggle even to keep the laboratory running . . . Researches on animals must be carried out as humanely in the laboratory as corresponding operations would be conducted in the clinic. Like all surgical procedures, they are therefore expensive. For this the School provides a most inadequate budget . . .

With all the emphasis that today is laid upon scientific and laboratory work, for anyone to imply that the clinicians and perhaps above all the surgeons, much of whose energy and time is expended in laborious hours at the operating table, should receive discouragements from the School in regard to the pursuit of knowledge is to me amazing . . .[23]

At the same time that this storm was breaking over his head, Edsall was writing to President Lowell on behalf of a raise for Cushing and negotiating for other funds for the surgical department budget. The amounts were small, but they were in line with what was being paid at the time.

"Dr. Cushing, as you know, is one of the most difficult men in the group to deal with," Edsall wrote to the president, "and is constantly looking for slights and constantly tending to stir up a sense of injustice. I do not think that he does any particular harm in this way . . . Dr. Cushing's brilliancy and accomplishments and his value in the School are beyond question." [24]

Edsall thought it a matter of simple justice that Cushing's salary, then $5000 from the hospital and $3500 from the school, should be $5000 from each, top salary for those times. It was common knowledge that Cushing was very well off, but as Edsall reminded the president, this should not be allowed to influence their decision. When Dr. John Warren had asked to be paid what his position warranted, Lowell had advised Edsall it ought to be done "on the basis of an established standard for certain services rather than with any consideration of a man's private income." [25]

In 1932 Harvey Cushing turned sixty-three and reached the retirement age he himself had set. He had written to Henry Christian in the days when the two of them were forming the policies of the Brigham, "I shall be very glad, for one, to have it legislated to stop active work at sixty . . . I have an idea that the surgeon's fingers are apt to get a little stiff and thus make him less competent before the physician's cerebral vessels do. However, as I told you, I would like to see the day when somebody would be appointed surgeon somewhere who had no hands, for the operative part is the least part of the work." [26]

Cushing and Edsall were in constant conflict throughout Edsall's tenure as dean at Harvard. Nevertheless, upon Cushing's retirement Edsall did his best to keep him at Harvard. He offered him a chair in the history of medicine and made provision so that Cushing could see his patients at the Deaconess Hospital, which was near the medical school. Edsall thought that he had arranged for Cushing to carry on his brilliant work in surgery and in addition to have the joy of teaching the history of medicine at Harvard, a field Cushing knew very well and could do well, for he was a great authority.

Cushing's response to this was to say that Edsall would do nothing for him and that in consequence he was going back to his old alma mater at Yale where they were offering him a professorship (Cushing chose to be Professor of Neurology rather than Neurosurgery.) He even made it sound as if Edsall had thrown him out of Harvard. I have always found that myth hard to take because the reverse was true.

It is possible that Edsall and President Lowell, who assumed that Dr. Cushing was a very rich man, did not realize that he was feeling very poor after certain reverses following the 1929 stock market crash. Cushing's friend Lucien Price wrote in a sketch of him:

> His patrimony had been swept away in the financial dither of a banking house whose repute had been thought to be above reproach. "I had supposed," he writes President Angell of Yale, "that the trust fund that I had established for my family was as secure as anything could legally be. I now learn that the people who had my affairs in hand

have so manhandled them that I may have to start afresh as a wage earner, which is not so easy at my time of life . . ." In this situation, Mr. Lowell, then President of Harvard, offered him a Professorship of the History of Medicine without salary. Yale did rather better, a chair in its Medical School and $15,000 a year for salary and office expenses. Cushing, in his turn, did rather better by Yale, carrying thither his immense abilities and prestige, a unique pathological collection, and one of the great libraries of our time which that university has fittingly enshrined. Harvard fumbled and Yale fell on the ball.[27]

When Cushing retired from Harvard he took copies of all his records from the Brigham Hospital and transported them to Yale. His remarkable collection of pathological material on brain tumors was offered to Harvard, but difficulties arose over plans to house it in the Warren Museum, and no one at Harvard seemed alert and enthusiastic enough to make a fight for it. This too went to Yale and became the core of the Brain Tumor Registry. Dr. Louise Eisenhardt was the pathologist who long had charge of the pathological specimens from Cushing's operations, and when this unique collection of brain tumors went to Yale Dr. Eisenhardt went with them.

* * *

In spite of the manifold difficulties of his responsible position, Edsall always had the whole-hearted and generous support of the Harvard administration. In May of 1929, President Lowell wrote Edsall: "Your letter of the fourteenth troubles me a little because I do not like your thinking even in a tentative way of giving up the Deanship in any future that I can foresee. You have done for the School what nobody else could have done. You have put it upon a plane that any school in the world may well envy, and I think your work is by no means finished. On the other hand, if you want assistance from the routine work and in the relations with the hospitals, you ought to have it."[28]

Harvard's support is really something special. A particular example of it in my own career came at this time when Edsall appointed me to the Huntington Hospital to replace George Minot after Minot took Peabody's place at the Thorndike Laboratory at the Boston City Hospital. I didn't want to go into cancer research in the way it was being done in those days, which seemed to me a hopeless pathological approach to a difficult problem. People spent their time adding to the catalog of substances which could produce tumors. The American Society for Cancer Research then was dominated by pathologists, who spent their time talking about what a cancer cell looked like and how a diagnosis of cancer could be made pathologically. This was certainly not my idea of how to study problems of growth. What interested me was the mechanism of normal cell division. Edsall thought that the approach to cancer from the metabolic point of view would be profitable. He thought I would do this well. I had to go to the Harvard Cancer Commission and put the choice to them, and they approved it. It was a Harvard approach and they were willing to gamble. They realized that this was a fundamental approach, that might never get anywhere near cancer, but it was a main line of development all the same and they approved of it. As I look back that was the sort of approach that Harvard would approve at that time.

The cancer problem was like a pyramid. You could either study the bottom of it and try to find the fundamental error of metabolism, or you could try for the Nobel Prize, scoop off the top of the pyramid — a proper therapy for cancer. You had to make up your mind which interested you, and it was perfectly clear from the beginning that I was interested in the bottom of the pyramid. I wanted to know why things happened, what the mechanisms were. It was hardly considered to be cancer research in that day, but after World War II it became the generally accepted approach.

Throughout his stay as dean, Edsall seemed to keep continuously in mind the coordination of different trainings so that the medical school seemed to blossom, and the viewpoint of teachers and of clinicians seemed to broaden away from the narrow interests of clinical medicine characteristic of the physicians of the early nineteen hundreds. An example was his advancement of Edward Churchill who became John Homans Professor of Surgery at the MGH in 1931 and chief of the general surgical services there some years later, a position similar to the one Edsall had held in the medical services when he was there. Edsall had chosen well again. Churchill had the necessary interests and equipment for an academic surgeon, a particularly rare thing in those days. He had the skill, patience, and judgment which were necessary to develop new fields effectively and skillfully. He was one of the first academic full-time surgeons who did much to open the field of surgery to academic improvement, and he made thoracic surgery possible, just as Harvey Cushing had made surgery of the brain a possible subject of academic pursuit and intellectual development.

The following letter from Churchill's files shows the kind of problem Edsall faced in building new or stronger units. Such problems would have been encountered in some degree wherever independent units existed in the same department, but they were exacerbated in this case by Cushing's assumption that he ruled the surgical department. Churchill had sent in a draft of his first budget, including sums which had been agreed on, and the changes Edsall made are a lesson in the tactics of budget-making.

> In the first place [wrote Edsall], I have put down your first item not as secretary, but as secretary and technician. That is merely a tactical matter because I think it may be less annoying to Dr. Cushing than to put down the full salary of a secretary. It is understood that if the secretary needs to be raised above $1,500. that we will go as high as $1,800., but it is desirable to start with a smaller figure.
>
> You will note also that I left out the second technician at $1,200. I hereby, however, guarantee to you that we will provide that second technician if you find that it is necessary. That again was merely a tactical measure at the start because the other surgical units are not yet provided with two technicians, and I do not want to seem to provide numerous extra things in a hurry.
>
> I would also note that you put down your running expenses at $3,000. I indicated in the other letter that you will have the $3,000., but that $1,500. will be written into the Surgical budget for your use and the other $1,500. will be written into the Tutorial expense budget, but will still be available for you . . .[29]

In 1930 Edsall was saddened by the loss of a close friend, Richard M. Pearce. He had known for some time that his friend was working too hard, and indeed had given him some sage medical advice about slowing down. Pearce had replied warmly: "I thank you for a very lovely letter. It is worth having to give up something I very much wanted to do, to have your heartfelt advice and counsel." He promised that his next trip abroad would include some loafing, and remarked: "For a year I have been trying to work into a 10 a.m. to 4:30 p.m. day and soon will have it. So you see I am not quite hopeless." and finished, "It was good of you to write me as you did and I love you for it. Thank you, dear Edsall, you have done me a lot of good. Affectionately, Pearce."[30]

It is not easy to describe Edsall's affectionate character which endeared him to his associates. But it can be seen in his words about Richard Pearce at the memorial meeting held at the Rockefeller Institute for Medical Research April 15, 1930.

Edsall said of Pearce:

> Few persons are endowed with the power to hold others to high standards and yet to rouse only feelings of gratitude and affection, but his was no coldly puritanical spirit. In persons who lead others on, qualities that reach the heart are no less important than those that stimulate the courage. A rarely deep and manifest loyalty, straightforward frankness, keenly affectionate interest in those whose lives touched his, considerate tenderness in times of trial, all these he had in abundance for his colleagues, and equally for his subordinates. They were given freely and sweetly to his intimates. And I cherish with special tenderness those rare times when the sweetness of his nature gave place to fiery indignation as evidence came forth that baser motives were blocking high endeavor, or that individuals were suffering from cold, unkind, or unjust treatment. Such traits are not easily found, and when found we lean upon them.

It may well be that Edsall's high regard and affection for Dr. Pearce stemmed from a close resemblance in their characters.

He paid tribute also to Pearce's statesmanship with the words:

> It fell to my province recently . . . to attempt to understand, to comment upon, and to recommend action regarding the manifold activities of his division of work. I venture to think that no one had ever had occasion to seek, and no one ever had gained, the knowledge of medical and public health activities in teaching and research in the various parts of the world that he had secured . . . As the astronomer comes to think in terms of universes, whereas others of us think of portions of a small planet, so he saw the world as we see individual institutions.[31]

* * *

Felix Frankfurter wrote to Edsall in 1929 about a proposed institute for the study of crime, and in outlining ideas for such an institute Edsall distilled much of his knowledge about the workings of his own school. He was aware, for instance, both of the necessity for specialization, and the danger of it, and described what had been done in the organization of the School of Public Health to minimize such dangers.

I am so much impressed with the increasing need of greater coordination of the specialized forms of work that I feel like emphasizing that point. Specialization has been of the utmost value, but it so strongly tends to very decided separation and to actual lack of knowledge of other people's work as it bears upon the work of the individual specialist . . . For instance, in public health there is a strong tendency for the atmosphere to develop that public health, which is a Siamese twin of medicine, is even a separate thing from Medicine. It goes far, here, to offset that tendency that a noteworthy group of our Medical Faculty are engaged in both the Medical and the Public Health Schools in teaching and in research, and those in the Public Health School who are not definitely in the Medical Faculty do offer work, especially in research or in advanced work, to students in the Medical School.

Edsall described to Frankfurter the pressure for immediate results which could interfere with the training of people who would become leaders:

I have had the experience here of being bombarded by a considerable group of Alumni and other persons interested in the Medical School who have the very sincere feeling that the great purpose of the Medical School is simply the training of doctors . . .

I have . . . spiked their guns by pointing out that by far the larger part of the endowment of the Medical School is given specifically for research and that we cannot morally use that money for any other purpose. It is quite an interesting thing, I think, that extremely little of the Medical School's money is given specifically for teaching . . .

That and other experience leads me to believe, that it is not true as is stated in the discussion you send with this memorandum, that it is easy to obtain money for a school, but hard to obtain it for research. In Medicine and Public Health, it is and has been for years far more easy to obtain money for research, for developing new knowledge, than to obtain it for the routine training of persons who apply that knowledge . . .

I have always felt — and I grow increasingly convinced and I find increasing numbers of persons becoming convinced — that not many persons dealing with the clinical aspects of medicine can be completely separated from the application of this in practice without suffering in their abilities as either teachers or research workers in the problems of medicine. Those that are engaged in the pure medical sciences must, I think, certainly devote themselves wholly to them. In the research aspects of the clinical branches, on the other hand, I believe that they must keep some touch with not merely hospital practice, but with practice outside. The difficulty is to find in what way and in what degree . . .

It is an interesting fact that with the very large development in the past twenty years, in this country, of full-time clinical departments in Medicine, and the expenditure of great sums of money in many places on this (including this School) there has been less development in the past decade or thereabouts of very able personnel in the clinical branches, and fewer men who show the qualities of leaders therein, than at any time in the thirty-five years or more that I have known Medicine. Whether that is the result of too decided a separation from the active problems of their profession, or not, I am not at all clear. I have been wholly in sympathy with the developments, in their not too rigid form, but it does seem clear that thus far throughout two decades this development has not led to producing just the kind of men that it was hoped it would lead to, and this at least makes one pause to think. What I mean in regard to your proposed institute is that I doubt if it can be completely cloistered from relations, in some way, with the practical aspects of problems of crime, and the handling of them, without running danger of becoming narrow and unserviceable in its activities.

Edsall indicated an awareness of another fact which plagues those who would accomplish things in medicine by spending money on them, which is that a certain amount of economic stringency acts as a stimulus to productive thought.

I would quite agree . . . that it would be better to start with a moderate amount of money . . . I think a very large amount of money in the beginning would lead to waste . . . it has tended in some institutions, I believe, to lead to less keen and ingenious work because of the ease in so many ways of purchasing what is wished, whereas a moderate restricting of resources and need of some economy really leads to a very considerable development of actual originality and ingenuity. I find that the most productive departments that I have to administer are, by and large, the most economical ones . . .[32]

Various large new gifts came in during these years. There was $100,000 for the revolving loan fund from Dr. Frederick Shattuck. Friends of Dr. Robert W. Lovett gave a fund of $120,000 for research in orthopedic problems. More than $60,000 came in for a memorial to Dr. Francis Peabody. The Rosenwald Fund of $100,000 was given to Dr. Bronson Crothers for study and treatment of handicapped children. Vanderbilt Hall, already outgrown, was enlarged. The Rockefeller Foundation made its grant to Cannon (Chapter 20).

In his 1931–1932 Dean's Report Edsall again took stock, this time at his fifteen-year mark. He described continuing work on the curriculum, freeing the schedule of the fourth year for all students, in the third year reducing the requirements from thirteen subjects and examinations to four, in the second year from nine to five.

The School of Public Health report laid particular stress on the many services provided by that school for government and private industry. Said Edsall:

> There is advice and guidance to other organizations and individuals that is of an extent and diversity that is scarcely recognized by any except those engaged in the activities of the School. A great part of the time of certain of the departments particularly is devoted to important consultations and advice with governmental, semi-public and other organizations, and a very large share of the value of the School lies and will lie in this . . .
>
> The direct teaching activities of the School . . . have been, from the beginning, relatively less in amount . . . In much of this country there has been but a very slowly growing demand for really well trained men in public health. This is largely due to undesirable political influence in the appointments and undesirable political control of health officers in many parts of the country. These conditions are slowly improving . . .
>
> Until very lately much the major portion of the students in the School have been foreigners sent here by foundations or governments.[33]

For Edsall the Harvard Medical School and the School of Public Health were two aspects of a unified field. He was quick to point out how much each side gained from their close cooperation, the saving in use of facilities and library in common, the cross-fertilization of ideas, the strengthening of special departments which would have had to be maintained in a rudimentary form if the schools had been separate. As for his double deanship, he kept two offices and two separate secretaries, but the two hats went on the same head. To the Associated Harvard Clubs he once described this as follows: "In one of his brilliant letters published after his death, Johnny Poe reported to his Princeton class that he was in the real estate business with a cousin also named Poe. He said, 'the firm name is Poe and Poe. I do not know whether I am Poe or Poe.' And so in regard to the Medical School and the School of Public Health, I do not know whether I am Poe or Poe. They are 'two souls with but a single thought, two hearts that beat as one.'"[34]

By the academic year 1932–33, even the carefully invested Harvard funds were feeling the strain of continuing depression. "Both Schools," Edsall wrote in his Dean's report, "have . . . suffered from forced reduction of expenditures . . . peculiarly severe as compared with other parts of the University . . . budgets of both Schools are so very largely dependent upon income from invested funds, while tuition fees and other rather stable sources of income, though significant, play a minor role."[35]

At the time Edsall was writing, 78% of the medical school funds came from investments, and nearly 96% of the School of Public Health funds; when these fell off, the budgets had to be cut back proportionately. Many departments had also enjoyed considerable support in the form of yearly gifts, which of course were sharply curtailed. Three years earlier the medical school budget had contained over $165,000 in the form of gifts for immediate use for research; by 1932–33 these had fallen to $91,000.

In November 1931, the university was getting 5¼% on its investments but foresaw a drop. By January 1932 the income rate had dropped to 4.8% and Edsall wrote, "All temporary grants, except those from the DeLamar Mobile Research Committee, will have to cease." In June 1932 he had to ask the departments to make plans to meet a further 10%, a 20%, or a 30% reduction in their budgets. Salaries, services, and wages were to be the last things cut. By wringing the last penny out of each little special fund, even counting "the prospective small surplus from rentals at the dormitory which have never yet been written into the budget," a cut in salaries was postponed.[36]

The dean even managed to make some economies in the budgets of Heat, Light, Power, and Refrigeration, Miscellaneous, and the Removal of Rubbish, and to shake something loose from Maintenance. These items were handled in Cambridge, and early in his administration Edsall had been involved in lengthy negotiations to get these items budgeted in an equitable manner. Once settled, they were difficult to change. The university authorities stated that the maintenance budget was not to be touched until the academic budget had been cut back to the level of 1922. The two schools had grown too much under Edsall's hand for that to be a reasonable position. He hurled at them some figures of 1931, showing that since he had been dean the budgets of the laboratory departments had increased 99%, those of the clinical departments (including increases from hospitals) 819%. "How could one argue," he wrote, "that the maintenance should be increased or decreased proportionately to the . . . budget of ten years ago?"[37] He and the comptroller finally agreed not to try to keep up the buildings "in handsome shape" but to settle for the prevention of deterioration.

One effort at economy involved the use of the elevators. Only specially privileged individuals were supposed to use them at all, and each elevator had its own key. In one department, a request was sent out to turn in the keys, since new ones were being issued. Seven keys had been issued originally, but thirty-five were returned, at which point the use of locks on the elevators was given up.[38]

In September 1932 the interest rate was 4.1% and Edsall was required to remove $100,000 from the budget. By February 1933 it had fallen to 3.85%. Still the schools managed to squeeze by. "Owing to the fact that some money is free in the budget and a number of members of the Faculty (amongst whom, Drs. Cutler, Churchill and Christian are particularly heavily involved) have generously consented to the use for the time being of unemployed funds in the general budget, it will be possible to write the new budget without asking the departments to make [further] reductions."[39]

As the plans for retrenchment were made, people came first. Remarkably enough, not only were the schools able to keep their personnel; they maintained salaries as well, and the teaching continued unimpaired. Funds were shifted from research and other less immediate needs. Money which had been earmarked for research or other projects, but not yet spent, was called back to help out. But even research suffered less than Edsall had feared. In his 1932–33 Dean's Report he wrote, "careful planning by the departments and additional effort has resulted in an amount of valuable research that is to a gratifying degree comparable with that carried out in the previous times of greater ease." [40]

CHAPTER 23

The Last Harvard Years, 1932–1935

In 1932 Edsall saw with regret that President Lowell's term of office was coming to an end. The letter that he wrote the President evoked a remarkable response.

17 Quincy Street
Cambridge
November 20th

Dear Dr. Edsall;

It was very kind of you to write me such a letter full of feeling about my resignation, and it touches me very deeply. That working with you has been a pleasure, I need hardly tell you. You have always been thoughtful and considerate, wise and just, the greatest of leaders of American Medical Schools of your day, and yet you have listened sympathetically to any suggestion I made. We have, in fact, never differed except on the number of men to be admitted to the School.

A pleasant time it has been to look back upon, and I believe the improvements in medical education have been highly important and will prove durable, for they have been framed on a definite rational plan toward a specific goal. I hope that you will long see their fruit, and contribute more and more thereto. Whatever you might have done by personal work as a teacher and investigator, I do not suppose could have equalled what your stimulus and guidance has effected through the direction you have given as Dean.

Of course I dislike to give up such associations; but for every man the time must come to yield to younger hands, and for me I think it has come. A letter like yours makes the occasion happier.

Yours very sincerely:
A. Lawrence Lowell [1]

In the spring of 1934 Edsall wrote his yearly reports for the last time. The success of the full-time, university-style professorship was much in his mind. This was one battle that had been fought successfully, and the old issues could now be laid aside. He looked forward to new developments, particularly in the field of psychiatry where he foresaw important advances. Never one to be satisfied with things as they were, he reviewed the state of the various departments in the medical school and noted the pressing needs of otology, dermatology, orthopedic and genito-urinary surgery. Nevertheless, the general state of the school was one that he could regard with confidence, particularly because of the promise of the younger men who were coming along. He wrote:

> About twenty years ago, a man with one of the most brilliant minds in academic medicine, thinking of this School from outside our own staff, said to the writer that the Harvard Medical School had collected a very distinguished group of primary men, but that it could not be considered to be in a strong and effective situation until it had very greatly increased the number of able and promising junior men who would serve to reenforce and carry on the work of the seniors . . .
>
> About that time the total number of men in our large staff of over 400 who could be considered to be doing serious work toward the progress of medicine was between 40 and 50 only, including senior men; whereas it was noted that ten years later the numbers had more than tripled and that four-fifths of them were junior men — a much more dependable situation.[2]

Finding men, developing men, giving them the necessary opportunities to do their work — this was one of the strengths of Edsall's administration, one of the goals he had seen most clearly from the beginning. He had early improved the system of selection. "There can be little doubt," he wrote in 1934, "that the School has acquired the standing of being the best place in the country and perhaps anywhere for advanced efforts in research and for advanced training of teaching and research personnel. There have been, at most times in recent years, for example, as many as twenty to thirty men resorting here on fellowships from outside organizations and coming from all parts of this country and most parts of the world . . . There were very few such twenty years ago." Edsall attributed this directly to the care used in selecting personnel.

"One of the most important causes," he wrote, "for the improvement of the personnel has been the development of committees on promotion and reappointments and special committees on each of the appointments to life positions. These committees carefully study the individuals when their advance is under consideration, give care to maintaining the standards of quality in the junior personnel, and in the case of life appointments go into detail as to all available individuals and their qualities. Earlier this most important of all questions . . . was often hastily and casually handled."[3]

Edsall gave about equal weight to three factors in choosing men: a) teaching ability; b) power to contribute to progress in knowledge; c) power to stimulate other men in advanced studies and in the continued progress of their particular subjects. The last point he said was often overlooked, being more subtle than the other points but no less important.

The fruition of these wise policies could be seen when Edsall came to list the outstanding departments of the two schools.

By retention of able personnel of diverse abilities and training, there has been built up here a "School" of Physiology and Biochemistry that has become more conspicuous than any elsewhere in this country and, perhaps, in recent years, than any in any other part of the world, chiefly due to the ability and accomplishments of the heads of the departments named, but in considerable part due to the retention of conspicuous men in other aspects of Physiology and Biochemistry than the heads of those departments represented. The prestige that has come to the University through this group is quite extraordinary. It is in part indicated by the fact that there are at present more than thirty men in professorships, chiefly in this country but also in various other parts of the world in Physiology or equivalent subjects who have obtained much of the essential parts of their training, or all of it, here.

Similar opportunities in training in the Pathological Medical Sciences (in Pathology, Bacteriology, and Protozoology) have gradually been developed in these departments and in Tropical Medicine, so that with relatively slight additions a situation comparable to that in Physiology would be in existence. In certain aspects of these subjects, especially in Bacteriology, already there is a large drift of men to this School for advanced training.

Similar opportunities for the training in the whole field of Neuropsychiatry and its fundamental scientific aspects have been gradually developed and are in the process of further development, so that there is a strong opportunity for making it a powerful center of teaching, research, and advanced training in all these lines. The physiological developments in this regard have been of conspicuous importance for a considerable period of years. The recent admirable developments in

Anatomy make it a real center for neuro anatomical training. The developments in the clinical branches in neuro-psychiatry and in the clinical facilities are such that it is apparently quite feasible to make it a thoroughly rounded opportunity for training from the fundamental to the practical and in the various aspects of research in this great field.

Conditions in Clinical Medicine and in Surgery have been vastly improved by the persons in charge of those departments and very greatly increased amounts of money provided for this, and the numbers of accomplishments of real importance and of able persons at work have greatly increased.

The Pediatric Department has, by the present incumbent, been changed from one chiefly doing forceful teaching into one that in its clinical, its research, its hygienic and its sociological relations is, I think, looked upon as the leading pediatric clinic.[4]

To mention only a few of the great group who were working at that time (1933–34):

In Physiology: Walter Cannon, Arturo Rosenblueth, Alexander Forbes, Hallowell Davis, Roy G. Hoskins, Magnus Gregersen, Milton O. Lee.

In Biological Chemistry: Otto Folin, Cyrus H. Fiske, Harry C. Trimble, Milan A. Logan, George Hitchings, Yellapragada Subba Row.

And in separate departments in the Laboratories of Physiology:

Physical Chemistry: Lawrence J. Henderson, Edwin J. Cohn, Ronald M. Ferry, John Edsall, Jesse P. Greenstein.

Physiology in the School of Public Health: Cecil Drinker, Lawrence T. Fairhall, Louis A. Shaw.

Pathology: S. Burt Wolbach, Frederick Parker, Tracy B. Mallory.

Bacteriology: Hans Zinsser, J. Howard Mueller, Hugh K. Ward, Benjamin White, William A. Hinton, John F. Enders, and in the School of Public Health, William L. Moss.

Protozoology (Comparative Pathology): Ernest C. Tyzzer, Donald L. Augustine, Marshall Hertig, Hans Theiler.

Tropical Medicine: Richard Strong, A. Watson Sellards, George C. Shattuck.

In Neuropsychiatry, Psychiatry under Macfie Campbell, Neurology under James B. Ayer and Stanley Cobb, Neuropathology under Stanley Cobb, a distinguished group of men which included Frank Fremont-Smith, Tracy J. Putnam, and Harry C. Solomon.

Clinical Medicine was a very large department with seventeen professors of various ranks, headed by Henry Christian, James H. Means, and George Minot. Surgery was equally large, leading off with Elliott Cutler, and Edward D. Churchill, and including David Cheever, William C. Quinby, and Ernest M. Daland.

Associated with Kenneth D. Blackfan in Pediatrics were such men as James Gamble, Bronson Crothers, Richard M. Smith, Allan M. Butler, Louis K. Diamond.

This was one important way in which Edsall had contributed to the growth of the two schools. Another was in increasing and regularizing the relations with the various teaching hospitals. It was an impressive list which he set out for Locke in 1934:

> The official relations with the Psychopathic Hospital were very inadequate, the Professor being practically a guest in the Hospital. With the change in personnel through Dr. Southard's death, the Commissioner of the Insane was prevailed upon to make the incoming Professor the actual head of the institution. Together with this, there was a large increase from the Medical School of the funds spent upon that subject.
>
> The Lying-In Hospital had no defined official relations, beyond tradition, with the Medical School. A large sum was obtained by the School to be used in the building of the new Lying-In Hospital, and with this the understanding was reached that the School should have continued facilities in teaching and should nominate the staff to the Trustees, and that the new Hospital should be built in the immediate neighborhood of the School.
>
> When the new Beth Israel Hospital was built, the Hospital officials directly sought an affiliation with the School, which has been already of marked advantage in certain ways, and may in future be of increasing importance.
>
> With the Massachusetts General Hospital our relations had been previously only traditional, and in many ways faulty and liable to disagreement. An arrangement has been reached for the most intimate conjoint securing of personnel and employment of funds and development of new activities.

With the Eye and Ear Infirmary there were no defined relations again but tradition, but largely through the cooperative efforts of Dr. Bradford, while he was President of that Board, arrangements similar to that with the Massachusetts General and an increasing degree of intimacy in the employment of funds and in the planning of new developments has been carried out.

At the City Hospital, largely through the efforts of Dr. Sears and Dr. Locke there has been developed a system whereby the School nominates to the Trustees for the headship, and hence for other positions, in that part of the medical and surgical services that is definitely assigned to this School. The important Thorndike Laboratory has been made a part of the responsibilities of the man at the head of the Harvard Medical service, and that service and Laboratory are at present among the most important, if not the most important, of the medical clinics in the country, as a consequence of the force of the personnel and the activities that have been in operation there. Further, the Neurological Service and laboratories have been greatly developed chiefly on special funds obtained by the School.

Through the aid of the School, extensive laboratories and a contagious ward have been developed at the Children's Hospital, largely increasing the activities in research in diseases of children, and greatly improving the facilities of the School in the contagious diseases.

Again through securing of funds conjointly, the laboratory and other investigative facilities of the Massachusetts General Hospital in Medicine and Surgery have been altogether transformed and made admirable, whereas previously they were very inadequate.

At the Eye and Ear Infirmary, the facilities and funds for ophthalmic research have been likewise transformed, and funds, in normal times amounting to more than three-quarters of a million dollars have been provided for this purpose through the School.

The relations with the Cancer Commission and the Huntington Hospital have been entirely altered and advanced. Previously there was no actual correlation with the School; the research and the Hospital, though maintained by the University, were run entirely separately. Now in the choice of personnel, in the administration, and in the teaching, research, and other activities of the Cancer Commission and the Huntington Hospital there is the most complete and intimate cooperation with the School.[5]

Relations with the university were also a subtle but important factor in improving the quality of the school's teaching. During

Edsall's and Lowell's terms, these relations had become much closer.

Edsall in particular had always cherished the ramifications of the wider Harvard world. In speaking of the extraordinary growth of medicine, particularly in the preceding forty years, he once said:

> We may not forget that most of this could not possibly have come about were it not for what the natural sciences had done previously, and often long before its bearing upon medical matters was apparent. The slow but secure development of scholarly thought, methods and information in other fields than Medicine, that had to come first, is the most fundamental factor in making Medicine what it is today; and the intimate associations with other fields that come from being a true part of the university family are the most precious of the influences that have led to what we have gained and will lead to what the future holds.[6]

The summary of his accomplishments that he sent to Locke showed the importance of his continued support of these relations:

> The general relations with the University have become very much closer in a variety of ways, and the School has become far more definitely a university school of medicine. This is, in my judgment, one of the most important things for the future progress of the School . . . There will need to be constantly increased relations with Economics, Sociology, Social Anthropology, Psychology, and various other parts of the University, including the Engineering School, the Law School, and the Business School, so that the utmost intimacy in relations with the general University is of fundamental importance in the whole conception of the future of the School.
>
> A large part of this has been due to the increasing numbers of men in our personnel who have intimate interests in, and dependence upon, fields in other parts of the University. A considerable part of it has been due to the establishment of the biochemical tutors in the College and the aid that has been given by the Medical School to the acquiring of these men, the providing of some part of their salaries, and the providing of facilities for them to continue their research. That field and the persons in the field have made for an increasingly close liaison between the Medical School and the departments of the Natural Sciences in Cambridge. There have, in addition, been increasing bonds with the Engineering School . . . especially in relation to public health . . .

Also with the Business School, and with other parts of the University, especially in certain of the biological activities . . .[7]

In looking to the future, as he told the Harvard medical alumni in 1931, Edsall wished to have three things borne in mind as important general principles. First in importance was the avoidance of complacency. Second, a consistent line of progress, carefully mapped out, without hasty or impulsive changes in policy. Third, constant care in the choice of personnel.[8]

Edsall would be sixty-five when he retired in the spring of 1935. He was willing to ease off. He wrote to Locke, "The labor, the nervous strain and the anxiety of securing large amounts of money is a thing that I shall be most happy to turn over soon to somebody else. I know nothing that is more exhausting than the care in detail, the tact, the diplomacy, and the worry that one has to go through in order to accomplish considerable things in that way."[9]

He would no longer have to deal with newspapers which made a heyday of an outbreak of food poisoning at the school, or with the awkward problem of power house chimneys which at times belched forth an embarrassing amount of smoke from an institution which pioneered in the problems of air pollution, or with such a problem as the following which Zinsser presented to the dean's office:

We are troubled in this building by an increasing pest of mice and rats against which our simple devices of trapping and poisoning now offer no appreciable effect. The thing has grown so annoying that we are beginning to find our postage stamps eaten into and as you will see from the upper edge of this sheet our stationery is also being damaged. We are beginning to get worried about the possibility of these animals attacking the feebler members of the laboratory staff.

There seems to be no sense in making a vigorous campaign against them on our own floor unless the thing is done throughout the building. We have tried this method both with mice and cockroaches and if we are unkind to them they merely migrate temporarily to Wolbach's department or down to Strong.

Seriously, the matter is one that is resulting in constant minor damages and soon is likely to cost the University considerable money . . .[10]

Edsall left the school early in 1935, although his resignation officially did not take effect until fall. Walter Cannon wrote him on February 7, 1935:

"I went to your office this morning, to say 'Farewell' to you before your departure, and found that you had already gone. It was with a feeling of sadness that I turned away, for I realized that when I return to the School next autumn I shall not find you in the Dean's office. Let me say again that in my opinion you have done more for the welfare of the Harvard Medical School than any one else, with the possible exception of President Eliot; and may I add that I look back upon my association with you during your illustrious administration as one of the choicest experiences of my life."[11]

As word of his retirement got about, the letters poured in. Good manners of course required some notice of the occasion, but the kind of tribute expressed went far beyond good manners. Zinsser wrote him in January, "If I have been able to accomplish anything for the school & for my subject it has been made possible, to a large extent, by your understanding of problems & difficulties & your support—financial and personal of every worthy effort."[12]

He wrote again in June, "thanking you for the innumerable things you have done during my ten years at Harvard to make my work pleasant and profitable. I record that, although by nature & temperament a 'minority' opponent of authority which gained me the soubriquet of 'pepper-box' at Columbia—I can recall no single instance in which we have clashed or had difficulties—a fortunate course which I attribute largely to your wisdom, cooperativeness & understanding of my problems and intentions."[13]

George Minot wrote at length in the warmest way:

It seems to me but yesterday that my father told me, shortly after I had passed my house officer examination, that the biggest man in medicine was to be my chief at the M.G.H. In these past 23 years you have always been the person who has served me as the finest example of how to do the correct thing in medicine. In my teaching I hear myself frequently saying Dr. Edsall taught me this or that about carbohydrate intestinal indigestion, headache, typhoid fever, etc., etc., and then when administrative problems arise I find I always try to think out how you would solve the problem and what are the fundamental basic aspects of the problem to be solved . . .

In accepting the honor conferred upon me by the Nobel Foundation it gave me great pleasure to mention you in my after dinner speech at the Town Hall in Stockholm as one with a few others (Thayer, Homer Wright, & Peabody,) who had inspired me . . .

It will be a long time before any Dean of the School has such a splendid record of accomplishments as you have.

I shall certainly never forget your kindness to me almost 14 years ago when I told you I had diabetes – in the days before insulin – To carry on at that time & since under this handicap was in large part possible because of your example & encouragement.[14]

President Conant wrote him:

It is impossible for me to express adequately the sincere appreciation of the Governing Boards and indeed of the whole Harvard community . . .

Personally I should like to tell you what a great pleasure it has been for me to work with you and how much I have learned from you not only about the difficult problems of medical education but about administration in general. I shall never forget the warm and kindly letter which you wrote me when I was elected President. It was a most encouraging note which was peculiarly welcome at a moment when the future seemed particularly forbidding . . .[15]

His close friend Alfred North Whitehead wrote for himself and his wife:

DEAREST DAVID

I am taking this opportunity to tell you what we think of you, and what you mean to us. It is difficult to write because words look conventional, and your worth is beyond all conventional phrases.

To us you have meant kindness, wisdom, and ever-active self-sacrifice. We have loved you dearly from the first beginning of our friendship. You have been to us a reliable force, ever mitigating the harshness in the nature of things – embodying in your person the reasons why America is so lovable . . .

You have made Harvard an outstanding example of the best side of medicine. Of course, there are other factors (human and material) which gave you the opportunity. But you took it.[16]

Louis Kirstein, chairman of the executive committee of the Beth Israel Hospital, and soon to become the hospital's president, sent his regrets that he could not attend the retirement dinner and added: "I want to particularly thank you for your cooperation in the development of the fine and harmonious relationship between the Medical School and the Beth Israel Hospital. I know that this has done much to improve the care of the patient at the institution and has been an excellent influence on every phase of its management and service."[17]

Plaudits came from far and near.

"You have had the opportunity . . . to have a determining influence upon the development of American medicine. It has been a fine job, consistently and thoroughly done throughout many years." (From Ray Lyman Wilbur, president of Stanford.[18])

"With many of his friends I regretted [Edsall's] decision to give up research work in order to devote himself to administrative and educational tasks, but that regret was not well founded and was relinquished very soon, for the life he chose proved eminently fruitful not only to Harvard Medical School but also to medical education throughout the land." (From Dr. David Riesman to Walter Cannon.[19])

"When you took over, the Medical School was a capably conducted, sound, but pretty provincial organization. Under your guidance it has taken on breadth and strength to a greater extent than any other school in the country . . . This is an extraordinary achievement in such a relatively short time. Your grasp of the importance of fundamental science tempered by realization of the importance of knitting clinical teaching with the fundamentals has been quite beyond that of any of your contemporary administrators." (From Dr. Hugh Cabot, surgeon at the Mayo Clinic and former dean of the University of Michigan Medical School.[20])

"Your retirement marks the climax of a phase in the development of the Medical School which will always remain as a brilliant period in its growth and in its accomplishment . . . of course none of these changes has taken place without you and your constant interest and thought and stimulus." (From Dr. Warfield T. Longcope, professor of medicine, Johns Hopkins.[21])

"Under your guidance and leadership, the Harvard Medical School has become the outstanding institution in the country . . . It has been a privilege to study under you and to have later been permitted to work with you on the Committee on the Training of Hospital Executives, and the Commission on Medical Education.

I have not known a man who has the breadth of vision, soundness of medical and educational ideas, and the effective manner of producing really great results." (From Willard C. Rappleye, dean of the Columbia Medical School.[22])

While we are summing up, it is worth quoting the verdict of Cecil Drinker, who stepped into Edsall's shoes as dean of the School of Public Health:

> The University benefited extraordinarily by the fact that Dr. Edsall was a man of national, indeed international outlook in medical scholarship, and that just as he became Dean of the Medical School we received the DeLamar bequest which gave latitude for expenditure in many desirable directions, and shortly after, the gift for the School of Public Health added further to the general resources of the medical establishments we now have. We were greatly lucky to have Dr. Edsall on the scene just as the first real opportunity for remaking the Medical School came to us . . .
>
> It would be fair to say that the interest of the [Rockefeller] Foundation in establishing a Health School at Harvard would never have materialized into principal had they not been able to count upon Dr. Edsall for the wise expenditure of the money given . . .[23]

At the same time that he retired as dean, Edsall ended his service as trustee of the Rockefeller Foundation. His colleagues there paid unusual recognition to their departing trustee. Alan Gregg, then director of the Division of Medical Sciences, wrote him:

> I wanted to thank you for the service and help you have given as a Trustee of the Foundation. It is warmly appreciated by others than myself but I think none of us is in a position more directly to know it and value it than I. In the search for new trustees your services are referred to as the paradigm to follow and it is the gap you create which it is hard to fill. There is not sufficient recognition of the services given by Trustees in general but in particular I don't want you to be without the warm thanks of at least one of the officers, and the one who has learned most from, and leaned most on, your advice. I am particularly grateful to you for the impersonality with which you dealt with everything and the candor with which you expressed your doubts or disagreement. It seems to me that your friendly patience and willingness to hear more and try things out has helped more than the contributions of any other Trustee and I shall miss you very greatly from the Board.[24]

From Raymond B. Fosdick, soon to become president of the Foundation, Edsall received an equally friendly letter:

"Your last paragraph gave me a feeling of deep and genuine regret. I had not realized that you were so close to the end of your term in the Rockefeller Foundation and, frankly, I do not know what we shall do without you . . . personally, I do not like the idea of board meetings in which your advice and counsel are no longer heard. I know I speak for all my associates in expressing this opinion."[25]

Drs. Locke, Means, Cutler, and Cecil Drinker were in charge of arranging a suitable celebration to mark Edsall's retirement. One of his friends declined to come and see him stuffed.[26] It was decided to have some speeches at an afternoon exercise, followed by a dinner and more speeches, and this took place on October 23, 1935.

Walter Cannon was in the chair at the afternoon meeting and spoke warmly of the many ways in which Edsall had gained distinction.

> We are gathered here this afternoon to recall and evaluate some of the services which our friend Dr. Edsall has performed for medicine. I say some of the services because he has had so varying a career, he has contributed to medicine in so many different ways, that a satisfactorily complete estimate of all he has done would be impossible in a short afternoon meeting.
>
> More than a quarter of a century ago his knowledge of pediatrics was recognized by his election as president of the American Pediatric Society, and by his being offered the chair of pediatrics at Johns Hopkins and later at the University of Texas. For some time he was professor of therapeutics and pharmacology at the University of Pennsylvania. Later he was professor of preventive medicine at Washington University, St. Louis. For eleven years he was Jackson professor of clinical medicine at Harvard and during that time he inspired many young men for their professional careers. As you probably know, he has long been recognized as a pioneer in the study of industrial disease in the United States. And he is known throughout the country as a great reformer of medical education. As dean he has done more to increase the resources and to promote productive scholarship in the Harvard Medical School than any other individual, in my opinion, except possibly President Eliot.[27]

Dr. Walter A. Jessup, president of the Carnegie Foundation

for the Advancement of Teaching, spoke on "Harvard in Medical Education." Eugene DuBois discussed "The Development of Clinical Subjects as Contributing to University Work," and had a personal as well as a general tribute to give:

It is a great personal pleasure to me that I am permitted to take part in this afternoon's tribute to Dr. Edsall. It so happens that my own interest in research was largely stimulated by Dr. Edsall, something of which he has been totally unaware. The particular medical school that I attended seemed to have little interest in anything except the preparation of men for hospital interneships. I do not think that any student was ever referred to an original article in the literature. Our diet was one hundred per cent. text-book. When I was interning at the Presbyterian Hospital, however, my chief, Dr. Francis Kinnicutt, for the first time in my career brought me face to face with an original article, and it was one of Dr. Edsall's papers dealing with typhoid fever. That started me on a study of typhoid that lasted for many years. Since that time I have frequently been indebted to Dr. Edsall's contributions and it was only a week ago that I had to consult and quote one of his articles on respiration.

Dr. Edsall was a leading member of the group of research clinicians who changed the whole atmosphere of clinical medicine in this country. His own studies were important, but I think the medical world is even more indebted to him for his share in the development of the Harvard Medical School . . . Harvard seems to have worked out a happy solution of the perplexing problem of full time versus part time, combining the advantages of both systems. The research work coming from the group of institutions that constitute the Harvard Medical School has been outstanding . . .[28]

Another aspect of Edsall's contribution to medical advance came under review in a speech on "The Relation of Medicine to the Fundamental Sciences" by Lawrence J. Henderson, who had himself shifted from the medical school to the university as Abbott and James Lawrence Professor of Chemistry.

The dinner was a mammoth affair held at the Harvard Club in Boston. After considering how he could invite some without others of the huge teaching staff, Dr. Locke decided to invite all of them, over 500. "I do this," he wrote to Edsall, "because it seems as though even the youngest of the teachers had a right to attend against men in the profession outside."[29]

Then there were men from various eastern medical schools, and men from the Harvard trustees, corporation, and overseers, as well as deans of other Harvard schools, and Edsall's own list of friends. Locke hoped to seat some of Edsall's family and friends, "including the ladies," on a small balcony over the dining room. It took a special dispensation to admit the ladies of the family, as the doors of the Harvard Club in those days had never been darkened by a woman. They were smuggled in a back door and up a back stairway, to seats on the balcony of the organ loft in the great hall. A near crisis arose when Mrs. Geoffrey Edsall dropped her handkerchief which threatened to flutter down to the head table. Fortunately it landed on the edge of the balcony and the ceremony was not disrupted.[30]

President Conant served as toastmaster at the dinner, and remarked, "I suppose that the committee that asked me to preside at this meeting had in mind the story of the colored gentleman who prayed, 'Oh, Lord, take me and use me, if only in an advisory capacity.'" Half seriously, he continued, "When two years ago I took over the task of being President, I must admit to this gathering that it was with considerable trepidation that I approached the deans of the various faculties, who knew far more than I could possibly know, and this was particularly true for the Medical School . . . Patiently and kindly [Dean Edsall] told me what a young president ought to know, not only in regard to the Medical School but in regard to the general problems of administration, and I turned to him for wise counsel . . ."[31]

Gracefully turning to the future, he then introduced the new dean of the medical school, Charles Sidney Burwell, who had been induced to leave his position as professor of medicine at Vanderbilt University in Tennessee, to return to Boston as dean.

Dr. Burwell surveyed Edsall's career and his contributions to Harvard and to medicine as a whole. In closing he said:

> Dr. Welch says in one of his charming discourses that one may expect the approval and appreciation only of those who are engaged in his own kind of work. This gathering, Dr. Edsall, is a gathering of people who have come to express their esteem for and admiration of you and the work which you have done. Most people would be very proud to have a group gather and make such an expression in regard to one kind

of endeavor. Here we have a gathering of your colleagues in several fields: your colleagues in clinical medicine and clinical investigation, your colleagues in the field of medical education, your colleagues in the field of public health, and your many colleagues in devotion and service to the Harvard Medical School.[32]

One president of Harvard in the chair, another president of Harvard on the speaker's rostrum, the diners were next addressed by President Lowell. He praised many things which Edsall had been able to do in relation to the medical school, seeing them all as "part of a much larger proposition, which was that of bringing all the aspects of medicine into relation, both in the minds of the student and also in the sentiment of their instructors."

So successful was that [Lowell continued], that he was able to entirely destroy the jealousy which existed at one time between the laboratory and the clinical subjects. I know when I first entered the School that jealousy was very, very marked. They sat more or less on opposite sides of the central table and they looked at one another as more or less natural enemies, and so strong was this feeling that it was very much like the sentiment which prevailed before the Civil War, where in the United States Senate it was said, "You mustn't add another new state to your side, unless you add another new state to our side." "And you mustn't add a professor in this body to the clinical side unless you add another to the laboratory side." Nobody could find any sign of an agreement. The difference between when I presided over that Faculty in the first years and the last years was enormous . . .

Now Dean Edsall went forward – the whole thing in his mind was one great thought of correlating all those things that had been separate – the correlation of courses, closer relation with clinical and laboratory branches, drawing together and at the same time expanding all the fields of medical education, building up the School of Public Health, trying experiments with tutors, instituting full-time clinical professors, and vastly increasing the number of men doing constructive research, always thinking of further improvement and extension.

Dreams, dreams. Was he a dreamer? Yes, I think he was, but there are various kinds of dreamers . . . the real dreamer is the buccaneer who goes out and finds [treasure-trove] or seizes it from a Spanish galleon. That is what he did. He found it. The great DeLamar bequest came. He found it, and then he lightened the coffers of the Rockefeller Foundation and other places, and he brought back the rich, golden treasure to the Harvard Medical School . . .[33]

Edsall was the next speaker of the evening. He began by contrasting conditions as he had found them elsewhere with those he had found in Boston and in the Harvard sphere of influence. "One of the characteristics in this community," he said, "is that element of divine discontent and the willingness to go on and do better things than had been done before." He was also struck by the loyal support for Harvard and its projects.

He then paid tribute to the Harvard administration:

> Before I came here I had had some unhappy experiences with university administrative officers, university trustees and hospital trustees. I even dallied at times in my thoughts, for my amusement, with the idea that some time I might persuade a benefactor to give a large gift in order that there might be established a school for training university presidents and trustees and hospital trustees. I came to the conclusion that that idea wouldn't work here because it had been so well done already. The corporations that I had known before were bodies that seemed to me – university and hospital trustees both – to be chiefly swayed by the feeling that their job was censorship and restraint. The Corporation of Harvard University, on the contrary, is thought by all the Faculty and especially by those of us who have, in administrative positions, had to make many hundreds of recommendations, to be a body that is there in order to encourage and help. The President that I served under most of the time I cannot speak of quite, I think, with calmness. I spent fifteen years serving with Mr. Lowell. Those fifteen years included, of course, many conferences and I never left any conference without a feeling that I had been helped, encouraged, and stimulated in a way that was one of the greatest gifts that I have had in all my active life . . . I think Mr. Lowell never knew, when he first asked me to be Dean and I quite flatly refused, one strong reason for my refusal was that I had still that complex about university administrators . . .
>
> I thought seriously of resigning when Mr. Lowell told me he had resigned. It seemed fairer to the new President, whomsoever he might be, not to precipitate that extra confusion and I decided to wait two years . . . The past two years with President Conant have been two very happy years of association with him. I have always had the comfortable feeling that Mr. Conant and I were brought up on the same bottle! [34]

Edsall then went on to speak of three of his guiding ideas in medical education, referring to the type of men who should be made the leaders in a medical faculty and of the responsibility of the medical school towards those superior students who would be the leaders of the future. He also spoke of the increasing necessity for ever closer relations between a medical school and its university:

> The Engineering School has become a blood brother of the School of Public Health, and Public Health and Medicine are Siamese twins. The Business School already has had many associations with the Medical Faculty. I have no doubt that in a very few years there will be active and interested cooperation between the Law School and the Medical School. Those are the practical aspects of it, but a larger aspect than that, I think, is that Medicine has become a scholarly pursuit in all its lines. It must exist in a scholarly atmosphere and not by itself if it is to progress adequately.

Lowell wrote to Edsall afterwards, "That was a beautiful speech of yours last night . . . You must have perceived that your hearers thought so, and one who owes you more than any other comrade likes to say so."[35] And Felix Frankfurter wrote to Cecil Drinker, "I don't know when I have heard anything equally impressive . . . Like all important utterances, it had much more than immediate technical application. The essence of his creed regarding medical education is equally applicable to legal education."[36]

CHAPTER 24

Epilogue

IN THE SPRING of 1935, the Edsalls bought some land in Tryon, North Carolina, which had all the advantages of mountainous Vermont combined with a more tolerable winter climate. Edsall had already had the pleasure of remodeling an old house — in his Vermont house he had uncovered some wainscotting 140 years old and pulled all the ceilings out of the downstairs rooms to expose the beams – and now he was going to build. In August he was in Greensboro waiting to go south as soon as the Tryon house had progressed far enough to need decisions about colors of walls and floors. By September 22nd they were staying at Thousand Pines, Tryon, to superintend the finishing touches, and they celebrated their first Christmas in the new home that year.

Edsall might have gone on rusticating at leisure except for the war. In 1939 he was made chairman of the Medical Advisory Committee of the American Red Cross, and the same year was appointed to the Science Committee of the National Resources Planning Board. The next year he became member at large of the Division of Foreign Relations of the National Research Council.

With so many Washington meetings to attend, travelling became onerous, and Edsall and Louisa moved to Washington the winter before the U. S. went to war. A letter from Louisa to her stepson Richard and his wife Katherine gives a picture of their life there:

> We have a nice little bright little apartment, which we were very lucky to get as Washington is very full indeed, and your father goes down to the Red Cross every morning and sometimes has to see people in the afternoon too . . .
>
> We came up by train, so we have our feet and the public conveyances only to get around in. Quite a change from just hopping into a car every time you feel like it. However, Washington is extremely fascinating to me to walk through and it is nice to be able to pause and stare whenever I feel like it . . . This part of Washington is right

among the Embassies. Latvia, the nearest, has been closed up since we were here before. That is just around the corner from us. The Greek Embassy of course has had its flag at half mast the last few days. What a beautiful blué it is. We went as far as the British Embassy this morning. They seem to be building on to that. We passed Chief Justice Hughes on the way. I think he must have been walking to Church . . . Yesterday we had tea with the Ben Joneses. They have such a lovely little girl named Pamela, about Louisa's * age, and a little boy of two, who had just succeeded in climbing out of his pen for the first time when we were there. Awful moment, and so much involved. However, it is perhaps as well for the parent to know when that point is reached.

Your father is well and much interested in all the questions that come up in relation to the Red Cross work. He is well stocked with puzzles for off moments, and there is a Lending Library just around the corner that is well supplied with detective stories . . .

We get two meals in and one out. There are heaps and heaps of restaurants as you may remember, and quite a lot right in this neighborhood. We are now on a system of trying a different one whenever we are not lunching down by the Red Cross. On those days we go to The Allies Inn, a superfine cafeteria, very crowded, very good food and very interesting faces to look at . . .

The Mellon Collection is not going to open till March 17th. I hope we shall be here. I don't suppose we should get a chance to go to the opening, but we could see it right afterwards. I am hoping to see something of the art exhibits and maybe get an idea what modern art is all about. I see Margaret Fitzhugh Browne of Boston is starting a society for Sanity in Art. I can't help wondering if it will get anywhere. However, it may be a good chance to get together and cuss.[1]

From the time Hitler came to power in Germany, the number of refugee physicians entering the United States rose year by year, hitting a peak of 1,384 in 1939.[2] Many of these men had world reputations, but distinguished or not, they ran up against a prime example of U.S. xenophobia. If state regulations did not forbid their practising, many state boards of medical examiners did, and this in the face of increasing shortages of physicians, especially in rural areas. In 1939 Edsall became chairman of the Boston Committee on Medical Emigrés, and Honorary Chairman of its parent

* This Louisa is Richard and Katherine's daughter, and named for the older Louisa. Louisa had no children of her own but she was a good grandmother, as she was a well-loved stepmother.

national organization, the National Committee for Resettlement of Foreign Physicians.

Edsall's plea for action, a letter appearing in the *Journal of the American Medical Association* in May 1939,[3] heralded a period of intense activity which enlisted such distinguished doctors as Lewellys Barker, Joseph A. Capps, Alfred E. Cohn, Alice Hamilton, D. B. Phemister, and many others. The training and credentials of the refugees were investigated, they were aided in taking the necessary examinations, in finding intern positions, in getting any necessary retraining. The country was combed for communities which needed what the foreign doctors had to give, and efforts were made to calm the fears of many officials, in and out of the medical profession, that the country would be flooded by ill-trained, incomprehensible foreign doctors. As Edsall pointed out in his third AMA article, it was not a flood, the men were not ill-trained, and the total number involved about equalled one year's graduates of American medical schools.[4]

In 1941 the committee could report that it had placed some 700 physicians in American hospitals, and had settled about 500 in practice. Less happy was the fact that they had failed to dent many barriers of prejudice:

> There is no doubt . . . that we are facing an increasing shortage of native physicians. Despite this and despite the unquestioned ability of most refugee physicians to fill it, there is a complex network of legal or semilegal restrictions which stands in the way. Unfortunately, discrimination against the licensure of legally admitted émigrés has increased in many states during the two and a half years of the committee's operation. This discrimination has contributed directly to their congregation in New York, in Boston and to a lesser degree in Chicago, San Francisco and Los Angeles — communities where they have had some chance to do useful medical work and maintain some kind of professional status.[5]

Wartime conditions were hard on Edsall, who was seventy-two the year of Pearl Harbor. He had what was probably a stroke in 1942, and two attacks of cardiac decompensation in the summer of 1943. For a month in the fall of 1943 he was a patient in his old hospital, the MGH. An elevette was installed in his Tryon house and in November he was allowed to go south. In January he was back in hospital in Asheville, and twice more in the MGH after he came north.

Louisa nursed him with every loving care. They had a little apartment on Memorial Drive in Cambridge the winter of 1944–45, and in February 1945 it was a good day when he could walk from his bed to the living room twice. Despite his physical troubles, or perhaps because he was forced to regard each day as a gift, the last two years of his life were extremely serene and happy. He probably got more joy out of his family and close friends than he had for a good many years.[6]

On August 12, 1945 he died and his ashes were taken to Greensboro to be buried among his beloved mountains.

The pathologist's report gave the usual definite answer to a complicated clinical picture. Paul Dudley White wrote the facts to Louisa, and there was a kind of comfort in his conclusion that "it is obvious there was absolutely no prospect of ultimate recovery and that Dr. Edsall probably survived his first acute heart failure of the summer of 1943 by twice the ordinary length of time."

The heart showed a "very high degree" of narrowing of the aortic valve due to deposit of calcium, Dr. White went on, and the heart muscle was something over twice normal weight. Signs of old inflammation of the peritoneum from diverticulitis and cystitis were found, as well as some congestion with fluid in the pleural cavities, lungs, liver, and feet, and traces of the final bronchopneumonia.

"There are two things in my mind that I want to tell you," Dr. White wrote, "first, my admiration of your own devoted, patient, and efficient care that I am sure was the most important reason for Dr. Edsall's long survival and relative comfort; and second, my own affectionate regard for Dr. Edsall himself. It was he incidentally who started me, as he has done for so many others, on my special professional career, and who always advised me after that start so helpfully. His photograph is on my wall as one of my choicest heroes."[7]

The memorial from his colleagues at the medical school was spread on the faculty minutes, December 7, 1945. In closing it said, "Edsall was preeminent as a founder of institutions. His imprint on Harvard and beyond was great.
It will endure."

DAVID L. EDSALL, M.D.

Published Papers

1897

"On the Estimation of Hydrochloric Acid in Gastric Contents." *U. of Penn. Med. Mag.*, 9 (Sept. 1897): 797–809.

"Tuberculous Occlusion of the Œsophagus, with Partial Cancerous Infiltration." (By William Pepper and D. L. Edsall.) *Am. Jnl. of Medical Sciences*, 114 (July 1897): 44–63.

1898

"Buttonhole Mitral and Tricuspid Regurgitation; Multiple Emboli of the Lung." *Trans.*, Path. Soc. Phila., 18 (1898): 232.

"A Case of Tuberculosis Pericarditis." (By F. S. Westcott and D. L. Edsall.) *Trans.*, Path. Soc. Phila., 18 (1898): 230–32.

"Dissociation of Sensation of the Syringomyelic Type: Occurring in Pott's Disease." *Jnl. Nervous & Mental Disorders*, 25 (April 1898): 257–63.

"A Note on the Gastric Conditions in Anemias." *Phila. Med. Jnl.*, 1 (Jan. 22, 1898): 159–61.

"Peculiar Bodies Found in the Central Nervous System." (By D. L. Edsall and J. Sailer.) *Proc. Path. Soc. Phila.*, n.s.1 (1897–98): 96–100.

"Perforation in Typhoid Fever." (By J. H. Lloyd and D. L. Edsall.) *Trans.*, Path. Soc. Phila., 18 (1898): 110–13.

"Tuberculosis of the Esophagus, with Cancerous Infiltration." *Trans.*, Path. Soc. Phila., 18 (1898): 87–93.

1900

"Absorption and Metabolism in Exclusive Rectal Alimentation." *U. of Penn. Med. Mag.*, 13 (March 1900): 23–30.

"A Critical Summary of the Literature on the Serum Diagnosis of Tuberculosis." *Am. Jnl. of Medical Sciences*, 120 (1900): 72–77.

"The Influence of Immoderate Water-Drinking upon Metabolism and Absorption." *Contributions, William Pepper Laboratory*, 4 (1900): 368–94.

1901

"The Carbohydrates of the Urine in Diabetes Insipidus." *Am. Jnl. of Medical Sciences*, 121 (May 1901): 545–51.

"A Contribution Concerning Creatinin Excretion." *Proc.*, Path. Soc. Phila., n.s.5 (1901–02): 35–43.

"A Critique of Certain Methods of Gastric Analysis." *U. of Penn. Med. Bull.*, 14 (April 1901): 46–50.

"The Estimation of the Urinary Sulphates and of the Fecal Fat in the Diagnosis of Pancreatic Disease." *Am. Jnl. of Medical Sciences*, 121 (April 1901): 401–410.

"General Metabolism in Diabetes Mellitus." *Phila. Med. Jnl.*, 7 (April 6, 1901): 673–80.

"Syphilis of the Liver with Large Gummata in Late Childhood." *Arch. Pediatrics*, 18 (June 1901): 425–32.

1902

"A Brief Report of the Clinical, Physiological and Chemical Study of Three Cases of Family Periodic Paralysis." (By John K. Mitchell, Simon Flexner, and D. L. Edsall.) *Brain*, Part 97 (Spring 1902): 109 ff.

"Concerning the Benzoyl Esters of the Urine in Diabetes Mellitus, and the Clinical Significance of an Excess of Glycuronic Acid." *U. of Penn. Med. Bull.*, 15 (April 1902): 34–37.

"A Contribution Concerning the Clinical Significance of the Readily Eliminable Sulphur of the Urine." *U. of Penn. Med. Bull.*, 15 (May 1902): 87–90.

"The Diagnosis of Diabetes Mellitus and the Urinary Findings in This Disease." *Medicine*, 8 (Oct. 1902): 793–806, and in *Proc.* Phila. County Med. Soc., n.s. 4 (1902): 112–25.

"The Relation of Uric Acid and Xanthin Bases to Gout and the So-called Uric Acid Diathesis." *Phila. Med. Jnl.*, 9 (May 3, 1902): 794–800.

"A Remarkable Case of Coma, Apparently Due to Acid Intoxication Sui Generis." *Phila. Med. Jnl.*, 9 (June 28, 1902): 1155–60.

"Some Recent Literature Concerning Ferments and Their Products." *Phila. Med. Jnl.*, 9 (1902): suppl. 5–8.

1903

"Concerning a Possible Etiological Factor in Tobacco-Alcohol Amblyopia, Revealed by an Analysis of the Urine of Cases of This Character." (By G. E. de Schweinitz and D. L. Edsall.) *Am. Jnl. of Medical Sciences*, 126 (Aug. 1903): 216–27, and in *Trans. Am. Ophth. Soc.*, Hartford, 10 (1903): 41–57.

"Concerning the Nature of Certain Cases of Chronic Polyarthritis." (By D. L. Edsall and Ralph S. Lavenson.) *Am. Jnl. of Medical Sciences*, 126 (Dec. 1903): 991–1004.

"A Contribution to the Chemical Pathology of Acromegaly." (By D. L. Edsall and Caspar W. Miller.) *U. of Penn. Med. Bull.*, 16 (June 1903): 143–50.

"A Preliminary Communication Concerning the Nature and Treatment of Recurrent Vomiting in Children." *Am. Jnl. of Medical Sciences*, n.s. 125 (April 1903): 629–35.

"Pyloric Carcinoma with Symptoms Resembling Gastrocolic Fistula." (By D. L. Edsall and Charles A. Fife.) *Am. Medicine*, 6 (Oct. 10, 1903): 584–87.

"Some Biological Differences Between the Natural and the Artificial Feeding of Infants." *Am. Medicine*, 6 (Sept. 26, 1903): 508–11, and in *Proc.* Phila. Co. Med. Soc., 24 (1903): 116–24, and in *St. Louis Med. & Surg. Jnl.*, 85 (1903): 23–31.

"A Study of Two Cases Nourished Exclusively Per Rectum, with a Determination of Absorption, Nitrogen-Metabolism, and Intestinal Putrefaction." (By D. L. Edsall and Caspar W. Miller.) *U. of Penn. Med. Bull.*, 15 (Jan. 1903): 414–24, and in *Wis. Med. Jnl.*, 1 (1903): 87–107.

1904

"A Case of Post-Pneumonia Endocarditis." (By D. L. Edsall and W. E. Robertson.) *Proc.* Path. Soc. Phila., n.s. 7 (1903–04): 199.

"The Clinical Behavior of the Lymph Glands in Typhoid Fever." *Am. Jnl. of Medical Sciences*, 27 (April 1904): 599–606.

"Concerning the Accuracy of Percentage Modification of Milk for Infants." (By D. L. Edsall and Charles A. Fife.) *N. Y. Med. Jnl.* and *Phila. Med. Jnl.*, 79 (Jan. 9 & 16, 1904): 58; 107.

"Concerning the Nature of Still's Type of Chronic Polyarthritis in Children." *Arch. Pediatrics*, 21 (March 1904): 175–83.

"The Clinical Chemistry of Disease of the Liver." *U. of Penn. Med. Bull.*, 16 (Feb. 1904): 427–36.

"The Physiological Chemistry of Diabetes." *Albany Med. Annals*, 25 (April 1904): 341–55, and in *Trans.* Med. Soc. N.Y., Albany 98 (1904): 89–101.

"A Small Hospital Epidemic of Pneumococcus Infections, with Remarks on the Transmission of Pneumonia and the Methods by Which It Can Be Prevented." (By D. L. Edsall and Albert A. Ghriskey.) *Therapeutic Gazette*, 20 (March 1904): 147–57, and in *Trans.* Coll. Phys. Phila., 26 (1904): 6–25.

"A Small Series of Cases of Peculiar Staphylococcic Infection of the Skin in Typhoid Fever Patients." *U. of Penn. Med. Bull.*, 17 (March 1904): 8–11.

"Two Cases of Bothriocephalus Infection, One of Them Showing Profound Anemia." *Am. Medicine*, 8 (Dec. 1904): 1087–91.

"Two Cases of Violent but Transitory Myokymia and Myotonia Apparently Due to Excessive Hot Weather." *Am. Jnl. of Medical Sciences*, 28 (Dec. 1904): 1003–11.

1905

"A Case of Acute Leukaemia, with Some Striking Clinical Features. Observations on Metabolism in This Case, and in a Case of Severe Purpura Hemorrhagica." *Am. Jnl. of Medical Sciences*, 130 (Oct. 1905): 589–600, and in *Trans.* Assn. Am. Phys., 20 (1905): 279–93.

"The Dietetic Use of Predigested Legume Flour, Particularly in Atrophic Infants: With a Study of Absorption and Metabolism." (By D. L. Edsall and Caspar W. Miller.) *Am. Jnl. of Medical Sciences*, 129 (April 1905): 663–84, and in *Buffalo Med. Jnl.*, 44 (1904–05): 455–57.

"Hodgkin's Disease with a Milky Non-Fatty Pleural Effusion." *N. Y. Med. Jnl.* and *Phila. Med. Jnl.*, 82 (Oct. 21 & 28, 1905): 838; 901.

"The Influence of Infected Milk in the Diet of the Sick — Particularly in Acute Infectious Diseases." *N. Y. Med. Jnl.* and *Phila. Med. Jnl.*, 81 (March 25 & April 1, 1905): 578; 644, and in *Penn. Med. Jnl.*, Pittsburgh, 8 (1904–05): 357–70, and in *Pediatrics*, 17 (1905): 562–78.

"The Influence of Mercury on Autolysis." (By D. L. Edsall and Caspar W. Miller.) *U. of Penn. Med. Bull.*, 17 (1904–05): 415.

"Some Further Experiments upon Rectal Alimentation." (By D. L. Edsall and Caspar W. Miller.) *Am. Medicine*, 9 (Feb. 4, 1905): 187–90.

"A Study of Metabolism in Leukaemia, under the Influence of the X-Ray: With a Consideration of the Manner of Action of the X-Ray and of Some Precautions Desirable in Its Therapeutic Use." (By John H. Musser and D. L. Edsall.) *U. of Penn. Med. Bull.*, 18 (Sept. 1905): 174–84, and *Trans.* Assn. Am. Phys., 20 (1905): 294–323.

"Typhoidal Insanity in Childhood, with Some Notes as to Its Character and Prognosis." *Am. Jnl. of Medical Sciences*, 129 (Feb. 1905): 327–39.

1906

"The Attitude of the Clinician in Regard to Exposing Patients to the X-Ray." *Jnl. Am. Med. Assn.*, 47 (Nov. 3, 1906): 1425–29.

"The Clinical Value of Blood Cultures." *Penn. Med. Jnl.*, 9 (1905–06): 844–53.

"Examination of the Urine in Relation to Surgical Measures." Chap. 55 in William Williams Keen, *Surgery: Its Principles and Practice*, vol. 4. Philadelphia: Saunders, 1906.

"Intolerance of Fats." *Boston Med. & Surg. Jnl.*, 155 (Dec. 27, 1906): 763–71.

"The Physiological Limitations of Rectal Feeding." *Am. Jnl. of Medical Sciences*, 132 (Nov. 1906): 679–86, and in *Trans.* Assn. Am. Phys., 21 (1906): 239–53.

"The Physiology of Glycolysis." *Proc.* Phila. Co. Med. Soc., 26 (1905–06): 313–15, and in *Med. Exam. & Pract.*, N. Y., 16 (1906): 134.

1907

"Diseases Due to Chemical Agents." In William Osler, ed., *Modern Medicine: Its Theory and Practice in Original Contributions by American and Foreign Authors.* Philadelphia: Lea Brothers, 1907.

"Observations Relating to the Nature of Atrophy of Intestinal Origin." *Jnl. Am. Med. Assn.*, 48 (May 4, 1907): 1469–76.

"Some Practical and Theoretical Considerations Concerning Dietetics." *International Clinics*, 3 (1907): 1–24.

"The Use of the X-Rays in Unresolved Pneumonia: Nature of the General Toxic Reaction Following Exposure to the X-Rays." (By D. L. Edsall and Ralph Pemberton.) *Trans.* Assn. Am. Phys., 21 (1906): 618–40, and in *Am. Jnl. of Medical Sciences*, 33 (Feb. & March 1907): 286–97 and 426–31.

1908

"The Bearing of Metabolism Studies on Clinical Medicine." Harvey Society lecture, Nov. 30, 1907. *Archives Internal Med.*, 1 (Feb. 1908): 154–74, and in *Trans.* Assn. Am. Phys., 22 (1907): 667–82.

"A Disorder Due to Exposure to Intense Heat: Characterized Clinically Chiefly by Violent Muscular Spasms and Excessive Irritability of the Muscles. Preliminary Note." *Jnl. Am. Med. Assn.*, 51 (Dec. 5, 1908): 1969–71.

"The Hygiene of Medical Cases, Particularly in Hospital Wards." *Am. Jnl.*

of Medical Sciences, 135 (April 1908): 469–508, and in *Trans.* Assn. Am. Phys., 22 (1907): 667–82.

1909

"Further Studies of the Muscular Spasms Produced by Exposure to Great Heat." *Trans.* Assn. Am. Phys., 24 (1909): 625–28.

"The Hygiene of Medical Diseases in Medical Wards." *Nat. Hosp. Record*, April 15, 1909.

"The Present Status of Organo-therapy." *Lancet Clinic*, Cincinnati, 100 (1908): 613–23, and in *Louisville Month. Jnl. Med. & Surg.*, 15 (1908–09): 193–206.

"Prophylaxis Against Infectious Diseases from the Standpoint of the Practitioner." *Jnl. Am. Med. Assn.*, 52 (Jan. 9, 1909): 123–28.

"Some of the Relations of Occupations to Medicine." *Jnl. Am. Med. Assn.*, 53 (Dec. 4, 1909): 1873, 1881, and in *Wisconsin Med. Jnl.*, 8 (1909–10): 425–47.

1910

Address of the President, American Pediatric Society, May 3, 1910, in "The Influence of David Edsall on Pediatrics in the United States" by Borden S. Veeder, *Jnl. Pediatrics*, 47 (Dec. 1955): 808–16.

"Some General Principles of Dietetics, With Special Remarks on Proprietary Foods." *Jnl. Am. Med. Assn.*, 54 (Jan. 15, 1910): 193–96.

"The Work of the Council on Pharmacy and Chemistry." *Jnl. Am. Med. Assn.*, 55 (Nov. 12, 1910): 1701–05.

1911

"Discussion on Acidosis." *British Med. Jnl.*, 2 (1910): 1033–38.

"The General Principles of Dietetics." In John H. Musser and A. O. J. Kelly, eds., *Handbook of Practical Treatment*, Vol. 1. Philadelphia: Saunders, 1911.

"Some Observations upon Metabolism in Relation to Muscular Disorders." *Bull. Manila Med. Soc.*, 3 (1911): 17.

1912

"The Clinical Study of Respiration." Shattuck Lecture, June 11, 1912. *Boston Med. & Surg. Jnl.*, 167 (Nov. 7, 1912): 639–51.

"The Clinician, the Hospital and the Medical School." *Boston Med. & Surg. Jnl.*, 166 (Feb. 29, 1912): 315–23.

"The Efficiency and Significance of Different Forms of Respiration." *Trans.* Assn. Am. Phys., 27 (1912): 560–70.

"Industrial Poisoning." *Am. Labor Legisl. Rev.*, 2 (1912): 231–34.

1913

"Medical Sepsis and Asepsis." Abstract. *Trans.* N. H. Med. Soc. (1913).

1914

"The Effect of Strychnin, Caffein, Atropin and Camphor on the Respiration and Respiratory Metabolism in Normal Human Subjects." (By D. L.

Edsall and J. H. Means.) *Arch. Internal Med.*, 14 (1914): 897–910, and in *Trans.* Assn. Am. Phys., 29 (1914): 69–80.

"The Relation of Industry to General Medicine." Symposium on Industrial Diseases, June 9, 1914. *Boston Med. & Surg. Jnl.*, 171 (Oct. 29, 1914): 659–62.

"The Relation of Syphilis to Internal Medicine." *Boston Med. & Surg. Jnl.*, 171 (Sept. 10, 1914): 412–15.

1915

"Observations on a Case of Family Periodic Paralysis." (By D. L. Edsall and J. H. Means.) *Am. Jnl. of Medical Sciences*, 150 (1915): 169–78, and in *Riforma Med.*, Napoli, 31 (1915): 1160.

"Relation of the Staff to the Administration of Hospitals." *Modern Hospital*, 4 (March 1915): 184–96.

1916

"Movements in Medicine." *Boston Med. & Surg. Jnl.*, 174 (June 22, 1916): 891–97.

1917

"A Study of Occupational Disease in Hospitals." *Monthly Review*, Bureau of Labor Statistics, U. S. Dept. of Labor, 5 (Dec. 1917): 169–85.

1918

"Medical-Industrial Relations of the War." *Johns Hopkins Hosp. Bull.*, 29 (Sept. 1918): 197–205.

"The Prevention of Disease in War Industries: Extent and Importance of the Problem." *Med. Rec.*, N. Y. 93 (1918): 611.

"Supposed Physical Effects of the Pneumatic Hammer on Workers in Indiana Limestone." *Public Health Reports*, 33 (March 22, 1918): 394–403.

1919

"The Clinical Aspects of Chronic Manganese Poisoning." (By D. L. Edsall and Cecil K. Drinker.) *Contributions to Medical and Biological Research*, Vol. 1. Dedicated to Sir William Osler, in honor of his seventieth birthday, July 12, 1919, by his pupils and co-workers. New York: Hoeber, 1919.

"Industrial Clinics in General Hospitals." *Jnl. Indust. Hygiene*, 1 (1919): 394, and in *Modern Med.*, 93 (1919): 575–77.

"The Occurrence, Course and Prevention of Chronic Manganese Poisoning." (By D. L. Edsall, F. P. Wilbur, and Cecil K. Drinker.) *Jnl. Indust. Hygiene*, 1 (Aug. 1919): 183–93.

1922

"Present Day Problems of the Medical School." *Harvard Alumni Bull.*, 25 (Dec. 7, 1922): 279–86.

1923

"The Prevention of Degenerative Diseases." (By D. L. Edsall and Paul D. White.) Chapter 27 in *Nelson Loose-Leaf Living Medicine*, Vol. 7. New York: Nelson, 1923.

1924

"The Product of Medical Education." *Boston Med. & Surg. Jnl.*, 191 (Aug. 14, 1924): 282–94.

1925

"An Adequate Examination at the End of the Clinical Courses." *Jnl. Am. Med. Assn.*, 84 (May 2, 1925): 1320–24.

"The Handling of the Superior Student." *Proc.* Assn. Am. Med. Colleges, 35 (1925): 114–24.

Speech to Associated Harvard Clubs, May 21, 1925. *Harvard Alumni Bull.*, 28 (Oct. 1, 1925): supplement 9–20.

1928

"A Decade of Progress in the Medical School Library." *Harvard Med. Alumni Bull.*, 3 (Oct. 1928): 4–6.

1929

"Some Features of Medical Education." *Southern Med. Jnl.*, 22 (Aug. 1929): 715–18.

"Some of the Human Relations of Doctor and Patient." Chapter 1 in L. E. Emerson, ed., *Physician and Patient: Personal Care.* Cambridge: Harvard, 1929.

1930

"Clinics in Preventive Medicine in Third-Year Teaching: Harvard Medical School." (By D. L. Edsall and Joseph C. Aub.) *Methods and Problems of Medical Education*, Eighteenth Series. New York: The Rockefeller Foundation, 1930.

"Certain Changes in the Harvard Medical School During the Last Twenty Years." *Harvard Med. Alumni Bull.*, 5 (June 1931): 6–9.

1931

"The Transformation in Medicine," address delivered at the Dedicatory Exercises of Duke University School of Medicine and Duke Hospital, Durham, N.C., April 20, 1931. *Southern Med. Jnl.*, 24 (Dec. 1931): 1103–13.

1932

"The Trend of Medical Research." *Jnl. Proceedings & Addresses*, Assn. Am. Universities, 34th Annual Conf., State U. of Iowa, Nov. 1932: 132–45.

1935

Address at a dinner held in his honor, Oct. 23, 1935. *New England Jnl. Med.*, 213 (Dec. 12, 1935): 1184–87.

1939

"A Program for the Refugee Physician," letter. *Jnl. Am. Med. Assn.*, 112 (May 13, 1939): 1986–87.

1940

"The Emigre Physician in American Medicine." *Jnl. Am. Med. Assn.*, 114 (March 23, 1940): 1068–73.

1941

"The Emigre Physician in America, 1941." (By D. L. Edsall and Tracy J. Putnam.) *Jnl. Am. Med. Assn.*, 117 (Nov. 29, 1941): 1881–88.

1942

Presentation of the George M. Kober Medal to Dr. D. D. Van Slyke. *Trans.* Assn. Am. Phys., 57 (1942).

"The Research Career in Public Health." Pamphlet. Washington, D.C.: National Research Council. No date.

Source Notes

Chief sources of material about David L. Edsall are the archives of the three medical schools where he taught, the archives of the Rockefeller Foundation, and the personal papers in the hands of his son, Dr. John T. Edsall. The abbreviation "HMS" in the following notes refers to the archives of the Harvard Medical School in Countway Medical Library, Boston, Mass.; "Harvard" refers to the Harvard University Archives, in particular the presidential papers of Charles W. Eliot and A. Lawrence Lowell, in Widener Library, Harvard University, Cambridge, Mass.

CHAPTER 1, HOME GROUND

1. James P. Snell, *History of Sussex & Warren Counties, New Jersey.* Philadelphia: Everts & Peck, 1881, p. 344.

2. George E. McCracken, "Samuel Edsall of Reading, Berks. and Some Early Descendants." *N. Y. Genealogical and Biographical Record,* 89 (July 1958): 129–145; Harriet Winfield Gibson, "Probable Originator of the Edsalls. A Story of the Life of Samuel Edsall, With Some of His Business Connections. His Connection with the Fish, DeKay and Other Families Prominent in New York and New Jersey. His Great Landed Interests." (MS. in possession of Dr. John Edsall.)

3. Jerome R. Reich, *Leisler's Rebellion: A Study of Democracy in New York, 1664–1720.* Chicago: U. of Chicago, 1953; "Documents Relating to the Administration of Leisler," in *Collections of the New-York Historical Society for the Year 1868,* Vol. 1; Samuel Eliot Morison, *The Oxford History of the American People.* New York: Oxford, 1965, pp. 118–122; Maud Wilder Goodwin, *Dutch and English on the Hudson: A Chronicle of Colonial New York.* The Chronicles of America Series, Vol. 7, Allen Johnson, ed. New Haven: Yale, 1920, pp. 150–164.

4. *New Jersey Herald,* July 30, 1964.
5. Snell, *op. cit.,* p. 344.
6. Geoffrey Edsall, personal communication to Aub, Sept. 1967.
7. Thomas L. Pellett to Aub, Oct. 5 and 11, 1963.
8. G. Edsall, *op. cit.*
9. *Ibid.*
10. Snell, *op. cit.,* p. 336.
11. G. Edsall, *op. cit.*
12. Snell, *op. cit.,* p. 336.
13. Dorothy L. Roberts to Aub, March 27, 1964.
14. Thomas L. Pellett, *op. cit.*
15. John T. Edsall, personal communication to Hapgood, 1963.

16. Geoffrey Edsall, "Notes and Recollections on D.L.E., Sept. 9, 1945." MS. (Edsall)

17. D. L. Edsall to Samuel Eliot Morison, Nov. 22, 1938. (HMS)

18. Mrs. W. B. Butterworth to Aub, Oct. 8, 1963; Mrs. W. B. Butterworth to Joseph W. Gardella, Sept. 30, 1963. (Aub)

19. D. L. Edsall to Cecil Drinker, Sept. 9, 1925. (HMS)

20. M. Halsey Thomas to Aub, Dec. 13, 1967.

21. Edsall to Drinker, *op. cit.*

22. D. L. Edsall, "The Trend of Medical Research," speech at 34th Annual Conference, Assoc. of Am. Universities, Iowa City, Iowa, Nov. 11, 1932. MS. (Edsall)

23. Edsall to Drinker, *op. cit.*

CHAPTER 2, MEDICAL STUDENT

1. William Osler, "The Medical Clinic: A Retrospect and a Forecast." *British Med. Jnl.* (Jan. 3, 1914).

2. D. L. Edsall to Thomas M. Woodward, Feb. 25, 1942. (Edsall)

3. *Ibid.*

4. William Osler, "In Memoriam — William Pepper." *Phila. Med. Jnl.*, 3 (1899): 607.

5. Francis Newton Thorpe, *William Pepper, M.D., LL.D. (1843–1898)*. Philadelphia: Lippincott, 1904.

6. James Tyson, "Dr. William Pepper." *Proceedings of the American Philosophical Society, Memorial Volume I.* Philadelphia: Am. Philos. Soc., 1900.

7. Geoffrey Edsall, personal communication to Aub, Sept. 1967.

8. U. of Penn. Dept. of Medicine, catalogue, 1890–91.

9. *Ibid.*, p. 159.

10. George W. Corner, *Two Centuries of Medicine: A History of the School of Medicine, University of Pennsylvania.* Philadelphia: Lippincott, 1965.

11. D. L. Edsall, untitled speech, "The Medical School has very especial reason for rejoicing." MS. No date. (Edsall)

12. Corner, *op. cit.*, pp. 158, 170–71.

13. Edsall, untitled speech, *op. cit.*

14. D. L. Edsall, "The Transformation in Medicine," address delivered at the Dedicatory Exercises of Duke University School of Medicine and Duke Hospital, Durham, N.C., April 20, 1931. *Southern Med. Jnl.*, 24 (Dec. 1931): 1103–13.

15. William Osler, "The Coming of Age of Internal Medicine in America." *International Clinics*, 4, 25th series (1915): 1.

16. *Ibid.*

17. D. L. Edsall, "Autobiographical Notes." MS. 1926. (Edsall)

18. The suggestion comes from Dr. T. Grier Miller of Philadelphia who was a student of Edsall's in 1910–11.

19. Edsall, "The Transformation in Medicine," *op. cit.*

20. D. L. Edsall to Cecil Drinker, Sept. 9, 1925.

21. Edsall, "Autobiographical Notes," *op. cit.*

22. *Ibid.*

23. Thomas L. Pellett to Aub, Oct. 5 and 11, 1963.
24. Edsall, "Autobiographical Notes," *op. cit.*
25. *Ibid.*
26. Edsall, "The Transformation in Medicine," *op. cit.*
27. D. L. Edsall to Emma Linn Edsall, June 30, 1895. (Edsall)
28. William Osler, "Letters to My House Physicians, Letter V." *N. Y. Med. Jnl.*, 52 (Sept. 20, 1890): 334.
29. Edsall, "Autobiographical Notes," *op. cit.*

CHAPTER 3, YOUNG DOCTOR EDSALL

1. Louisa R. Edsall, "Notes for D.L.E.'s children." MS. (Edsall)
2. D. L. Edsall, "Autobiographical Notes." MS. 1926. (Edsall)
3. D. L. Edsall, "Some of the Human Relations of Doctor and Patient," in L. Eugene Emerson, ed., *Physician and Patient: Personal Care.* Cambridge: Harvard, 1929, pp. 26–27.
4. Geoffrey Edsall, personal communication to Aub, Sept. 1967.
5. *Ibid.*
6. Edsall, "Autobiographical Notes," *op. cit.*
7. N. I. Bowditch, *A History of the Massachusetts General Hospital.* Second edition, with a continuation to 1872. Boston: Mass. General Hospital, 1872, p. 493.
8. D. L. Edsall, "The Transformation in Medicine," address delivered at the Dedicatory Exercises of Duke University School of Medicine and Duke Hospital, Durham, N.C., April 20, 1931. *Southern Med. Jnl.*, 24 (Dec. 1931): 1103–13.
9. Edsall, "Autobiographical Notes," *op. cit.*
10. *Ten Year Book of the Class of 1890 of Princeton.*
11. John Irwin Bright, ed., *Twenty Years After, A Record of the Class of 1890.* Princeton, 1910.
12. Edsall, "Autobiographical Notes," *op. cit.*
13. *Ibid.*
14. *Ibid.*
15. *Ibid.*
16. *Ibid.*
17. Geoffrey Edsall, "Notes and Recollections on D.L.E., Sept. 9, 1945." MS. (Edsall)
18. G. Edsall, personal communication to Aub, Sept. 1967.
19. Samuel P. Orth, *The Boss and the Machine: A Chronicle of the Politicians and Party Organization. The Chronicles of America series,* Vol. 43, Allen Johnson, ed. New Haven: Yale, 1920, graduates' edition, pp. 93 ff.
20. Owen Wister, quoted in Orth, *op. cit.*, p. 98.
21. Orth, *op. cit.*, pp. 93 ff.
22. J. T. Salter, *Boss Rule: Portraits in City Politics.* New York: Whittlesey, 1935, p. 216.
23. Robert C. Brooks, "William Scott Vare," in Dumas Malone, ed., *Dictionary of American Biography.* New York: Scribners, 1936.
24. G. Edsall, personal communication to Aub, Sept. 1967.

25. "Margaret Harding (Tileston) Edsall." *Radcliffe Magazine*, 15 (June 1913): 172.
26. Harriet Mixter to Mrs. John B. Tileston, January 1927. (Edsall)
27. Margaret Tileston to John B. Tileston, March 10, 1897. (Edsall)
28. Margaret Tileston, diary, 1898. (Edsall)
29. Margaret Tileston to Mrs. John B. Tileston, March 17, 1898. (Edsall)
30. D. L. Edsall to Margaret Tileston, May 19, 1899. (Edsall)
31. D. L. Edsall to Henry A. Christian, April 27, 1923. (HMS)
32. Margaret T. Edsall, diary, April 15, 1900. (Edsall)
33. Margaret T. Edsall, diary, April 18, 1900. (Edsall)
34. *Ten Year Book of the Class of 1890 of Princeton.*

CHAPTER 4, FRUITFUL YEARS IN PHILADELPHIA

1. Interview with Wilder Tileston and John Edsall, May 15, 1957. (Aub)
2. Margaret T. Edsall, diary, 1902. (Edsall)
3. Simon Flexner to William Welch, draft of letter, 1901. (Copy in possession of Aub)
4. Irwin Bright, ed., *Twenty Years After: A Record of the Class of 1890*. Princeton, 1910.
5. D. L. Edsall, "The Work of the Council on Pharmacy and Chemistry." *Jnl. Am. Med. Assn.*, 55 (Nov. 12, 1910): 1701–05.
6. *Ibid.*
7. *Ibid.*
8. Thomas L. Pellett to Aub, Oct. 5 and 11, 1963.
9. *Ibid.*
10. James H. Means, *The Association of American Physicians: Its First Seventy-five Years*. New York: McGraw-Hill, 1961.
11. N. C. Gilbert, "Archives of Internal Medicine." In Morris Fishbein, *A History of the AMA: 1847 to 1947*. Philadelphia, Saunders, 1947, pp. 1111 ff.
12. Walter B. Cannon, *The Way of an Investigator: A Scientist's Experiences in Medical Research*. New York: Norton, 1945, pp. 154–57.
13. Edsall, "The Work of the Council," *op. cit.*
14. *Ibid.*
15. *Ibid.*
16. *Ibid.*
17. *Ibid.*
18. *Ibid.*
19. D. L. Edsall, "Examination of the Urine in Relation to Surgical Measures." In Wm. Wms. Keen, *Surgery*. Philadelphia: Saunders, 1906, p. 168.
20. John K. Mitchell, Simon Flexner, and D. L. Edsall, "A Brief Report of the Clinical, Physiological and Chemical Study of Three Cases of Family Periodic Paralysis." *Brain*, London, Part 97 (Spring 1902): 109 ff.
21. D. L. Edsall and Caspar W. Miller, "A Contribution to the Chemical Pathology of Acromegaly." *U. of Penn. Med. Bull.*, 16 (June 1903): 143–50.

22. D. L. Edsall, "Some General Principles of Dietetics, with Special Remarks on Proprietary Foods." *Jnl. Am. Med. Assn.*, 54 (Jan. 15, 1910): 193–96.

23. John H. Musser and D. L. Edsall, "A Study of Metabolism in Leukemia, Under the Influence of the X-Ray: With a Consideration of the Manner of Action of the X-Ray and of Some Precautions Desirable in its Therapeutic Use." *U. of Penn. Med. Bull.*, 18 (Sept. 1905): 174–84.

24. D. L. Edsall, "The Attitude of the Clinician in Regard to Exposing Patients to the X-Ray." *Jnl. Am. Med. Assn.*, 47 (Nov. 3, 1906): 1425–29.

25. D. L. Edsall, "Some of the Relations of Occupations to Medicine," address to the Wisconsin State Medical Society. *Jnl. Am. Med. Assn.*, 53 (Dec. 4, 1909): 1873–81.

26. D. L. Edsall, "The Hygiene of Medical Cases, Particularly in Hospital Wards." *Am. Jnl. of Medical Sciences*, 135 (April 1908): 469–508.

27. D. L. Edsall, "The Trend of Medical Research." *Jnl. Proceedings & Addresses*, Assn. Am. Universities, 34th Annual Conf., State U. of Iowa, Nov. 1932: 132–45.

28. D. L. Edsall, "The Transformation in Medicine," address delivered at the Dedicatory Exercises of Duke U. School of Medicine and Duke Hospital, Durham, N.C., April 20, 1931. *Southern Med. Jnl.*, 24 (Dec. 1931): 1103–13.

29. Geoffrey Edsall, personal communication to Aub, Sept. 1967.

30. D. L. Edsall, Address of the President, American Pediatric Society, May 3, 1910. *Jnl. Pediatrics*, 47 (Dec. 1955): 808–16.

31. D. L. Edsall, Address of the President, Phila. Pediatric Society, Jan. 14, 1908. (Edsall)

32. D. L. Edsall, "A Disorder Due to Exposure to Intense Heat, characterized clinically chiefly by violent muscular spasms and excessive irritability of the muscles. Preliminary Note." *Jnl. Am. Med. Assn.*, 51 (Dec. 5, 1908): 1969–71.

33. Interview with Wilder Tileston and John Edsall, *op. cit.*

34. D. L. Edsall, "Further Studies of the Muscular Spasms Produced by Exposure to Great Heat." *Transactions*, Assn. Am. Physicians, 24 (1908): 625–28.

35. D. M. Glover, "Heat Cramps in Industry: Their Treatment and Prevention by Means of Sodium Chloride." *Jnl. Industrial Hygiene*, 13 (Dec. 1931): 347–60.

36. J. H. Talbott, "Heat Cramps." *Medicine*, 14 (Sept. 1935): 323–76.

37. D. L. Edsall, "Medical-Industrial Relations of the War." *Johns Hopkins Hosp. Bull.*, 29 (Sept. 1918): 197. © The Johns Hopkins Press.

38. William Osler, ed. *Modern Medicine: Its Theory and Practice in Original Contributions by American and Foreign Authors.* Philadelphia: Lea Bros., 1907.

39. "Memorial on Occupational Diseases," prepared by a committee of experts and presented to the President of the United States. Am. Assoc. for Labor Legislation, pub. no. 12, 1910.

40. Edsall, "Some of the Relations of Occupations to Medicine," *op. cit.*

41. G. Edsall, *op. cit.*

Chapter 5, Early Relations with Harvard

1. Starr J. Murphy to John D. Rockefeller, Jr., Dec. 19, 1901, quoted in Thomas Francis Harrington, *The Harvard Medical School: A History, Narrative and Documentary, 1782–1905*. Vol. 3. New York: Lewis Pub. Co., 1905, pp. 1178 ff.
2. *Ibid.*
3. This and the following quotations from D. L. Edsall, "Report on a visit to the Medical Department of Harvard University, in February 1904." MS. (Library, Coll. Phys. Phila.)
4. Edwin A. Locke to Frederic Washburn, March 19, 1938. (Aub)
5. D. L. Edsall to Edwin A. Locke, April 22, 1908. (Aub)
6. Edsall to Locke, May 24, 1909. (Aub)

Chapter 6, Invitation to St. Louis

1. Marjorie E. Fox, *History of the Washington University School of Medicine.* (An authorized history, manuscript, unpaginated.) Ch. 2. (Wash.)
2. *Ibid.*
3. *Ibid*, Ch. 3.
4. Abraham Flexner, *I Remember*. New York: S & S, 1940, pp. 124–25. Copyright 1940 by Abraham Flexner.
5. *Ibid.*
6. *Ibid*, p. 126.
7. Fox, *op. cit.*, Ch. 4.
8. D. L. Edsall to David F. Houston, Dec. 24, 1909. (Wash.)
9. Fox, *op. cit.*
10. *Ibid.*
11. *Ibid.*
12. John F. Fulton, *Harvey Cushing: A Biography*. Springfield, Ill.: Thomas, 1946, p. 335.
13. *Ibid.*, p. 336.
14. *Ibid.*, p. 306.
15. Margaret T. Edsall to Edith Johnson, Feb. 23, 1910. (Edsall)
16. D. L. Edsall to Edwin A. Locke, May 24, 1909. (Aub)
17. Margaret Edsall to Johnson, *op. cit.*
18. Margaret T. Edsall to Mary Kirkbride, Jan. 22, 1910. (Edsall)
19. Margaret T. Edsall to Mary Kirkbride, March 14, 1910. (Edsall)
20. Elizabeth Thomson, *Harvey Cushing: Surgeon, author, artist.* New York: Abelard, 1950, p. 169.
21. Harvey Cushing to Henry A. Christian, Feb. 2, 1911. (HMS)
22. Fulton, *op. cit.*, p. 307.

Chapter 7, Explosion in Philadelphia

1. George W. Corner, *Two Centuries of Medicine: A History of the School of Medicine, University of Pennsylvania.* Philadelphia: Lippincott, 1965, pp. 221–22.
2. Edgar Fahs Smith, quoted in Corner, *op. cit.*, p. 222.

3. Petition to Dr. Charles C. Harrison, Provost, Jan. 1910. (Edsall)
4. Margaret T. Edsall to Mary Kirkbride, Feb. 3, 1910. (Edsall)
5. *Ibid.*
6. *Ibid.*
7. D. L. Edsall to Simon Flexner, Feb. 3, 1910. (Copy in possession of Aub)
8. D. L. Edsall to Edwin A. Locke, Feb. 25, 1910. (Aub)
9. D. L. Edsall to Henry A. Christian, March 3, 1910. (HMS)
10. Philadelphia *Ledger*, March 31, 1910.
11. George Dock, "Address Made at the Opening of the New Medical Amphitheatre & Clinical Laboratory of the Hospital of the University of Pennsylvania." *U. of Penn. Med. Bull.*, 22 (Aug. 1909): 186.
12. A. N. Richards, conversation with John Edsall, June 30, 1955.
13. Corner, *op. cit.*, pp. 226–27.
14. "Report of committee of nomination for the position of Research Professor of Medicine," Dec. 13, 1909. MS. (Penn.)
15. D. L. Edsall to Edward B. Robinette, June 11, 1910. (Penn.)
16. "Upheaval in U. of P. Medical Teaching." Philadelphia *Record,* undated clipping after May 4, 1910.
17. D. L. Edsall to A. Lawrence Lowell, Dec. 11, 1913. (Harvard)
18. Edsall to S. Flexner, Feb. 15, 1911. (Aub)
19. S. Flexner to Edsall, Feb. 23, 1911. (Aub)
20. Edsall to Charles C. Harrison, Sept. 3, 1910. (Penn.)
21. Corner, *op. cit.*, p. 234.
22. Richards, *op. cit.*
23. Margaret T. Edsall, diary, 1910. (Edsall)
24. Edsall to Locke, Nov. 19, 1910. (Aub)
25. John Howland to Eugene Opie, Dec. 6, 1910. (Opie)
26. Howland to Opie, Dec. 29, 1910. (Opie)
27. Margaret T. Edsall to Mrs. John B. Tileston, Jan. 3, 1911. (Edsall)
28. Richards, *op. cit.*
29. Edsall to S. Flexner, Feb. 15, 1911. (Aub)
30. Edsall to Christian, Jan. 14, 1911. (HMS)
31. Edsall to Edgar Fahs Smith, Jan. 6, 1911. (Penn.)
32. Henry W. Cattell, quoted in *North American*, Jan. 9, 1911.
33. Edsall to Thomas Woodward, Feb. 25, 1942. (Edsall)
34. George W. Corner to Hapgood, June 27, 1968. (Aub)
35. Edsall to S. Flexner, June 18, 1912. (Aub)

CHAPTER 8, "A SECOND TIME AMBUSHED"

1. D. L. Edsall, "The Clinician, The Hospital and the Medical School." *Boston Med. & Surg. Jnl.*, 167 (Feb. 29, 1912): 315–23.
2. Marjorie E. Fox, *History of the Washington University School of Medicine.* (An authorized history, manuscript, unpaginated.) Ch. 4. (Wash.)
3. Fox, *op. cit.*
4. Fox, *op. cit.*, Ch. 4 & 5.
5. Fox, *op. cit.*, Ch. 4.
6. *Who's Who in America*, 1966–67.

7. H. Schuck et al., *Nobel: The Man and His Prizes.* Norman: U. of Okla., 1950.

8. Howard A. Kelly, "John Howland," in Howard A. Kelly & Walter L. Burrage, *Dictionary of American Medical Biography.* New York: Appleton, 1928, pp. 610–11.

9. Wilburt C. Davison, "John Howland," in Borden S. Veeder, ed., *Pediatric Profiles,* reprinted from the *Jnl. of Pediatrics,* Nov. 1954–Nov. 1957. St. Louis: Mosby, 1957, pp. 161 ff.

10. J. C. Schwarz, ed., *Who's Who Among Physicians and Surgeons, 1938.* New York, 1938.

11. Fox, *op. cit.*

12. *Who's Who in America,* 1932–33.

13. Fox, *op. cit.*

14. Eugene L. Opie, "The Reorganization of the Washington University School of Medicine." MS. No date. (Opie)

15. Fox, *op. cit.*

16. Robert J. Terry, conversation with Aub, Nov. 1, 1963.

17. Opie, *op. cit.*

18. Fox, *op. cit.,* Ch. 4.

19. *Who's Who in America,* 1932–33.

20. Fox, *op. cit.*

21. D. L. Edsall to Edwin A. Locke, no date. (Aub)

22. Edward W. Dempsey to Ethan A. H. Shepley, Feb. 11, 1964. (Aub)

23. St. Louis *Republic,* Jan. 8, 1911.

24. Opie, *op. cit.*

25. *Ibid.*

26. D. L. Edsall to Simon Flexner, Jan. 23, 1912. (Copy in possession of Aub)

27. D. L. Edsall to Eugene L. Opie, Jan. 11 [1912]. (Opie)

28. Edsall to S. Flexner, Jan. 23, 1912. (Aub)

29. Edsall to S. Flexner, July 25, 1912. (Aub)

30. Edsall to S. Flexner, Jan. 23, 1912. (Aub)

31. Edsall to Locke, no date. (Aub)

32. Margaret T. Edsall, diary, 1912. (Edsall)

33. D. L. Edsall to Henry A. Christian, April 21, 1912. (HMS)

34. Edsall to S. Flexner, no date. (Aub)

35. Edsall to S. Flexner, Feb. 10, 1912. (Aub)

36. Robert J. Terry, conversation with J. C. Aub, Nov. 1, 1963.

37. Edsall to Christian, July 14, 1912. (HMS)

38. Edsall to Locke, Feb. 22, 1912. (Aub)

39. Edsall to Christian, July 14, 1912. (HMS)

40. *Ibid.*

41. James J. Minot to Henry A. Christian, April 21, 1933. (HMS)

42. Edwin A. Locke to Frederic Washburn, March 19, 1938. (Aub)

43. Edsall to Locke, April 7, 1912. (Aub)

44. Edsall to Locke, April 15, 1912. (Aub)

45. Joint letter to Chancellor Houston, April 16, 1912. (Copy in possession of Aub)

46. Edsall to Locke, April 15, 1912. (Aub)
47. David F. Houston to D. L. Edsall, May 3, 1912. (Wash.)
48. Edsall to Locke, April 15, 1912. (Aub)
49. Edsall to S. Flexner, no date. (Aub)
50. Edsall to Christian, July 14, 1912. (HMS)
51. *Ibid.*
52. S. Flexner to Edsall, July 20, 1912. (Aub)
53. S. Flexner to Edsall, July 29, 1912. (Aub)
54. Edsall to S. Flexner, July 29, 1912. (Aub)
55. Edsall to Christian, April 21, 1912. (HMS)
56. Fox, *op. cit.*, Ch. 4.
57. Fox, *op. cit.*, Ch. 4 and 5.
58. Fox, *op. cit.*, Ch. 5.

CHAPTER 9, EDSALL ARRIVES IN BOSTON

1. Richard Cabot, article in *Boston Transcript*, May 11, 1912.
2. Richard Cabot, "Dr. Edsall Elected Jackson Professor." *Harvard Alumni Bull.*, 14 (May 15, 1912): 523–24. Copyright 1912, Harvard Bulletin, Inc.
3. *Boston Med. & Surg. Jnl.*, 166 (June 13, 1912): 898.
4. D. L. Edsall, "The Clinical Study of Respiration." Shattuck lecture, Mass. Medical Society, June 11, 1912. *Boston Med. & Surg. Jnl.*, 167 (Nov. 7, 1912): 639–51.
5. Leonard K. Eaton, *New England Hospitals: 1790–1833*. Ann Arbor: U. of Mich., 1957, p. 98.
6. Francis M. Rackemann, *The Inquisitive Physician: The Life and Times of George Richards Minot*. Cambridge: Harvard, 1956, pp. 57–58.
7. James Howard Means, conversation with Aub, June 1963.
8. *Ibid.*
9. *Ibid.*
10. *Ibid.*
11. D. L. Edsall, "Address." *New England Jnl. Med.*, 213 (Dec. 12, 1935): 1184–87.
12. D. L. Edsall to Simon Flexner, June 18, 1912. (Copy in possession of Aub)
13. D. L. Edsall to Henry A. Christian, July 14, 1912. (HMS)
14. D. L. Edsall to Charles Scudder, July 1919. (HMS)
15. Edwin A. Locke to Frederic Washburn, March 19, 1938. (Aub)

CHAPTER 10, THE HARVARD MEDICAL SCHOOL AS EDSALL FOUND IT

1. Edward D. Churchill, ed. *To Work in the Vineyard of Surgery: The Reminiscences of J. Collins Warren (1842–1927)*. Cambridge: Harvard, 1958.
2. Harvard University, Announcement of the Medical School, 1912–13.
3. Abraham Flexner, *Medical Education in the United States and Canada*. Bulletin 10 of the Carnegie Foundation for the Advancement of Teaching, p. 240.
4. *Ibid.*, p. 109.
5. A. Lawrence Lowell, first Annual Report, Harvard University, 1909–1910, p. 16–17.

6. Oliver Wendell Holmes, "Scholastic and Bedside Teaching," in *Medical Essays 1842–1882*. Boston: Houghton Mifflin, 1891, p. 303.
7. *Ibid.*, p. 308.
8. Leonard K. Eaton, *New England Hospitals: 1790–1833*. Ann Arbor: U. of Mich., 1957, pp. 209–10.
9. Churchill, *op. cit.*, p. 52.
10. *Ibid.*, p. 81n.
11. James Jackson, *Memoir of James Jackson Jr. With extracts from his letters, and reminiscences of him, by a fellow student*. Boston: Hilliard, 1836.
12. Pierre Charles Alexandre Louis, *Researches on the Effects of Bloodletting in some inflammatory diseases*. With preface and appendix by James Jackson. Boston: Hilliard, 1836.
13. Frederick C. Shattuck and J. Lewis Bremer, "The Medical School, 1869–1929," in Samuel Eliot Morison, ed. *The Development of Harvard University since the inauguration of President Eliot 1869–1929*. Cambridge: Harvard, 1930.
14. Henry James, *Charles W. Eliot: President of Harvard University, 1869–1909*. Boston: Houghton Mifflin, Vol. 2, pp. 170–71.
15. Shattuck & Bremer, *op. cit.*, p. 557.
16. *Ibid.*, pp. 557–58.
17. Churchill, *op. cit.*, p. 182.
18. D. L. Edsall, "The Medical School has a very especial reason for rejoicing." (Untitled and undated speech, MS.) (Edsall)
19. *Ibid.*
20. James, *op. cit.*, pp. 160–61.
21. Charles A. Eliot, speech before Medical Faculty, May 1, 1909. *Harvard Graduates' Mag.*, 27 (June 1909): 778–81.
22. Flexner, *op. cit.*, p. 240.
23. Abraham Flexner, *I Remember*. New York: S. & S., 1940, pp. 310–11. Copyright 1940 by Abraham Flexner.
24. D. L. Edsall to Thomas M. Woodward, Feb. 25, 1942. (Edsall)
25. Solomon R. Kagan, *The Modern Medical World*. Boston: Medico-Historical Press, 1945, p. 186.
26. Churchill, *op. cit.*, p. 241.
27. D. L. Edsall to Henry A. Christian, July 14, 1912. (HMS)
28. Edsall to Christian, Oct. 11, 1918. (HMS)
29. A. J. McLaughlin, "The Eradication of Typhoid Fever." *Boston Med. & Surg. Jnl.*, 166 (May 23, 1912): 766.
30. D. L. Edsall, "Medical Sepsis and Asepsis." *N. H. State Med. Soc. Transactions*, 1913.
31. John Edsall, personal communication to Aub, May 1968.
32. Geoffrey Edsall, personal communication to Aub, May 1968.

CHAPTER 11, TRAGEDY

1. Margaret Edsall to Mary Kirkbride, April 30, 1912. (Edsall)
2. Margaret Edsall to Mary Kirkbride, Oct. 30, 1912. (Edsall)
3. Samuel Hamill to Edwin A. Locke, Nov. 17, 1912. (Aub)
4. Edwin A. Locke, conversation with Aub, July 1956.

5. D. L. Edsall to Mrs. Edwin Locke, Nov. 27, 1912. (Aub)
6. Samuel Hamill to Edwin A. Locke, Dec. 30, 1912. (Aub)

CHAPTER 12, THE JACKSON PROFESSOR AT WORK

1. D. L. Edsall, "The Clinical Study of Respiration." Shattuck lecture, Mass. Medical Society, June 11, 1912. *Boston Med. & Surg. Jnl.*, 167 (Nov. 7, 1912): 639–51.
2. D. L. Edsall, Address as president of the American Pediatric Society, May 3, 1910, quoted in Borden S. Veeder, "The Influence of David Edsall on Pediatrics in the United States." *Jnl. Pediatrics*, 47 (Dec. 1955): 808–16.
3. Harvard University, Announcement of the Medical School, 1914–15, p. 54.
4. Harvard University, Announcement of the Medical School, 1919–20, p. 65.
5. D. L. Edsall, "Relation of the Staff to the Administration of Hospitals." *Modern Hospital*, 4 (March 1915): 9.
6. Frederic A. Washburn, *The Massachusetts General Hospital: Its Development, 1900–1935*. Boston: Houghton Mifflin, 1939, p. 86.
7. D. L. Edsall, "The Trend of Medical Research." *Jnl. Proceedings & Addresses*, Assn. Am. Universities, 34th Annual Conf., State U. of Iowa, Nov. 1932: 132–45.
8. D. L. Edsall, "The Medical School has a very especial reason for rejoicing." (Untitled and undated speech, MS.) (Edsall)
9. Richard Cabot to Henry Christian, March 1 [1919]. (HMS)
10. Washburn, *op. cit.*, pp. 123–31.
11. D. L. Edsall to Abraham Flexner, May 22, 1922. (HMS)
12. "You and Your Hospital." Pamphlet. Mass. General Hospital, Boston, Mass.
13. D. L. Edsall, "Statement of the Dean to the Faculty of Medicine," April 7, 1919. (Edsall)
14. Harvey Cushing, quoted in John F. Fulton, *Harvey Cushing: A Biography*. Springfield, Ill.: Thomas, 1946, pp. 383.
15. *Ibid*, p. 384.
16. Harvey Cushing to A. Lawrence Lowell, March 30, 1914, quoted in Fulton, *op. cit.*, pp. 379–80.
17. A. Lawrence Lowell to Harvey Cushing, April 4, 1914, quoted in Fulton, *op. cit.*, p. 380.
18. D. L. Edsall, "The following seem to me to be worth making clear . . ." (Undated and untitled memorandum, MS.) (Harvard)
19. Abraham Flexner to A. Lawrence Lowell, May 1, 1914. (Harvard)
20. Lowell to A. Flexner, May 26, 1914. (Harvard)
21. Lowell to A. Flexner, April 23, 1914. (Harvard)
22. Flexner to Lowell, May 1, 1914. (Harvard)
23. Lowell to Henry P. Walcott, Sept. 15, 1914 (Harvard)
24. Jerome Greene, quoted in Lowell to Edsall, Nov. 14, 1914. (Harvard)
25. Wallace Buttrick, Memorandum, Jan. 30, 1917. (Harvard)
26. Charles W. Eliot to Wallace Buttrick, April 23, 1917. (HMS)

27. Francis M. Rackemann, *The Inquisitive Physician: The Life and Times of George Richards Minot.* Cambridge: Harvard, 1956, p. 94.

28. J. H. Means, "The Medical Laboratory." Massachusetts General Hospital. Memorial & Historical Volume: Together with the Proceedings of the Centennial of the Opening of the Hospital. 1921.

29. J. H. Means, *Ward 4: The Mallinckrodt Research Ward of the Massachusetts General Hospital.* Cambridge: Harvard, 1958.

30. Rackemann, *op. cit.*, p. 95.

31. J. H. Means, " 'Massachusetts General' — Teaching Hospital." *New England Jnl. Med.*, 264 (Jan. 26, 1961): 177.

32. 110th Annual Report of the Massachusetts General Hospital, 1923.

CHAPTER 13, THE INDUSTRIAL DISEASE CLINIC AND OTHER RESEARCH

1. D. L. Edsall, "Medical-Industrial Relations of the War." *Johns Hopkins Hosp. Bull.*, 29 (Sept. 1918): 197. © The Johns Hopkins Press.

2. D. L. Edsall, "The Relations of Industry to General Medicine." *Boston Med. & Surg. Jnl.*, 171 (Oct. 29, 1914): 659 ff.

3. *Ibid.*

4. *Ibid.*

5. *Ibid.*

6. Edsall, "Medical-Industrial Relations," *op. cit.*

7. D. L. Edsall, "The Bearings of Industry on Medical Practice," abstract of speech before Mass. Medical Society, June 13, 1917. *Boston Med. & Surg. Jnl.*, 177 (Oct. 25, 1917): 575.

8. Ida M. Cannon, *On the Social Frontier of Medicine: Pioneering in Medical Social Service.* Cambridge: Harvard, 1952, p. 185.

9. D. L. Edsall, "The Study of Occupational Disease in Hospitals." *Monthly Review*, Bureau of Labor Statistics, U. S. Dept. of Labor, 5 (Dec. 1917): 169–85.

10. Wade Wright, "An Industrial Clinic." *Monthly Review*, Bureau of Labor Statistics, U. S. Dept. of Labor, 5 (Dec. 1917): 185–93.

11. Edsall, "The Relation of Industry," *op. cit.*

12. *Ibid.*

13. Edsall, "The Study of Occupational Disease in Hospitals," *op. cit.*

14. D. L. Edsall, Exhibit C 17, in Dean Edward H. Bradford's report to Abraham Flexner, Dec. 16, 1915. (HMS)

15. Alice Hamilton to Aub, May 18, 1957.

16. Edsall, "Medical-Industrial Relations," *op. cit.*

17. David L. Edsall, F. P. Wilbur, Cecil K. Drinker, "The Occurrence, Course and Prevention of Chronic Manganese Poisoning." *Jnl. Industrial Hygiene*, 1 (Aug. 1910): 183 ff.

18. Edsall, "Medical-Industrial Relations," *op. cit.*

19. *Ibid.*

20. *Ibid.*

21. Edsall et al., "The Occurrence, Course and Prevention of Chronic Manganese Poisoning," *op. cit.*

22. D. L. Edsall and Cecil K. Drinker, "The Clinical Aspects of Chronic Manganese Poisoning," in *Contributions to Medical and Biological Research*; Dedicated to Sir William Osler; In Honor of His Seventieth Birth-

day, July 12, 1919, by his pupils and co-workers. Vol. 1. New York: Hoeber, 1919.

23. Cecil K. Drinker, quoted in George Cheever Shattuck, *Industrial Medicine at Harvard*, MS. p. 107. (HMS)

24. D. L. Edsall, "The Efficiency and Significance of Different Forms of Respiration." *Transactions*, Assn. Am. Physicians, 27 (1912): 560–71.

25. James Howard Means, *The Association of American Physicians: Its First Seventy-Five Years*. New York: McGraw-Hill, 1961, p. 108.

26. D. L. Edsall, "The Clinical Study of Respiration." Shattuck Lecture, June 11, 1912. *Boston Med. & Surg. Jnl.*, 167 (Nov. 7, 1912): 639–51.

27. D. L. Edsall and J. H. Means, "The Effect of Strychnin, Caffein, Atropin and Camphor on the Respiration and Respiratory Metabolism in Normal Human Subjects." *Archives Internal Med.*, 14 (Dec. 1914): 897–910.

28. D. L. Edsall and J. H. Means, "Observations on a Case of Family Periodic Paralysis." *Am. Jnl. of Medical Sciences*, 150 (1915): 169–78.

29. D. L. Edsall, "Relation of the Staff to the Administration of Hospitals." *Modern Hospital*, 4 (March 1915): 184 ff.

30. D. L. Edsall, "Movements in Medicine." *Boston Med. & Surg. Jnl.*, 174 (June 22, 1916): 891–97.

31. William Lawrence, *Memories of a Happy Life*. Boston: Houghton Mifflin, 1926, p. 297.

CHAPTER 14, EDSALL MAKES A NEW HOME

1. Harriet T. Mixter to Mrs. John B. Tileston, Jan. 1927. (Edsall)

2. *Amelia Peabody Tileston and her Canteens for the Serbs*. Privately printed. Boston: Atlantic Monthly Press, 1920, p. 169.

3. Geoffrey Edsall, personal communication to Aub, Sept. 1967.

4. D. L. Edsall to F. C. Shattuck, July 22 (no year date). (HMS)

5. D. L. Edsall to Geoffrey Edsall, Sept. 4, 1926.

6. Geoffrey Edsall, personal communication, *op. cit.*

7. D. L. Edsall to John Edsall, Sept. 13, 1926.

8. D. L. Edsall, notes on "Bill of fare of a luncheon 'feast.' Oct. 10, at Dr. Wu Hsien's." MS. (Edsall)

9. Francis R. Dieuaide to Aub, April 25, 1963.

10. *Ibid.*

CHAPTER 15, THE NEW DEAN

1. Mass. General Hospital, *Memorial & Historical Volume: Together with the Proceedings of the Centennial of the Opening of the Hospital.* 1921.

2. Henry A. Christian to Charles W. Eliot, April 25, 1918. (HMS)

3. Charles W. Eliot to Henry A. Christian, April 27, 1918. (HMS)

4. D. L. Edsall, "Address." *New England Jnl. Med.*, 213 (Dec. 12, 1935): 1184–87.

5. D. L. Edsall to A. Lawrence Lowell, June 3, 1918. (Harvard)

6. Robert M. Green, "Dr. Worth Hale." *Harvard Med. Alumni Bull.*, (June 1943).

7. D. L. Edsall, "Statement of the Dean to the Faculty of Medicine, April 7, 1919." (Edsall)

8. Faculty of Medicine Minutes, HMS, April 7, 1919.

CHAPTER 16, A STORMY BEGINNING

1. D. L. Edsall to A. Lawrence Lowell, May 26, 1921. (HMS)
2. D. L. Edsall, Dean's Report, Harvard Medical School, 1927–28.
3. W. B. Cannon, quoted in Cornelia James Cannon, *A Servant of Science: Walter Bradford Cannon*. MS., pp. 91–92.
4. Mrs. Walter Cannon to Aub, no date.
5. J. C. Aub, "Walter Bradford Cannon, the Professor," in *Walter Bradford Cannon, 1871–1945*. A Memorial Exercise. Held at the Harvard Medical School, Nov. 5, 1945.
6. D. L. Edsall speech, in *Walter Bradford Cannon, Exercises celebrating twenty-five years as George Higginson Professor of Physiology, Oct. 15, 1931*. Cambridge: Harvard, 1932.
7. D. L. Edsall to Charles Scudder, July 2, 1919. (HMS)
8. D. L. Edsall to Algernon Coolidge, Jan. 31, 1921. (Churchill)
9. Edsall to Scudder, July 2, 1919. (HMS)
10. Abraham Flexner, *I Remember*. New York: S. and S., 1940, p. 313. Copyright 1940 by Abraham Flexner.
11. "Report of Committee on the Medical School," chairman Dr. F. C. Shattuck. *Harvard Alumni Bull.*, 18 (May 31, 1916). Copyright 1916, Harvard Bulletin Inc.
12. D. L. Edsall, "Present Day Problems of the Medical School." *Harvard Alumni Bull.*, 25 (Dec. 7, 1922): 279–86. Copyright 1922, Harvard Bulletin Inc.
13. Harvard University, *Endowment Funds of Harvard University*, June 30, 1947, pp. 256–57.
14. D. L. Edsall to Henry A. Christian, March 30, 1922. (HMS)
15. Frederick C. Shattuck and J. Lewis Bremer, "The Medical School," in Samuel Eliot Morison, ed., *The Development of Harvard University*. Cambridge: Harvard, 1930, pp. 579 ff.
16. Charles W. Eliot to Henry A. Christian, Jan. 21, 1920. (HMS)
17. Henry A. Christian to A. Lawrence Lowell, Jan. 16, 1920. (HMS)
18. D. L. Edsall to Charles W. Eliot, Jan. 23, 1920. (HMS)
19. D. L. Edsall, Dean's Report, Harvard Medical School, 1923–24.
20. D. L. Edsall, Speech to Associated Harvard Clubs. *Harvard Alumni Bull.*, 28 (Oct. 1, 1925): sup. 9–20. Copyright 1925, Harvard Bulletin Inc.
21. D. L. Edsall to Henry A. Christian, March 25, 1920. (HMS)
22. *Ibid.*
23. Francis B. Harrington to Harvey Cushing, May 1, 1910. (Harvard)
24. Harvey Cushing, in Ninth Annual Report of the Peter Bent Brigham Hospital for the Year 1922.
25. Edsall, "Present Day Problems of the Medical School," *op. cit.*
26. D. L. Edsall to A. Lawrence Lowell, June 5, 1923. (HMS)
27. D. L. Edsall to Henry A. Christian, Oct. 15, 1919. (HMS)

CHAPTER 17, REFORMS IN TEACHING

1. D. L. Edsall to Abraham Flexner, Nov. 4, 1921. (HMS)
2. D. L. Edsall, Dean's Report, Harvard Medical School, 1918–19.

3. Faculty of Medicine Minutes, Harvard Medical School, May 2, 1913.
4. D. L. Edsall, Dean's Report, Harvard Medical School, 1921–22.
5. D. L. Edsall, remarks in discussion of Hugh Cabot, "Report of Committee on Curriculum of the Association of American Medical Colleges." *Proceedings*, Assn. Am. Med. Colleges, 1922: 73–91.
6. Edsall, Dean's Report, HMS, 1921–22, *op. cit.*
7. D. L. Edsall, "The Handling of the Superior Student." *Proceedings*, Assn. Am. Med. Colleges, 1925: 114–124.
8. D. L. Edsall to Edwin A. Locke, May 25, 1934. (Aub)
9. D. L. Edsall to Walter Cannon, Feb. 24, 1922. (HMS)
10. Edsall, Dean's Report, HMS, 1921–22, *op. cit.*
11. D. L. Edsall, "The Product of Medical Education." *Boston Med. & Surg. Jnl.*, 191 (Aug. 14, 1924): 283–94.
12. Edsall, Dean's Report, HMS, 1921–22, *op. cit.*
13. Edsall, "The Product of Medical Education," *op. cit.*
14. Edsall, Dean's Report, HMS, 1918–19, *op. cit.*
15. D. L. Edsall, Dean's Report, Harvard Medical School, 1927–28.
16. Edsall to Locke, May 25, 1934. (Aub)
17. *Ibid.*
18. D. L. Edsall, "An Adequate Examination at the End of the Clinical Courses." 1925. *Jnl. Am. Med. Assn.*, 84 (May 2, 1925): 1320–24.
19. Edsall to Locke, May 25, 1934. (Aub)
20. Committee on Examinations, "A Critical Review of the Methods of Promotion, Examination and Grading in the Harvard Medical School." Cambridge: Harvard University, 1920.
21. Edsall, "An Adequate Examination," *op. cit.*
22. D. L. Edsall, Dean's Report, Harvard Medical School, 1923–24.
23. Edsall, "An Adequate Examination," *op. cit.*
24. Edsall to Locke, May 25, 1934. (Aub)
25. Edsall, "An Adequate Examination," *op. cit.*
26. D. L. Edsall to Abraham Flexner, Nov. 4, 1921. (HMS)
27. David Riesman, *History of the Interurban Clinical Club, 1905–1937*. Philadelphia: Winston, no date, pp. 53–54.
28. Edsall, "An Adequate Examination," *op. cit.*
29. D. L. Edsall to Abraham Flexner, Nov. 4, 1921. (HMS)
30. *Ibid.*
31. Walter Bauer, "The Tutorial System in the Harvard Medical School." *Jnl., Assn. Am. Med. Colleges*, March 1940.
32. Edsall to Locke, May 25, 1934 (Aub)
33. *Ibid.*
34. *Ibid.*
35. Arlie V. Bock to Edsall, Oct. 29, 1924. (HMS)
36. Edsall to A. Flexner, Jan. 23, 1925. (HMS)
37. Edsall, "The Handling of the Superior Student," *op. cit.*
38. Edsall to Locke, May 25, 1934. (Aub)
39. *Ibid.*
40. *Ibid.*
41. Edsall, "The Product of Medical Education," *op. cit.*

42. D. L. Edsall, "Some Features of Medical Education." *Southern Med. Jnl.*, 22 (Aug. 1929).

CHAPTER 18, THE SCHOOL OF PUBLIC HEALTH

A basic source for this period is: Jean A. Curran, *Founders of the Harvard School of Public Health.* New York: Josiah Macy, Jr. Foundation, 1970.

1. D. L. Edsall, Dean's Report, Harvard Medical School, 1920–21.
2. D. L. Edsall, Dean's Report, School of Public Health, 1922–23.
3. Cecil K. Drinker, "Industrial Medicine at Harvard." *Harvard Alumni Bull.*, 21 (April 24, 1919). Copyright 1919, Harvard Bulletin Inc.
4. Cecil K. Drinker, quoted in George Cheever Shattuck, *Industrial Medicine at Harvard.* MS., 1954, pp. 9–10. (HMS)
5. Steven Jonas, "From Journal of Industrial Hygiene to Archives of Environmental Health." *Arch. Environ. Health*, 14 (April 1967): 634–39.
6. D. L. Edsall to Raymond B. Fosdick, April 8 [1938]. (Edsall)
7. D. L. Edsall to A. Lawrence Lowell, Dec. 20, 1918. (Harvard)
8. A. Lawrence Lowell to Henry P. Walcott, Dec. 21, 1918. (Harvard)
9. D. L. Edsall to Alice Hamilton, Dec. 27, 1918. (HMS)
10. Alice Hamilton endorsement on Edsall's letter of Jan. 24, 1919 (from the Schlesinger Library, Radcliffe College).
11. Alice Hamilton to Aub, May 18, 1957.
12. Alice Hamilton, *Exploring the Dangerous Trades.* Boston: Little, Brown, 1943, p. 264.
13. D. L. Edsall to Wickliffe Rose, Jan. 3, 1921. (HMS)
14. Edsall to Rose, Jan. 25, 1921. (HMS)
15. Milton J. Rosenau to D. L. Edsall, Feb. 17, 1921. (HMS)
16. Edsall to Rose, March 23, 1921. (HMS)
17. Rockefeller Foundation, Information Service, release dated Aug. 3, 1921.
18. Robert Shaplen, *Toward the Well-Being of Mankind: Fifty Years of the Rockefeller Foundation.* Garden City, N.Y.: Doubleday, 1965, p. 53.
19. Minutes of Meetings of Organization Committee of Harvard School of Public Health, Dec. 16, 1921. (SPH)
20. Obituary of Henry P. Walcott, *New England Jnl. Med.*, 209 (Nov. 9, 1933).
21. Frederick C. Shattuck, quoted in *Boston Transcript* (no date on clipping, HMS).
22. Edsall, Dean's Report, HMS, 1920–21, *op. cit.*
23. D. L. Edsall, Dean's Report, SPH, 1922–23, *op. cit.*
24. D. L. Edsall to Raymond B. Fosdick, April 8 [1938]. (Edsall)
25. Alice Hamilton to D. L. Edsall, May 1, 1920. (HMS)
26. E. J. Cornish to D. L. Edsall, March 29, 1923. (HMS)
27. Joseph C. Aub, Lawrence T. Fairhall, A. S. Minot, & Paul Reznikoff, *Lead Poisoning.* Baltimore: William & Wilkins, 1926; Donald Hunter & Joseph C. Aub, "Lead Studies. XV. The Effect of the Parathyroid Hormone on the Excretion of Lead and of Calcium in Patients Suffering from Lead Poisoning." *Quarterly Jnl. of Medicine*, 20 (Jan. 1927).

28. D. L. Edsall to Abraham Flexner, May 22, 1922. (HMS)

29. Cecil K. Drinker, "David Linn Edsall: 1869–1945." *Harvard Public Health Alumni Bull.*, 2 (Nov. 1945): 20–21.

CHAPTER 19, THE DEANSHIP BECOMES FULL TIME

1. D. L. Edsall to Frederic A. Washburn, May 19, 1919.
2. *Ibid.*
3. Frederic A. Washburn, *The Massachusetts General Hospital: Its Development, 1900–1935*. Boston: Houghton Mifflin, 1939, p. 96.
4. D. L. Edsall to Henry A. Christian, March 25, 1920. (HMS)
5. Christian to Edsall, April 16, 1920. (HMS)
6. D. L. Edsall to Edwin A. Locke, April 1922. (Aub)
7. D. L. Edsall to Abraham Flexner, May 22, 1922. (HMS)
8. Edsall to Locke, March 22 (1923?). (Aub)
9. Edsall to A. Lawrence Lowell, April 11, 1923. (HMS)
10. Edsall to Washburn, May 12, 1923. (HMS)
11. Lowell to Edsall, April 3, 1923. (Harvard)
12. Edsall to Washburn, May 12, 1923. (HMS)
13. *Ibid.*
14. Washburn to Edsall, May 18, 1923. (HMS)

CHAPTER 20, ADVANCES IN THE MEDICAL SCHOOL

1. D. L. Edsall, Dean's Report, Harvard Medical School and School of Public Health, 1927–28.
2. Raymond B. Fosdick, *Adventure in Giving: The Story of the General Education Board: A Foundation Established by John D. Rockefeller*. Based on an unfinished MS. prepared by the late Henry F. Pringle and Katharine Douglas Pringle. New York: Harper, 1962, pp. 114–15.
3. D. L. Edsall to Abraham Flexner, Aug. 3, 1919. (HMS)
4. *Ibid.*
5. Flexner to Edsall, Aug. 14, 1919. (HMS)
6. Edsall to Flexner, Sept. 13, 1919. (HMS)
7. *Ibid.*
8. D. L. Edsall, "The Product of Medical Education." *Boston Med. & Surg. Jnl.*, 191 (Aug. 14, 1924): 283–94.
9. D. L. Edsall to Edwin A. Locke, May 25, 1934. (Aub)
10. "Full-time accepted by President and Fellows of Harvard College." Nov. 24, 1924. (Churchill)
11. Edsall, Dean's Report, HMS & SPH, 1927–28, *op. cit.*
12. *Ibid.*
13. D. L. Edsall to Henry A. Christian, Aug. 26, 1919 (?). (HMS)
14. Edsall, Dean's Report, Harvard Medical School and School of Public Health, 1933–34.
15. George Dock to Henry A. Christian, May 29, 1922. (HMS)
16. Christian to Dock, June 1, 1922. (HMS)
17. Fosdick, *Adventure in Giving, op. cit.*, p. 171.
18. *Ibid.*, p. 119.
19. Harvard Medical School, dean's office memo, Dec. 11, 1929. (HMS)
20. Rockefeller Foundation, *Annual Report*, 1931, pp. 156–57.

21. Fosdick, *Adventure in Giving, op. cit.*, p. 328.

22. D. L. Edsall to Walter Cannon, quoted in Cornelia James Cannon, *A Servant of Science: Walter Bradford Cannon.* MS. pp. 84–86. (Cannon)

23. Cornelia James Cannon, *op. cit.*

24. D. L. Edsall, "The Handling of the Superior Student." *Proc.* Assn. Am. Med. Colleges 35 (1925): 114–24.

25. Abraham Flexner to Charles W. Eliot, April 6, 1923. (HMS)

26. Geoffrey Edsall, personal communication to Aub, Sept. 1967.

27. *Ibid.*

28. Alan Gregg to Aub, Feb. 1, 1957.

29. Roger I. Lee, "Health Examinations." MS. No date. (HMS)

30. Reginald Fitz to D. L. Edsall, April 7, 1928. (HMS)

31. Edsall to Locke, May 25, 1934. (Aub)

32. Frederick C. Shattuck, quoted in Harvard University, *Endowment Funds of Harvard University, June 30, 1947.* Cambridge, Mass.: 1948, p. 420.

33. D. L. Edsall to Henry A. Christian, Dec. 17, 1929. (HMS)

34. Frederic Sharf and Louis Stein, *The Beth Israel Hospital of Boston, 1916–1966: A Half Century of Community Service.* MS. in preparation, pp. 23–26.

35. Edsall, Dean's Report, HMS & SPH, 1927–28, *op. cit.*

36. *Ibid.*

37. D. L. Edsall to Hans Zinsser, June 16, 1924. (HMS)

CHAPTER 21, SECOND SPRING

1. Radcliffe College, Class of 1906, *Fiftieth Anniversary Record*, June, 1956.

2. D. L. Edsall to Louisa Richardson, letters collected for the children by Louisa Richardson. (Edsall)

3. Louisa R. Edsall to Mrs. John Richardson, July 14, 1930. (Edsall)

4. D. L. Edsall to Mrs. John Richardson, June 14, 1930. (Edsall)

5. Edsall to Mrs. Richardson, June 22, 1930. (Edsall)

6. Edsall to Mrs. Richardson, Aug. 16, 1930. (Edsall)

7. D. L. Edsall to Edwin A. Locke, Aug. 31, 1930. (Aub)

8. Mary Lee, personal communication to Hapgood, June 17, 1966. (Aub)

9. John Edsall to Aub, April 10, 1968.

CHAPTER 22, RENEWED CRISES

1. D. L. Edsall to A. Lawrence Lowell, May 17, 1929. (HMS)

2. *Ibid.*

3. *Ibid.*

4. *Ibid.*

5. *Ibid.*

6. *Ibid.*

7. *Ibid.*

8. A. Flexner to D. L. Edsall, April 7, 1930. (HMS)

9. Edsall to Flexner, April 3, 1930. (HMS)

10. D. L. Edsall to Edwin A. Locke, May 25, 1934. (Aub)

11. D. L. Edsall and J. C. Aub, "Clinics in Preventive Medicine in Third-Year Teaching: Harvard Medical School," in *Methods and Problems of Medical Education*, Eighteenth Series. New York: The Rockefeller Foundation, 1930.

12. Harvard University, *Synopsis of the Practice of Preventive Medicine.* 1929.

13. Kenelm Winslow, review of *Synopsis of the Practice of Preventive Medicine.* MS. (HMS)

14. D. L. Edsall to Henry A. Christian, March 18, 1930. (HMS)

15. Harvey Cushing to D. L. Edsall, April 1, 1930. (HMS)

16. Cushing to Edsall, March 15, 1924. (HMS)

17. Edwin A. Locke, interview with Aub, July 1956.

18. John F. Fulton, *Harvey Cushing: A Biography*. Springfield, Ill.: Thomas, 1946, pp. 557–58.

19. Edsall to Locke, May 25, 1934. (Aub)

20. D. L. Edsall to Charles P. Curtis, April 24, 1929. (HMS)

21. Cushing to Edsall, March 30, 1923. (HMS)

22. Edsall to Cushing, Jan. 9, 1925. (HMS)

23. Cushing to Edsall, Jan. 30, 1925. (HMS)

24. D. L. Edsall to A. Lawrence Lowell, Jan. 28, 1925. (HMS)

25. *Ibid.*

26. Harvey Cushing to Henry A. Christian, Nov. 20, 1911.

27. Lucien Price, "Harvey Cushing," in Edward W. Forbes and John H. Finley, Jr., eds., *The Saturday Club*. Boston: Houghton Mifflin, 1958, p. 183.

28. A. Lawrence Lowell to D. L. Edsall, May 15, 1929. (HMS)

29. D. L. Edsall to Edward Churchill, June 25, 1928. (Churchill)

30. Richard Pearce to D. L. Edsall, Oct. 13, 192 (?). (Edsall)

31. D. L. Edsall, "Richard M. Pearce, Jr. from the Personal Side." Memorial Meeting held April 15, 1930 at the Rockefeller Institute for Medical Research, New York City.

32. D. L. Edsall to Felix Frankfurter, Nov. 2, 1929. (HMS)

33. D. L. Edsall, Dean's Report, School of Public Health, 1931–32.

34. D. L. Edsall, speech to Associated Harvard Clubs. *Harvard Alumni Bull.*, 28 Oct. 1, 1925: sup. 9–20. Copyright 1925, Harvard Bulletin Inc.

35. D. L. Edsall, Dean's Report, Harvard Medical School and School of Public Health, 1932–33.

36. D. L. Edsall to Edward Churchill (form letter), Nov. 19, 1931, Jan. 19, 1932, June 20, 1932, June 23, 1932. (Churchill)

37. D. L. Edsall to A. L. Endicott, Feb. 4, 1932. (HMS)

38. Geoffrey Edsall, personal communication to Aub, Sept. 1967.

39. D. L. Edsall to Edward Churchill, Sept. 6, 1932; Edsall to Churchill (form letter), February 11, 1933. (Churchill)

40. D. L. Edsall, Dean's Report, HMS & SPH, 1932–33, *op. cit.*

Chapter 23, The Last Harvard Years

1. D. L. Edsall, Dean's Report, Harvard Medical School and the School of Public Health, 1933–34.

2. D. L. Edsall to Edwin A. Locke, May 25, 1934. (Aub)
3. *Ibid.*
4. *Ibid.*
5. *Ibid.*
6. D. L. Edsall, untitled speech, "The Medical School has a very especial reason for rejoicing." MS. (Edsall)
7. Edsall to Locke, May 25, 1934.
8. D. L. Edsall, "Certain Changes in the Harvard Medical School During the Last Twenty Years." *Harvard Med. Alumni Bull.*, June, 1931.
9. Edsall to Locke, May 25, 1934.
10. Hans Zinsser to Worth Hale, Dec. 18, 1929. (HMS)
11. Walter Cannon to D. L. Edsall, February 7, 1935. (Edsall)
12. Hans Zinsser to D. L. Edsall, Jan. 3, 1935 (HMS)
13. Hans Zinsser to D. L. Edsall, June 4, 1935. (HMS)
14. George R. Minot to D. L. Edsall, June 16, 1935. (Edsall)
15. James B. Conant to D. L. Edsall, Dec. 6, 1934. (Edsall)
16. Alfred North Whitehead to D. L. Edsall, Oct. 6, 1935. (Edsall)
17. Louis Kirstein to D. L. Edsall, Oct. 22, 1935. (Edsall)
18. Ray Lyman Wilbur to D. L. Edsall, Oct. 12, 1935. (Edsall)
19. David Riesman to Walter Cannon, Oct. 19, 1935. (Edsall)
20. Hugh Cabot to D. L. Edsall, Dec. 24, 1934. (Edsall)
21. Warfield T. Longcope to D. L. Edsall, Oct. 28, 1935. (Edsall)
22. Willard C. Rappleye to D. L. Edsall, July 10, 1935. (Edsall)
23. Cecil Drinker to Edwin J. Cohn, Oct. 17, 1945. (HMS)
24. Alan Gregg to D. L. Edsall, April 24, 1936. (Edsall)
25. Raymond B. Fosdick to D. L. Edsall, Jan. 5, 1935. (Edsall)
26. J. L. Yates to D. L. Edsall, Oct. 3, 1935. (Edsall)
27. Walter Cannon, in "Addresses in Honor of Dean Edsall," *Science,* 82 (Nov. 22, 1935).
28. Eugene J. DuBois, "The Development of Clinical Subjects as Contributing to University Work," in "Addresses in Honor of Dean Edsall," *Science, op. cit.*
29. Edwin A. Locke to D. L. Edsall, August 9, 1935. (Edsall)
30. Geoffrey Edsall, personal communication to Aub, Sept. 1967.
31. James B. Conant, in "Notes of after-dinner speeches, Oct. 23, 1935," taken by a student for Marian Dale. MS. (Edsall)
32. Charles Sidney Burwell, in "Notes of after-dinner speeches," *op. cit.*
33. A. Lawrence Lowell, in "Notes of after-dinner speeches," *op. cit.*
34. D. L. Edsall, "Address of Dr. David L. Edsall." *New England Jnl. Med.*, 213 (Dec. 12, 1935): 1184–87.
35. A. Lawrence Lowell to D. L. Edsall, Oct. 24, 1935. (Edsall)
36. Felix Frankfurter to Cecil Drinker, Dec. 17, 1935. (HMS)

CHAPTER 24, EPILOGUE

1. Louisa R. Edsall to Mr. & Mrs. Richard Edsall, Feb. 2, 1941. (Edsall)
2. David L. Edsall and Tracy J. Putnam, "The Emigré Physician in America, 1941: A Report of the National Committee for Resettlement of

Foreign Physicians." *Jnl. Am. Med. Assn.*, 117 (Nov. 29, 1941): 1881–88.

3. D. L. Edsall letter, "A Program for the Refugee Physician." *Jnl. Am. Med. Assn.*, 112 (May 13, 1939): 1986–87.
4. Edsall and Putnam, *op. cit.*
5. *Ibid.*
6. Geoffrey Edsall, personal communication to Aub, Sept. 1967.
7. Paul Dudley White to Louisa R. Edsall, Aug. 13, 1945. (Edsall)

Index

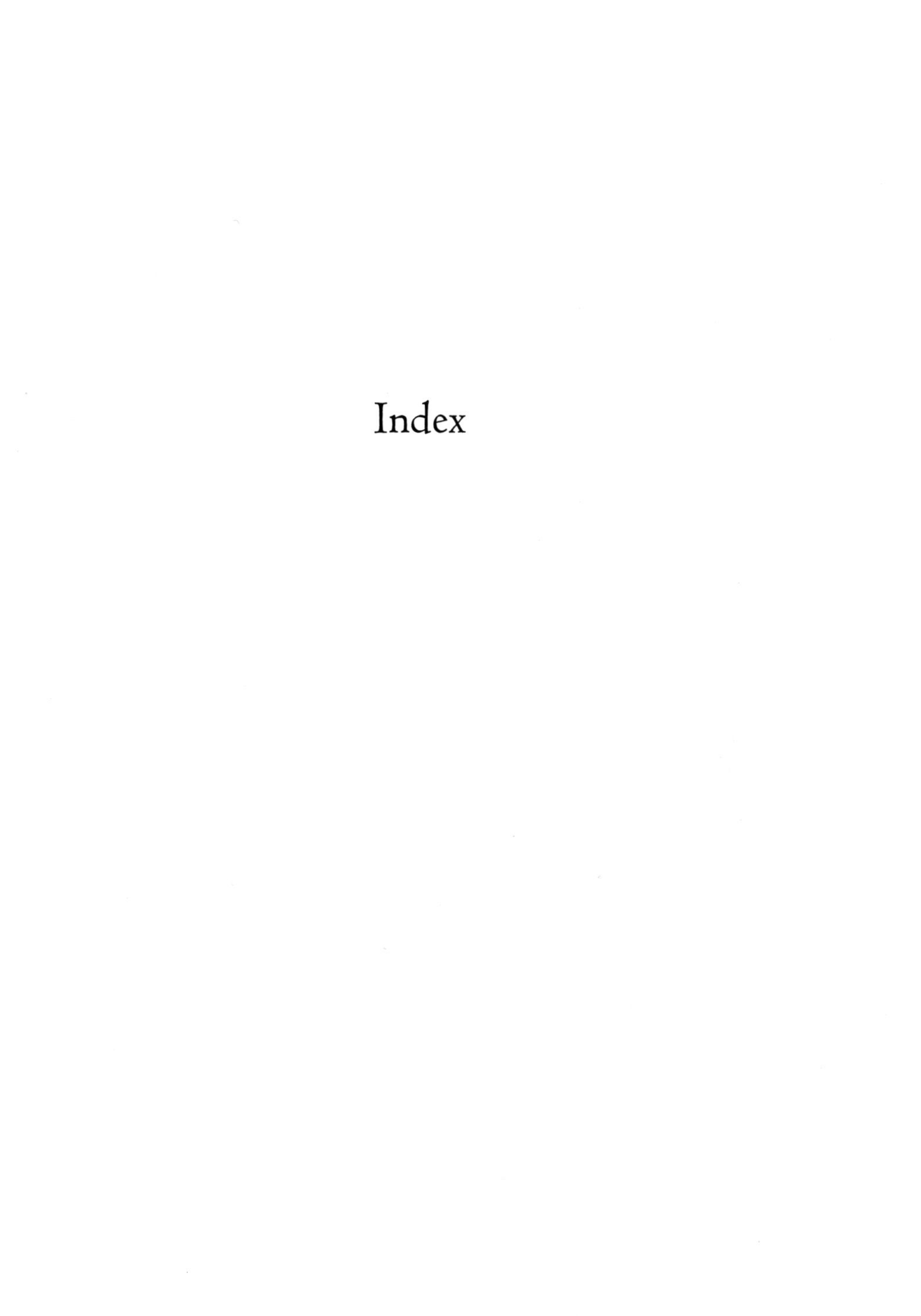

Index